How to Build a Small Budget Recording Studio

How to Build a Small Budget Recording Studio

From Scratch
...with 12 Tested Designs

THIRD EDITION

**Mike Shea
and
F. Alton Everest**

McGraw-Hill

New York   Chicago   San Francisco   Lisbon   London   Madrid
Mexico City   Milan   New Delhi   San Juan   Seoul
Singapore   Sydney   Toronto

Cataloging-in-Publication Data is on file with the Library of Congress

McGraw-Hill

A Division of The McGraw·Hill Companies

1 2 3 4 5 6 7 8 9 0   AGM/AGM   0 9 8 7 6 5 4 3 2

ISBN 0-07-138700-5

The sponsoring editor of this book was Scott Grillo. The editing supervisor was Steven Melvin, and the production supervisor was Pamela Pelton. It was set in Times New Roman per the TAB4 Design by Deirdre Sheean of McGraw-Hill Professional's Hightstown, N. J. composition unit.

Printed and bound by Quebecor/Martinsburg.

 This book is printed on recycled, acid-free paper containing a minimum of 50% recycled, de-inked fiber.

McGraw-Hill books are available at special quantity discounts to use as premiums and sales promotions, or for use in corporate training programs. For more information, please write to the Director of Special Sales, Professional Publishing, McGraw-Hill, Two Penn Plaza, New York, NY 10121-2298. Or contact your local bookstore.

CONTENTS

PREFACE

This book is about small studios: How to build them and how to treat them acoustically. Details of design, construction, and treatment of twelve actual studio suites are included. Acoustical principles are discussed along the way in the context of real life projects and problems rather than as blue sky theory.

The emphasis of this book is on budget studios which are eminently suited to efficient production of radio, audiovisual, video, film, and television program material on a day-to-day, routine basis and for training students in these fields. These studios operate in a different world from the flamboyant recording studios with gold records on the wall designed to catch the eye of the well-heeled client. The operators of the studios described in this book stress function and economy over glamour yet strive for good (natural, faithful) sound quality.

The word *studio* is often used in this book in the inclusive sense to include control and monitoring rooms. Acoustical defects in the recording studio quite obviously affect the recorded sound. If the control room acoustics are poor, the processing changes made by the operator may actually degrade the signal because they are correcting for acoustical defects. For this reason the same care is given to control room acoustics as to those of the studio.

The studios described in this book (with the exception of Chapter 7) follow closely, but not exactly, designs I prepared in my consulting practice for many different types of organizations in various parts of the world. I am greatly indebted to the following for permission to share the design of their studios with the readers of this book: Missionary TECH Team, Longview, Texas (Chapter 3); proposal submitted to the Christian Church in Zaire (Chapter 4); Hong Kong Baptist College, Kowloon, Hong Kong (Chapter 5); The Russ Reid Company, Pasadena, California (Chapter 6); Centro Bautista de Communicaciones, Montevideo, Uruguay (Chapter 8); Golden West Broadcasting Ltd., Altona, Manitoba, Canada (Chapter 9); The Paraguay Mission, Southern Baptist Convention, Asuncion, Paraguay (Chapter 10); Medios Educativos, A.C., Mexico City, Mexico (Chapter 11); Baptist Caribbean Media Center, Nassau, Bahamas (Chapter 12); Cathedral Films, Westwood Village, California (Chapter 13); and Far East Broadcasting Company, LaMirada, California (Chapter 14).

F. Alton Everest

INTRODUCTION

I began my career as a recording studio engineer in the 1960s, at a time when one was not only designated as the audio mixer but also the studio designer/builder along with part of the crew that built the studio's console. By 1970 I was attending lectures at MIT by Leo L. Beranek, a partner with R. H. Bolt of Bolt, Beranek and Newman, and a leading auditorium acoustician.

Attending college close to full time, attending seminars and lectures and working in a recording studio running my own four-track "portable" recording business, you'd think that all kinds of needed information would be pouring in my direction. While all of the above gave me a great deal of insight into the field of acoustics it seeemed there wasn't anybody talking about the things that were most directly important to the pop music recording artist. I, like many others at this point in time, was the first to deal acoustically with a whole new situation. Take two electric guitarists each running through mega-power stacks comprised of "Hi-Watt" amps on top of dual 4 × 12 inch speaker cabinets, a bass player using dual Ampeg folded horn cabinets with 18-inch speakers driven by a 250-watt Plush amp, a drummer with a giant drum kit played with sticks held backwards so as to have the thick end out and a keyboardist using a Hammond organ attached to two Leslie speakers (with variable motor speed drives) and a synthesizer capable of producing sub-audible tones plugged directly into the console. Place all of this in a 20 × 25 foot room, hit record and you'll have an idea of what I'm referring to.

Then along came Mr. F. Alton Everest's groundbreaking book in the field of recording studio acoustics and his generosity in not only making available these findings, but doing so in an accurate yet easily understood manner. You'd never understand what a big deal this was unless you've tried to explain to a professional acoustician/noise control engineer that it would not be possible to "simply" decrease the electric guitarist's amplifier level or remove one of two 4 × 12 speaker cabinets that made up his stack, and yes, the electric bass guitarist did feel it was "appropriate to thump" his fingers against the instrument's strings thereby causing those low frequency shock waves of vibration.

My introduction to the previous edition of Mr. Everest's book did not adequately put forth the feeling of gratitude that I hold for him. To further illustrate this point, I will explain what it was like back in 1971 when I was a student of the science of acoustics. At that time one of the better publications in the field was the Soviet acoustician, V.S. Mankovsky's *Acoustics of Studios and Auditoria.* The drill with this book was to read a short amount of text, then labor through a given mathematical equation. Read another line or two of text which varied the acoustical situation slightly then work your way through the resulting more complex equation. The section covering Helmholtz resonators is only 15 pages in length, however, it contains no less than 37 mathematical equations!

Not only did it take me close to two full semesters to work through this book's 360 pages, but being caught, even in the campus library surrounded by pieces of paper covered with mathematics was a serious blow to one's image, what little of it there was. Now do not get me wrong; I have never regretted this arduous undertaking because at its completion, Mr. Mankovsky's teaching method had been entrenched in my mind for good, a deep understanding of the fundamental principles of the art and science of acoustics. Yet what a tremendous joy it was for me and I'll bet a lot of other "recording studio types" when not 8 years later in 1979, F. Alton Everest, building upon the work of a great many previous acousticians including that of V.S. Mankovsky, undertook the simplification of the process of acoustically correct recording studio construction with the first publication of *How to Build a Small Budget Recording Studio* in which he gave an unheard of 12 room designs that he had actually built and thoroughly tested to boot! In his explanation of Helmholtz resonators Mr. Everest gives all details including construction blueprints using only one very simple equation, with an easily understood explanation and specific examples of how they are to be used all in only 6 pages. Later, in 1981, with the publication of *The Master Handbook of Acoustics* he went further letting everyone in on even more by giving a full description of his experiments along with details of the scientific tests he used to confirm the results. Here, he also expands on his coverage of slot-type Helmholtz resonators and sheds additional light on construction methods with the use of photographs. This ends up adding another 9 pages (half of which is made up of the photographs) on this subject to the original 6, ending up with the same amount of coverage as Mankovsky but far more in tune (pun intended) with the specific requirements of a popular music recording facility, all made easily understandable and compiled in a manner that leaves the reader with a much more useful reference work. I would therefore like to take this opportunity to finally say "Thank you very much Mr. F. Alton Everest" for myself and, I'm sure, many others.

Since the second edition of this book was published 15 years ago, more than a little has changed in the field of small budget recording. The number of studios has multiplied beyond all imagination. Back then, it was surprising to find even a full-time professional musician who had his or her own small recording studio. Now it seems like everyone involved in music has home recording facilities. So much has changed that it made an update to this publication a necessity so that it would retain it's original intent of providing the most accurate and up-to-date information to the small recording studio owner/builder dealing with the constraints of a tight budget.

From my work in consulting during the design stage of various studios whose owners had and were using this publication, I became acutely aware that apparently many people today are not as well versed as I had assumed in the reading and interpretation of construction blueprints. The added chapter (Chapter 20) explains in detail the process of correctly reading blueprints and utilizes as examples drawings used in building the components of the room designs that are actually covered in this text.

The chapter covering acoustical materials which describes them, their functioning, and the criteria by which they are selected has been greatly expanded. Due to the vastly increased demand for acoustic materials, many new and enhanced products that are useful in a variety of acoustical work are now available. Take for example the uncomplicated glass edge trims that can be used to surround studio windows to aid in vibration isolation and avoid post-installation breakage caused by structural shifting. They have now become complete multichore handling systems within themselves, with several layers of shock and vibration isolation along with internal spring-like self-attachment cores and even decorative finishes. These are now covered in depth with this edition.

The reader is given examples of some of the latest in new materials that enable the use of cost-easing methods in terms of construction labor and the fastening techniques that they make possible. Included is the use of acoustic "kits" that aim their sights on achieving the complete correction of almost every audio problem or imbalance within the studio control room. The list of manufacturers who make materials used for acoustical work, which had already been *the* most extensive anywhere, had to be increased to include many of the newest manufacturers and the latest offerings available.

Over the years I've noticed that there is a dearth of information available to the small budget studio owner concerning the often very difficult task of dealing with vibration. Therefore, the section explaining vibrations and how to control them has been greatly expanded. It now goes even further in giving information covering isolation as achieved by both static materials and dynamic isolators, along with a full in-depth explanation of the methodology used to tackle these problems professionally. Obviously the list of manufacturers also had to be extended to provide greater depth of coverage as far as materials and systems used for vibration isolation.

The electroacoustics chapter has been enlarged to include the latest in modern computer-controlled acoustical test equipment. The chapter illustrates how these tools can be utilized by even those not part of the scientific community. It offers a window into what the near future holds for the studio owner concerning this highly important field where personal computer– and even laptop-based acoustic and vibration measurement via PCMCIA cards is now possible. Consequently, a whole new appendix (D) had to be added just for listing the manufacturers of acoustical test equipment and the products they make available. As part of this chapter on electronic testing, the internal workings of all types of microphones are now explained in an easily understood manner, along with the advantages and shortcomings of each type. Examples are included of the latest, low-cost, yet highly accurate, acoustical test microphones which have now been made affordable to the even budget studio owner/builder and are suitable for both test *and* recording usage.

MY STUDIO—HOW BIG
AND WHAT SHAPE?

Feature: Understanding the room resonance problem.

CONTENTS AT A GLANCE

How Big Should a Studio Be?

Distribution of Modes

Deciding on Best Studio
Proportions

Studio Size and Low Frequency
Response

Room Cutoff Frequency

Summary of Room Mode Effects

The radio broadcasting industry nourished the "talk booth" concept in the early years. The size of the speech studio only needed to be large enough to accommodate one person (or possibly two for an interview), a table, a few chairs, and a microphone. This sanctified telephone booth-sized studios with very serious built-in acoustical problems. To understand the reason for these problems, we must realize that a roomful of air is a very complex acoustical vibrating system. In fact, it is a series of many resonant systems superimposed upon each other forming a super complex problem.

Let us consider a rectangular studio 12 feet high, 16 feet wide, and 20 feet long (ratios 3:4:5) as sketched in Fig. 1-1. First we shall pay attention to the two opposite and parallel N-S walls, neglecting for the moment the effects of all the other surfaces. Even though acoustically treated to some extent, some sound is reflected from these surfaces. For sound to travel a distance of one round trip between the two walls, or 2L feet, takes a certain, finite length of time. This time is determined by the velocity of sound which is about 1130 feet per second (about 770 miles per hour). At a frequency of 1130/2L Hz, this pair of opposing, parallel walls L feet apart comes into a resonance condition and a standing wave is set up. For example, in the N-S pair of walls in Fig. 1-1, L = 20 feet and the frequency of resonance is 1130/40 or approximately 28 Hz.

This resonance effect also appears at every multiple of 2L. Harmonics of 28 Hz appear at 56, 84, 112, 140 ..... etc. Hz. Although the term "harmonics" is not precisely accurate in this context, we'll use it for convenience. The single pair of walls then gives a fundamental frequency and a train of harmonics at which resonance effects also occur, called axial mode room resonances. The E-W walls (Fig. 1-1) give another fundamental frequency of 35 Hz and a train of harmonics. The floor-ceiling combination gives a fundamental frequency of about 47 Hz and a third series of harmonics. These modal frequencies determine the *sound* of a room. They yield bad effects only if they pile up at certain frequencies or are spaced too far apart, as we shall see later.

We have considered only axial modes of this room involving a single pair of surfaces. There are also tangential modes involving two pairs of surfaces and oblique modes involving three pairs of surfaces which have still different fundamental frequencies and harmonics (Fig. 1-2). Taken all together, these make the sound field of an enclosed space extremely complex. Fortunately, the effect of tangential and oblique modes is less than that of the axial

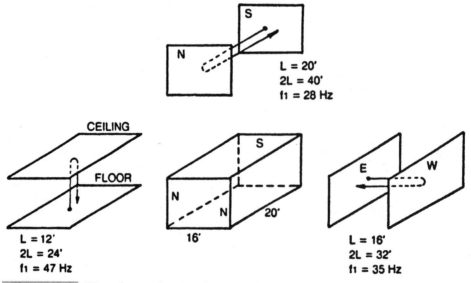

Figure 1-1 The six surfaces of a rectangular room are broken down into three pairs of opposite and parallel surfaces when considering axial mode room resonances. Each pair has its fundamental resonance frequency and train of harmonics.

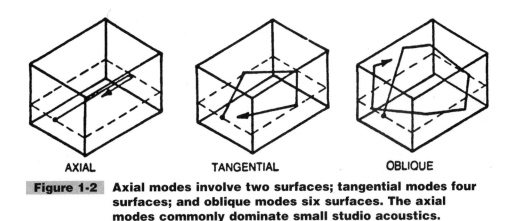

AXIAL   TANGENTIAL   OBLIQUE

Figure 1-2 **Axial modes involve two surfaces; tangential modes four surfaces; and oblique modes six surfaces. The axial modes commonly dominate small studio acoustics.**

modes. Basically, you can get by quite easily and reasonably accurately in designing a studio by considering only the axial modes, although you must depend on the resonance effects of tangential and oblique modes to do a certain amount of filling in between the axial modes.

Because of the standing wave effect the sound pressure at 28 Hz is far from uniform across the room, being very high near the surface of the N and S walls and zero at the center of the room. The situation is very much like the inside of an organ pipe closed at both ends. At the second harmonic near 56 Hz, however, the distribution of sound pressure in the standing wave is quite different, with two null points and a maximum in the center of the room as well as at each wall surface. When complex sounds of speech or music excite the fundamental and harmonics of a single series there is extreme complexity between a single pair of wall surfaces as the standing wave patterns shift. Adding the effects of the E-W and vertical modal series and adding the tangential and oblique modes results in a constantly shifting music or voice sound field, the complexity of which defies description.[1]

How Big Should a Studio Be?

Each room resonance frequency has a certain bandwidth (or Q).[6] The ideal situation is to have adjacent resonances (fundamentals and harmonics) locked arm in arm with neighbors through these resonance skirts. This results in signal components of constantly fluctuating frequency being treated uniformly. If the spacing of these resonances is too great, some of the precious signal energy is boosted by resonances, and some which "fall in the cracks" is discriminated against. On the other hand, if three or four room resonances occur at the same frequency or are very close together, signal energy in this part of the spectrum receives an abnormal boost. Such pileups are inevitably accompanied by gaps elsewhere in the spectrum. Good studio sound requires careful attention to these resonance frequencies which are, in turn, controlled by room dimensions and proportions.

Room size determines how the low frequencies are treated. The larger the room, the lower the frequency components the room can support. Small rooms result in great spacing of

[1]All references are found at the conclusion of this book.

room modes. A talk booth of 6 feet × 8 feet cannot support sound lower than about 70 Hz. Even though there is little energy in voice below 150 Hz, such a small room is unsuitable for recording because of excessive mode spacing. The BBC has concluded that any studio of less than 1500 cubic feet is not practical. Any saving in construction cost is outweighed by cost of correcting acoustical deficiencies—and usually successful correction of deficiencies is not feasible.

Distribution Of Modes

We have considered the three axial frequencies and the three series of harmonics of the studio of Fig. 1-1. Now we must ask the question, "Are these frequencies properly distributed?" To answer this, each frequency must be computed and examined. This may be a bit tedious, but it involves only the simplest mathematics.

A convenient approach is to tabulate the series for each of the three room dimensions, such as in Table 1-1. The 141.3 Hz and 282.5 Hz occur in each column, and they bear a 2:1 relationship to each other. These coincident frequencies are called *degeneracies*. How well the other modal frequencies are distributed overall is difficult to see from columns of figures.

In Fig. 1-3 each modal frequency is plotted on a linear frequency scale. Each mode is represented by a vertical line, although actually each one has an average bandwidth of about 5 Hz (as shown for the lowest mode of Fig. 1-3). The triple coincident frequencies at 141.3 and 282.5 Hz are greatly spaced from their neighbors. The piling up at 141.3 and 282.5 Hz means that signal energy near these frequencies will be unnaturally boosted.

TABLE 1-1 AXIAL RESONANCE FREQUENCIES			
	N-S WALLS (2L = 40 FT.)	E-W WALLS (2L = 32 FT.)	FLOOR-CEILING (2L = 24 FT.)
f1 (fundamental)	28.3 Hz	35.3 Hz	47.1 Hz
f2 (2nd harmonic)	56.5	70.6	94.2
f3 (3rd harmonic, etc.)	84.8	105.9	(141.3)
f4	113.0	(141.3)	188.3
f5	(141.3)	176.6	235.4
f6	169.5	211.9	(282.5)
f7	197.8	247.2	329.6
f8	226.0	(282.5)	
f9	254.3	317.8	
f10	(282.5)		
f11	310.8		

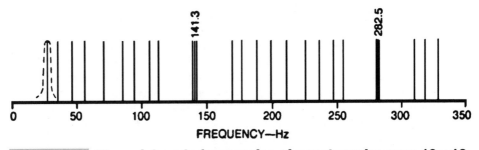

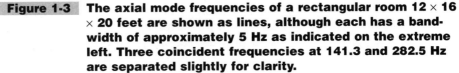

Figure 1-3 The axial mode frequencies of a rectangular room 12 × 16 × 20 feet are shown as lines, although each has a bandwidth of approximately 5 Hz as indicated on the extreme left. Three coincident frequencies at 141.3 and 282.5 Hz are separated slightly for clarity.

Also, the great separation from neighboring modal frequencies guarantees that signal components in these gaps will be unnaturally depressed. This adds up to almost certain audible colorations at these two frequencies that are monotonous, and repetitive blasts of energy which distort music and are particularly obnoxious in speech.

Deciding on Best Studio Proportions

The 3:4:5 proportion of Figs. 1-1 and 1-3 were, at one time, highly recommended but they are ill-suited for studio construction because of poor modal distribution. What room proportions should be used? Numerous studies have been made on this subject. Three suggestions from each of two authors are presented in Table 1-2. In this table, studio dimensions following the suggested ratios are included based on a ceiling height of 10 feet. The studio volume, which varies from example to example, is included for each ratio.

For a visual comparison of these ratios, the fundamental and harmonic series is plotted for each in Fig. 1-4. Although there are scattered double coincidences, all six show better distribution than the unfortunate 3:4:5 choice of Fig. 1-1. Yet none shows the equally spaced modal frequencies desired in our ideal studio. Nor should this bother you too much. It is just as bad to place too much emphasis on room proportions as it is to neglect them completely.

The presence of people and furnishings in a room so affects the actual modal frequencies that it is futile to worry over minor deviations from some assumed optimum condition. A practical approach is to be alert to problems of coincidence and mode spacing; to do what can be done to optimize them and then relax. If an existing space is being considered as a studio or control room, check the modal frequencies after the fashion of Table 1-1 and plot them as in Fig. 1-4 to see if serious problems exist. For new construction, do the same for proposed dimensions of any sound-sensitive rooms. A liberal education in axial modes awaits the persevering student who varies one dimension of a room (on paper) while holding the others constant and noting how the modal distribution is affected. You'll soon conclude that eliminating coincidences is a major victory—beyond that, there is little to be gained.

In Table 1-2, the ceiling heights of the room are held constant and volume is allowed to vary. If the several proportions were adjusted so that the volume remained constant, the ratios would draw closer together. This is illustrated in Table 1-3 in which the ratios of Table 1-2 are adjusted so that all volumes are the same as that corresponding to the first ratio. In other words, when rooms of the same volume but of different proportions are considered, the six ratio examples we have been studying are not as different from each other as they first appeared to be. However, they are different, relatively speaking, and you cannot escape the fact that smaller dimensions yield higher fundamental frequencies.

The fundamental resonance frequencies and harmonics of the ratios of Table 1-3 adjusted for the same room volume are plotted in Fig. 1-5. Figures 1-4A and 1-5A are, of course, identical as they represent the common point of comparison in the two cases. Although frequencies are shifted, a family resemblance can be seen when comparing Figs. 1-4B and 1-5B, Figs. 1-4C and 1-5C, and other corresponding pairs. The coincident or nearly coincident pairs of Fig. 1-4 still exist in Fig. 1-5. In general, however, the larger rooms of Fig. 1-4 have the advantage of yielding closer average spacings.

Studio Size and Low Frequency Response

An inspection of the left edge of the six plots of Fig. 1-4 reveals that the larger rooms have lower fundamental frequencies than the smaller ones. In Fig. 1-6 the fundamental room resonances corresponding to the longest dimension of the six cases are plotted against room volume. For the smallest studio having a volume of 1585 cubic feet (Fig. 1-4A), the lowest signal frequency which would have the advantage of resonance assistance is about 41 Hz. The studio of Fig. 1-4C, with its volume of 3728 cubic feet can handle signal components down to 24 Hz. For speech purposes, the smaller studio is quite adequate, because speech energy below 40 Hz is extremely low. The 3728 cubic foot studio adds almost another octave in the lows, which is quite advantageous for music recording.

TABLE 1-2 STUDIO PROPORTIONS					
	RATIOS	**CEILING HEIGHT, FT.**	**LENGTH, FT.**	**VOLUME WIDTH, FT.**	**CUBIC FT.**
(A)[7]	1:1.14:1.39	10.0	13.9	11.4	1585
(B)[7]	1:1.28:1.54	10.0	15.4	12.8	1971
(C)[7]	1:1.60:2.33	10.0	23.3	16.0	3728
(D)[8]	1:1.90:1.40	10.0	19.0	14.0	2660
(E)[8]	1:1.90:1.30	10.0	19.0	13.0	2470
(F)[8]	1:1.50:2.10	10.0	21.0	15.0	3150

TABLE 1-3 VARYING PROPORTIONS WITH THE SAME VOLUME

	RATIOS	CEILING HEIGHT, FT.	LENGTH FT.	WIDTH FT.	VOLUME CUBIC FT.
(A)	1:1.14:1.39	10.0	13.9	11.4	1585
(B)	0.93:1.19:1.43	9.3	14.3	11.9	1585
(C)	0.75:1.20:1.75	7.5	17.5	12.0	1585
(D)	0.84:1.60:1.18	8.4	16.0	11.8	1585
(E)	0.86:1.64:1.12	8.6	16.4	11.2	1585
(F)	0.80:1191.67	8.0	16.7	11.9	1585

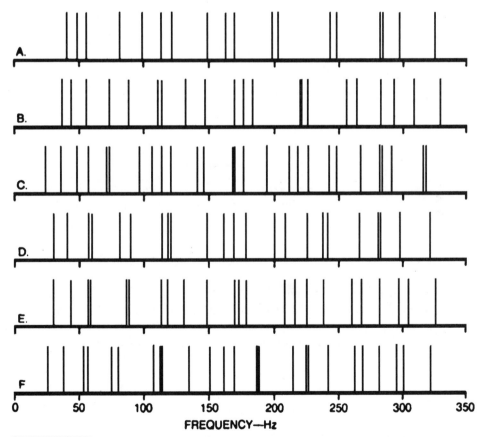

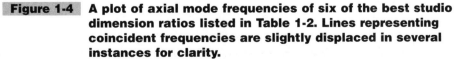

Figure 1-4 A plot of axial mode frequencies of six of the best studio dimension ratios listed in Table 1-2. Lines representing coincident frequencies are slightly displaced in several instances for clarity.

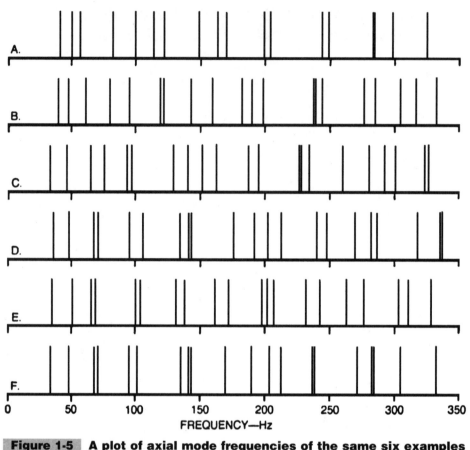

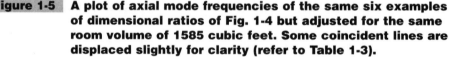

Figure 1-5 A plot of axial mode frequencies of the same six examples of dimensional ratios of Fig. 1-4 but adjusted for the same room volume of 1585 cubic feet. Some coincident lines are displaced slightly for clarity (refer to Table 1-3).

This discussion has been based on the axial modes of a room. The tangential and oblique modes have somewhat longer paths and hence would tend to extend the low frequency limit of a room. While axial modes have two reflections per round trip, tangential modes have four and oblique modes have six (refer again to Fig. 1-2). As energy is lost at each reflection, the reason for the lower amplitudes of tangential and oblique modal resonances is apparent.

However, axial reflections are perpendicular to the surfaces, the angle of incidence which gives the most efficient absorption. Tangential and oblique modal paths, while involving more reflections, are usually at smaller angles of incidence. This results in less absorption at each reflection. The limit is grazing incidence at which absorption is very small. Thus, the number of reflections for tangential and oblique is greater than for axial modes, but, acting in the other direction, there is the fact that the smaller angles of incidence result in less loss per reflection than the 90° incidence of axial modes. How do these opposing factors add up? Since the axial modes are more dominant than the tangential and

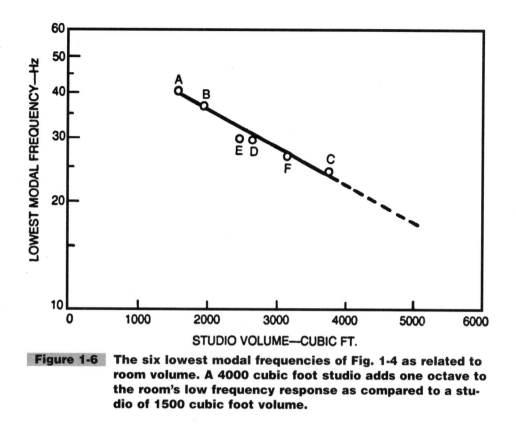

Figure 1-6 The six lowest modal frequencies of Fig. 1-4 as related to room volume. A 4000 cubic foot studio adds one octave to the room's low frequency response as compared to a studio of 1500 cubic foot volume.

oblique modes, the practical extension of a room's low frequency response due to tangential and oblique modes is limited.

ROOM CUTOFF FREQUENCY

Every studio has some frequency above which the modal frequencies are close enough together to merge into a statistical continuum. This is called the cutoff frequency of the room. At frequencies higher than the cutoff frequency, various components of the signal will be treated more or less uniformly and the room will act more like a large auditorium. At frequencies below cutoff, excessive spacing of modes exists with resulting uneven treatment of signal components.

The cutoff frequency of a room is a function only of its reverberation time and volume and may be computed approximately from the following statement[9]:

$$\text{Room cutoff frequency} = 20{,}000 \sqrt{\frac{T}{V}}$$

where

T = reverberation time, seconds

V = volume of room, cubic feet

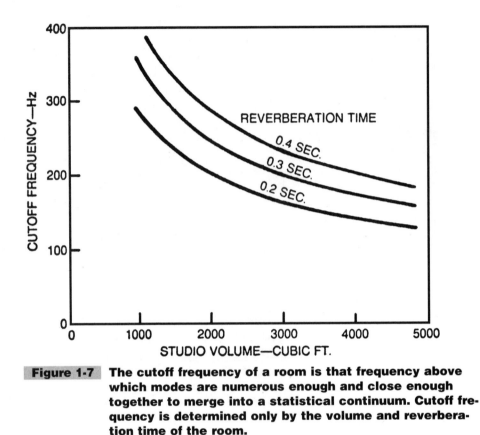

Figure 1-7 The cutoff frequency of a room is that frequency above which modes are numerous enough and close enough together to merge into a statistical continuum. Cutoff frequency is determined only by the volume and reverberation time of the room.

Figure 1-7 is computed from this basic statement for volumes and reverberation times common in small studios. The larger and the more dead a studio, the lower the cutoff frequency and the less difficulty in handling the low end of the audible spectrum. In Figs. 1-3, 1-4, and 1-5 the plots of modal frequencies were terminated at about 300 Hz, which approximates the cutoff frequency of the average small studio above which modes are increasingly closer together.

Summary of Room Mode Effects

- Rooms smaller than 1500 cubic feet are subject to insurmountable room mode problems and should be avoided for quality recording studios and control rooms. The larger the room, the closer the average spacing of modal frequencies and the more uniform the treatment of the various components of the signal.
- In choosing studio proportions, try to eliminate coincident frequencies below 300 Hz. Multiple coincidences in the region below 200 Hz are more apt to cause audible colorations than those between 200 and 300 Hz.

■ Once multiple coincident frequencies are reduced or eliminated it is futile to carry modal analyses to extremes because occupants, furnishings, and other room irregularities result in great deviations from the idealized paper condition.

■ The splaying of studio walls may be helpful in reducing flutter echo, but modal frequencies are only shifted to some unknown values. Splaying may also tend to break up degeneracies (coincident frequencies) in an otherwise symmetrical room, but keeping things under control by choosing proper proportions of a rectangular room is a satisfactory approach if flutter echo and diffusion are controlled by proper placement of absorbing material.

■ What can be done to treat an unavoidable coloration in a studio? A tuned Helmholtz resonator can be introduced to tame the coincident frequency. The sharpness of tuning may have to be controlled to avoid the slow decay of sound in a high-Q structure. The construction and tuning of such absorbers is discussed in later sections.

ELEMENTS COMMON
TO ALL STUDIOS

*Features: Sound lock treatment, doors and sealing, combatting air condi-
tioner noise, wall constructions, floor-ceiling constructions,
wiring precautions, lighting, observation windows, the permit.*

Each of the 12 studio plans studied in this book has certain elements in common with all the others. For example, all require protection from interfering noise, whether it originates outside or inside the studio. All studios require an observation window for visual contact between the control operator and those in the studio. It would be ridiculous to repeat descriptions of each of these common elements a dozen times, so this chapter discusses them all.

Sound Lock Acoustical Treatment

A typical sound lock corridor places two doors in series and two walls in series so that external noise must traverse both to penetrate to the quiet rooms. Many examples of sound locks sharing common principles and differing only in details are seen in the studio plans.

Functionally, the sound lock both isolates and absorbs. Wall construction (one way of achieving isolation), and the treatment of doors and door seals are considered later in this chapter.

As you enter a sound lock from the outside, the exterior noise momentarily floods the corridor. A sound level reading outside the open door is higher than one inside the sound lock, even with the door open, if the interior surfaces are highly absorbent. The sound transmitted through a sound lock corridor is significantly reduced if the corridor surfaces are properly treated.

Consider an untreated corridor (average absorption coefficient, say, 0.1) and the same corridor with surfaces treated (average absorption coefficient 0.9). By adding absorbing material the noise level in the sound lock is reduced as much as 9.5 dB from the bare condition. Even if both the exterior door and the studio door were open at the same time (an unusual occurrence), the exterior noise level in the studio would be reduced by about this amount due to the inner treatment of the sound lock alone. This is why it is so important to make sound lock surfaces as absorbent as possible.

Although sound locks come in all shapes and sizes, the treatment of Fig. 2-1 is representative. Heavy carpet and pad is the almost universal solution for the sound lock floor. Of the numerous approaches for the ceiling, the ubiquitous suspended ceiling supported by a T-bar grid offers many advantages. The better grade of lay-in panels offer absorption coefficients averaging 0.75 to 0.85 throughout the 125 Hz–4 kHz band. They give excellent low frequency absorption which is further increased by introducing a thick layer of household insulation into the cavity above the lay-in panels. Manufacturers specify coefficients for the standard Mounting 7 (or E-405 in the new system) in which the lay-in ceiling is 16 inches below the structural ceiling. Other distances may be used, of course, with some modest change in absorption. A lay-in ceiling of this type also provides an ideal hidden location for air conditioning ducts and electrical service runs.

Another approach is covering ceilings and upper walls with common acoustical tile or 2 to 4 inches of dense glass fiber boards. The latter require some sort of protective cover such as expanded metal, wire screen, or loosely-woven cloth. Cloth has the advantage of controlling the sloughing off of tiny irritating glass fibers.

Instead of old-fashioned acoustical tile or glass fiber with a protective cover, one of the new foam products such as Sonex (Illbruck), Sound-Sorber (Discrete Technology), or Acoustafoam (FM Tubecraft) may be used (see Appendix D for manufacturers' addresses). They are more expensive, but labor-saving. Their dramatic and colorful appearance may provide an intangible advantage in the reactions of clients or visitors.

The lower walls (wainscoting) must withstand considerable mechanical abrasion. There are dozens of proprietary panels which serve well in this location, but all are expensive. One straightforward and relatively inexpensive approach is to mount panels of Tectum (see Tectum and Gold Bond Products, Appendix D), a structural board of wood fiber with a cement binder. It stands up well under abuse and can be painted with a nonbridging paint

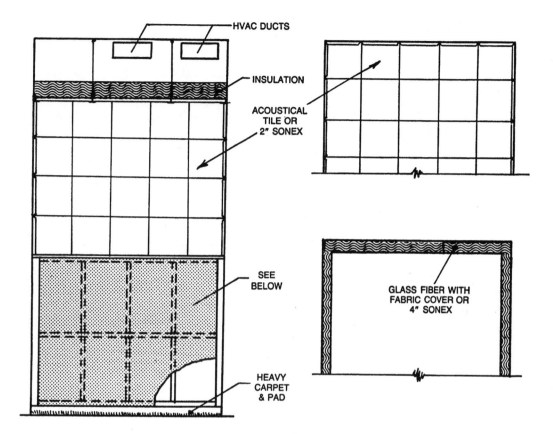

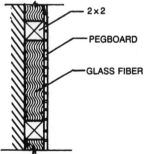

WAINSCOT DETAIL

Figure 2-1 Suggestions for the acoustical treatment of the sound lock corridor. To minimize noise being transmitted through the sound lock it should be very absorbent, although the requirements otherwise are not critical.

with minimum effect on absorbing properties. A 1-inch layer of Tectum over a glass fiber-filled air space obtained by furring out on 2 × 2s gives good midband absorption as shown in Fig. 2-2. The panel acts as a diaphragmatic absorber.

Figure 2-1 shows another approach utilizing 2 × 2 furring and ordinary pegboard. The sound absorption of pegboard backed by glass fiber is shown in Fig. 2-2. The Tectum arrangement is less sharply tuned than the pegboard, which is really a Helmholtz absorber. Although the Tectum absorbs less at low frequencies, above 500 Hz it is reasonably uniform throughout the important audible frequencies. The absorption coefficients of Fig. 2-2 are only for the purposes of comparison. There is little justification for spending time on calculating the absorption of a sound lock corridor.

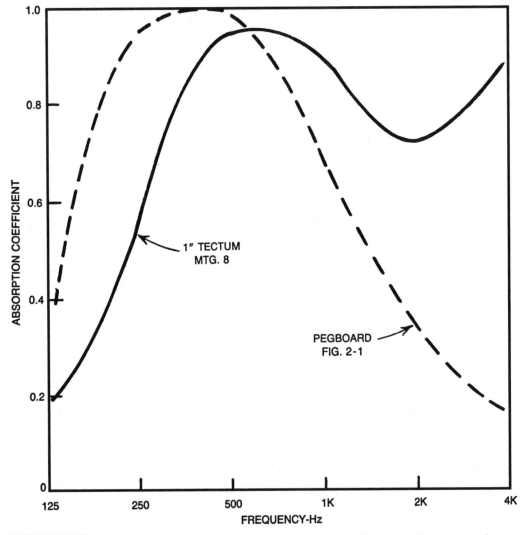

Figure 2-2 **Approximate absorption characteristics of 1-inch Tectum and pegboard wainscot treatments.**

Acoustical Doors

The access to each studio-control room suite should be via a sound lock. All doors should open into the sound lock. The use of a sound lock relaxes door requirements since heavy loaded and laminated doors with awkward ice box type of clamping hardware are eliminated. However, at the opposite extreme, ordinary household doors are far too thin acoustically to be acceptable. A reasonably inexpensive intermediate solution is 1¾ inch solid core doors which are readily available. Such doors have solid wood or particleboard cores which provide the density necessary to impede the flow of sound through them. A typical particleboard core faced with ⅛ inch hardboard has a density of 5.3 pounds per square foot of surface. This gives a transmission loss of 33 dB at 500 Hz on a mass law basis. In common with other barriers, its transmission loss is less than this amount below 500 Hz and greater above 500 Hz. This would be insufficient alone, but the sound lock principle places two such doors in series between a high noise area (exterior or control room monitor loudspeaker) and the studio.

A 10 inch × 10 inch window in each door (Fig. 2-3) reduces the possibility of injury by a door opened suddenly. It also allows visual appraisal of the situation in a room before entering. Such a window of ¼-inch plate glass, adequately sealed, does not seriously deteriorate the acoustical quality of the door but adds materially to functional efficiency. Some people think that upholstering a door makes it "more acoustical." As far as transmission loss through the door is concerned, vinyl cloth covering over a sheet of foam rubber and studded with large-headed tacks is a waste of time and effort. However, such a covering does offer sufficient high-frequency absorption to reduce unwanted reflections or flutter echoes.

Weatherstripping Doors

A hermetically sealed door is the ideal type from an acoustical standpoint. Such a condition can be approached in sound lock doors by careful installation of common weatherstripping materials. These are available in a host of different configurations. Figure 2-4 illustrates representative types. Strips of foam or felt in the form of metal backed strips, rolled beading or gummed strips are shown in Fig. 2-4A. Strip magnets, enclosed in flexible plastic as shown in Fig. 2-4B, are attracted to mild steel strips imbedded in the door, effectively sealing the opening. These are commonly used in refrigerator doors and they can provide an excellent acoustical seal.

The sides and top of a door are easier to seal than the bottom. A wiping rubber seal is held in place by a shaped metal or plastic retainer in Fig. 2-4C. In Fig. 2-4D, a mechanical threshold closer automatically lifts as the door is opened and drops on closure. The types of weatherstripping in Fig. 2-4A are usually considered impractical for door bottoms because in such a position they take a beating from passing shoes. For this reason, one of the types shown in B, C, or D are far more effective at the threshold.

All of the door sealing methods pictured in Fig. 2-4 require periodic inspection to assure a continuing tight seal. Occasional adjustments are required by some and complete

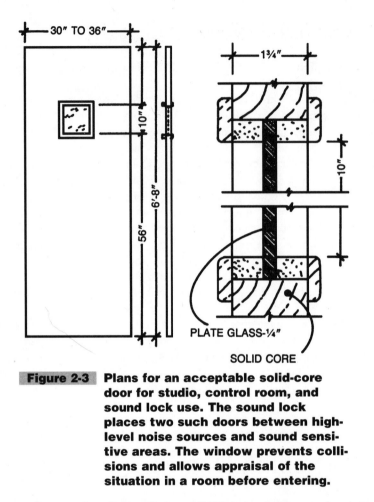

Figure 2-3 Plans for an acceptable solid-core door for studio, control room, and sound lock use. The sound lock places two such doors between high-level noise sources and sound sensitive areas. The window prevents collisions and allows appraisal of the situation in a room before entering.

replacement by others as foam and felt deteriorate. Door seal maintenance should be added to the studio maintenance reminder, list which normally stops at such things as amplifier response, recorder head alignment and cleaning, etc.

HVAC Noise

It is a difficult word to pronounce, but HVAC stands for heating, ventilating, and air conditioning facilities. In this book the term *air conditioning* (A/C) includes heating and ventilating. With proper advance precautions, A/C equipment noise in the sound-sensitive studio and control room areas need not be a problem. In practice, it is very often a problem because the higher standards required in recording studios are infrequently encountered by architects, building contractors, and air conditioning equipment suppliers and installers.

What background noise level is acceptable in recording studios? Single figure noise levels, lumping the entire audible frequency range together, are of very limited value. For this

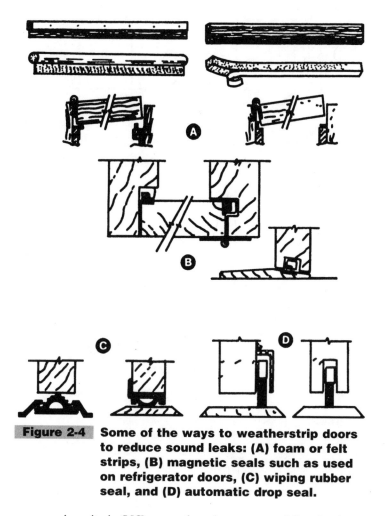

Figure 2-4 **Some of the ways to weatherstrip doors
to reduce sound leaks: (A) foam or felt
strips, (B) magnetic seals such as used
on refrigerator doors, (C) wiping rubber
seal, and (D) automatic drop seal.**

reason noise criteria (NC) curves have been proposed for adoption as standards. Figure 2-5
illustrates a family of such curves.

The downward slope of these curves reflects both the ear's increasing sensitivity with
increasing frequency and the spectral shape of common noises decreasing with increasing
frequency.

An NC-15 contour is a reasonably stringent, though usually quite attainable, noise specification and goal for recording studios. An NC-20 contour is a more relaxed specification
for less critical recording. The sound analysis contours of Fig. 2-5 are for octave bands of
noise. Sometimes single frequency hums or whines associated with motors or fans stand
out prominently from the general background of distributed noise. The peaks of such single frequency components determine the NC rating contour applicable. Thus, if the noise
levels in all bands are below the NC-15 contour except one which reached the NC-30 contour, the NC-30 applies to that noise rating. The contours of Fig. 2-5 offer a means of specifying the maximum permissible noise from air conditioning facilities or from other
sources.

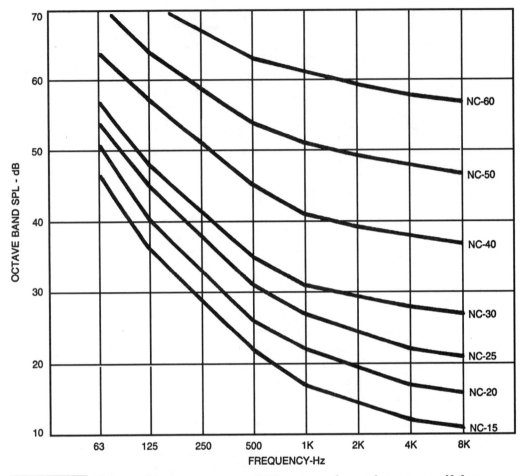

Figure 2-5 **Noise criteria curves useful in assessing noise or specifying maximum allowable noise levels in studios and other sound-sensitive areas.**

The ducting arrangement of Fig. 2-6A illustrates a method of reducing crosstalk between adjacent rooms through the duct from grille to grille. By avoiding serving adjacent rooms directly from the same duct (supply or exhaust), the duct path is lengthened and attenuation via the duct path increased. The greater the length of lined duct and the more lined bends between one grille and the next, the greater the attenuation and the less the crosstalk.

Figure 2-6B isolates the effect of absorptive duct lining on the attenuation of sound through the duct. Note that high-frequency noise is absorbed better than low-frequency noise. The 2-inch thick liner attenuates sound down the duct better than 1-inch thickness.

Figure 2-6C pictures the effect of a lined duct bend. The attenuation is greater at high frequencies than low, unfortunately similar to the straight lined duct. With an arbitrary rule that a minimum of 15 feet of lined duct and two lined bends must be installed between any two adjacent rooms, what kind of attenuation of crosstalk from room to room can you expect? The curves help a rough estimate of the following:

	ATTENUATION		
	250 HZ	**1 KHZ**	**4 KHZ**
15' lined duct (1" lining)	15 dB	33 dB	27 dB
2 lined bends	4	10	20
Total	19 dB	43 dB	47 db

This dramatically illustrates the fact that duct attenuation is easier to achieve at high frequencies than in the low-frequency region. Placing 30 feet of ducting between rooms increases the total 250 Hz attenuation to 34 dB. Hums and whines often occur in the 125 or 250 Hz bands. Another possibility not detailed here is to use a sharply tuned duct stub (acoustical band reject filter) to attenuate a troublesome single-frequency noise component.

It is usually easier to combat noises at their source than to introduce adequate attenuation downstream. The lined plenum of Fig. 2-6D, which offers excellent attenuation across the band, can be inserted in the duct between the A/C machinery and the sound sensitive areas. Baffles in the plenum increase the high-frequency attenuation, although as noted above, adequate high-frequency attenuation is usually available elsewhere. In some installations a large step-inside plenum or air chest is part of the basic equipment. If this plenum is lined with acoustically absorbent material, the noise traveling down the ducts is greatly reduced and the larger the plenum, the greater the low-frequency attenuation.

Air velocity is commonly higher in the budget A/C installations than in the more professional jobs. The air velocity should never exceed 500 feet per minute at the grilles to avoid generating excessive hissing noise due to air turbulence.

Wall Construction

The common 2×4 frame wall construction with a single layer of gypsum drywall on both sides is shown in Fig. 2-7A. It is included as a point of comparison only; its sound transmission class rating of STC 34-38 is too low for walls contingent with studios and control rooms. However, it could be used for sound lock walls not touching studios and control rooms. This wall can teach us several things. (Sound transmission loss of a wall varies with frequency. To achieve a convenient rating scheme a standard contour has been adopted which, when applied to the actual characteristic of a wall, results in a single number rating called the *Sound Transmission Class*, STC. Think of the STC rating number as a sort of average midband transmission loss in dB.)

For example, insulation in the cavity of this type of wall, with both faces closely coupled by being nailed to the same 2×4s, increases transmission loss only modestly. Another problem is that both faces vibrate as diaphragms. As they are identical they resonate at the same frequency and at this frequency there is a sort of acoustical hole in the wall. By making the faces of different thicknesses, of different densities, or supporting them in different ways, the two resonances are made to occur at different frequencies, improving the wall performance. The caulking of the perimeter, of at least one face but preferably both, seals the tiny cracks that are inevitable in normal construction.

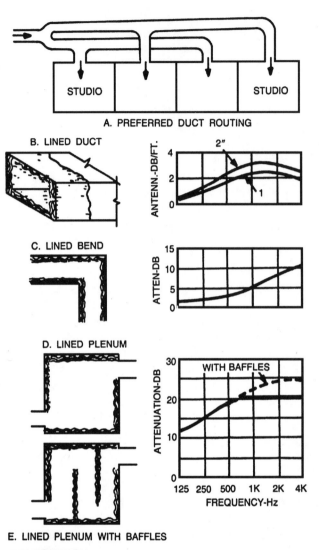

A. PREFERRED DUCT ROUTING

B. LINED DUCT

C. LINED BEND

D. LINED PLENUM

E. LINED PLENUM WITH BAFFLES

Figure 2-6 **Means of reduction of air conditioner noise in studios and the reduction of crosstalk between adjacent rooms via a duct path.**

Resilient Mounting

There is a very substantial improvement in transmission loss if gypsum board on one face is screwed to a resilient channel as in Fig. 2-7B. This must be done carefully, with screws designed for the purpose, or else the resilient channel might be "shorted out."

For example, if a screw that is too long hits a stud, the resilience is lost. Acoustical elements, cabinets or shelves attached to such walls must be mounted carefully lest the extra

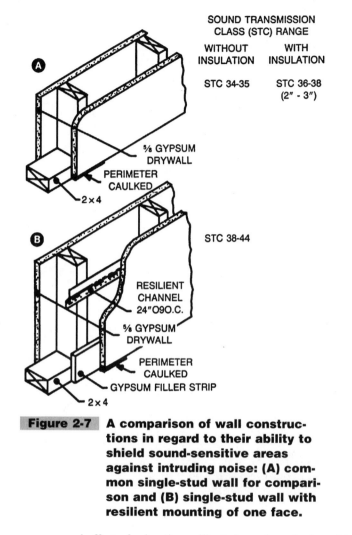

SOUND TRANSMISSION
CLASS (STC) RANGE

WITHOUT INSULATION	WITH INSULATION
STC 34-35	STC 36-38 (2″ - 3″)

⅝ GYPSUM DRYWALL

PERIMETER CAULKED

2 × 4

STC 38-44

RESILIENT CHANNEL 24″ O9 O.C.

⅝ GYPSUM DRYWALL

PERIMETER CAULKED

GYPSUM FILLER STRIP

2 × 4

Figure 2-7   **A comparison of wall constructions in regard to their ability to
shield sound-sensitive areas
against intruding noise: (A) common single-stud wall for comparison and (B) single-stud wall with
resilient mounting of one face.**

expense and effort of using the resilient channel are destroyed by solid contact of the gypsum board with the studs.

If a double layer of gypsum board is required on a resiliently mounted wall, the base
layer is attached vertically with screws and the face layer cemented in place following the
manufacturer's recommendations.

Staggered Stud Construction

Figure 2-8A shows typical staggered 2 × 4 studs with 2 × 6 plate. This eliminates the solid
connection of one wall diaphragm with the other except around the periphery. This does
essentially the same thing as mounting one of the wall diaphragms resiliently as in Fig. 2-7B,
and the STC results are somewhat comparable. .

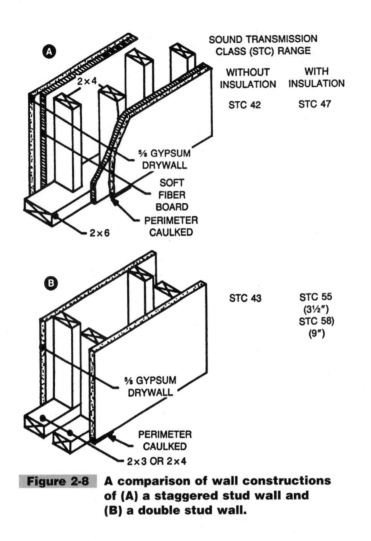

SOUND TRANSMISSION
CLASS (STC) RANGE

WITHOUT INSULATION	WITH INSULATION
STC 42	STC 47

Ⓐ

2 × 4

⅝ GYPSUM DRYWALL

SOFT FIBER BOARD

PERIMETER CAULKED

2 × 6

Ⓑ

STC 43	STC 55 (3½") STC 58) (9")

⅝ GYPSUM DRYWALL

PERIMETER CAULKED

2 × 3 OR 2 × 4

Figure 2-8 **A comparison of wall constructions of (A) a staggered stud wall and (B) a double stud wall.**

An additional feature of Fig. 2-8A is the use of soft fiberboards under each gypsum layer as sound deadeners. Because of their low density such soft boards contribute little to transmission loss directly, but they do serve as frictional elements in dampening vibrations of the gypsum diaphragms.

Nailing is the common method of supporting both the soft base layer and the gypsum face layer. In Fig. 2-8A both sides are identical which, as you have seen, is less desirable than making one side differ from the other. One satisfactory combination is a layer of ⅝-inch gypsum board over a ½-inch layer of soft sound deadening board on one side and a double layer of ⅝-inch gypsum on the other side. With ample insulation fill, a staggered stud wall with such facings adequately caulked comes close to STC-50—a good value for normal studio walls. The effectiveness of the filler insulation depends on thickness, but is independent of density. Therefore, the cheaper household thermal type of insulation is quite adequate for filling acoustical walls.

Double Walls

Double wall frame construction is shown in Fig. 2-8B. There is only a minor difference between walls framed of double 2×3s and 2×4s. The two wall diaphragms are still connected at the periphery through a common foundation (concrete floor?) which is somewhat less coupling than that provided by a common plate in staggered stud construction. The double 2×3 wall, if carefully constructed and sealed, can reach STC-55 to 58 with proper insulation fill.

Concrete and Masonry Walls

In new construction and in some cases of renovation, concrete or masonry walls are a viable choice. Ratings that apply to several common walls are found in Table 2-1. Practical concrete and masonry walls are quite comparable to framed walls in their STC ratings. The hard walls are somewhat inferior in another way—that of efficiently conducting structure-borne impulse noises from afar and reradiating them into sound-sensitive areas.

Floor-Ceiling Construction

If the space above a studio is occupied and people are stalking around with hardheeled shoes, the situation calls for careful attention. Impulsive sounds of this type penetrate

TABLE 2-1 TRANSMISSION LOSS IN CONCRETE AND MASONRY WALLS	
WALL	**SOUND TRANSMISSION CLASS**
Concrete—4 inches	STC-48
Concrete—8 inches	STC-52
Concrete blocks—4 inches	STC-40
Painted both sides	STC-44
Plastered both sides	STC-44
Concrete block—8 inches	STC-45
Painted both sides	STC-46-48
Plastered one side	STC-52
Concrete block—8 inches	
Voids filled with well-rodded concrete and plastered both sides	STC-56

to an extent that noises of other types seem tame by comparison. Floor-ceiling construction becomes very important in such cases. The construction in Fig. 2-9A is very common and the carpet helps to reduce footfall noise. The STC-42 rating, however, is marginal for most budget studios. The construction in Fig. 2-9B yields STC-51 by adding a resilient ceiling below and some insulation in the cavity between the floor joists.

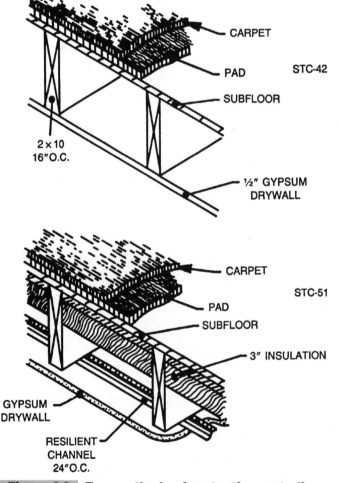

Figure 2-9 **Two methods of protecting a studio from noise from the floor above with frame construction: (A) with normal gypsum board ceiling and (B) with resiliently mounted ceiling and insulation in the air space.**

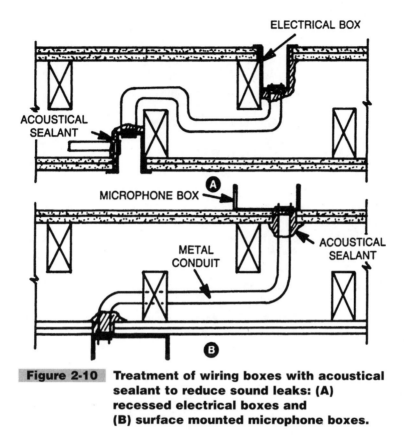

ELECTRICAL BOX

ACOUSTICAL
SEALANT

MICROPHONE BOX

Ⓐ

METAL
CONDUIT

ACOUSTICAL
SEALANT

Ⓑ

Figure 2-10 **Treatment of wiring boxes with acoustical sealant to reduce sound leaks: (A) recessed electrical boxes and (B) surface mounted microphone boxes.**

Electrical Wiring

Building a 50 dB wall and then loosely mounting electrical boxes back to back is an exercise in futility. A surprising amount of sound can leak through a very tiny opening and through small areas of thin spots in a wall. Electrical boxes are necessary, however, and Fig. 2-10A suggests staggering them and using copious quantities of acoustical sealant to seal openings and beef up the boxes. Surface boxes for microphone connectors reduce compromising the wall and they may be handled as shown in Fig. 2-10B. In addition to sealant at the ends of the metal conduit, it is a good idea to also pack glass fiber tightly around the audio pairs to avoid sound traveling through the conduit itself.

Lighting

If fluorescent lighting is considered, the ballast reactors should be removed from the fixtures and mounted in a metal box in the sound lock or completely outside the suite. Although this takes more wire, it removes the electrical and acoustical buzzes these reactors are famous for generating outside the recording and sound evaluating areas.

Track lighting fixtures have the advantage of flexibility in concentrating the light where it is needed and hiding the light source from the eyes of those in the other room. This is the proper way to eliminate troublesome reflections in the observation window glass.

Light dimmers of the selenium controlled rectifier type create electrical noises which are likely to give trouble in the low-level microphone circuits.

Observation Window

An observation window plan for staggered stud and double wall construction is shown in Fig. 2-11A. The window frame is in two parts, one nailed to the studs on the control room side and the other to the studs on the studio side. In this way the glass on each side is an extension of its own wall and has no solid connection to the other side. Positioning felt or sponge strips in this gap between the two frames prevents accidental solid contact between them. A comparable wall for single-leaf construction is shown in Fig. 2-11B. Beads of nonhardening acoustical sealant seal off tiny cracks between the window frame and the walls in both types.

Each glass plate is a resonant system as is the air cavity between. By using glass plates of different thicknesses the effects of plate resonance are minimized by preventing them

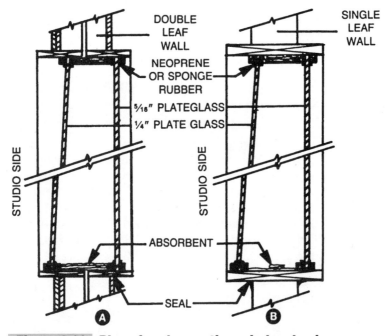

Figure 2-11 Plans for observation window having a reasonably high transmission loss: (A) for a double-leaf wall (staggered stud, double-stud, or double-masonry walls), and (B) for single-leaf wall.

from occurring at the same frequency. The cavity resonance is controlled by utilizing an absorbent reveal periphery. This can be acoustical fiberboard or glass fiber of the 703 type (Owens-Corning Fiberglas Corporation Type 703 industrial glass fiber semirigid boards of 3 pounds per cubic foot density, used widely in the studio designs to be described) with a cloth cover, or even strips of heavy carpet.

In any event, the periphery between the glass plates should be black or of a dark color to avoid attracting undue attention to it. Each glass plate is isolated from its retaining stops and the frame by strips of neoprene or sponge rubber. The strips bearing the weight of the glass plates should not compress more than about 20 percent under load, but the side strips can be much more pliant. Fasten the stops on one side at least with screws so that the inside surfaces of the glass plates can be reached for cleaning if necessary.

Absorber Mountings

All of the absorption coefficients listed in the appendix have been measured in reverberation rooms following standard procedures (ASTM:C-423-84a or the equivalent). A 72 square foot patch of the material to be measured is arranged on the floor of the reverberation room. The specimen is mounted to reflect the requirements of the test. For example, the absorbence of the material mounted on a hard surface may be desired, or a 16-inch air space may be needed in the case of material for suspended ceilings.

Mounting designations have been recently changed (ASTM: E-795-83) and the industry is now in a transition between the old and the new. Fortunately, the old and the new are quite comparable; only the designations have been changed significantly. A comparison of old and new mountings is given in Table 2-2. Drawings of the six most commonly used

TABLE 2-2 RELATIONSHIP BETWEEN OLD AND NEW MOUNTING DESIGNATIONS		
RELATIONSHIP BETWEEN OLD AND NEW MOUNTING DESIGNATIONS		
NEW MOUNTING DESIGNATION *		**OLD MOUNTING DESIGNATION ****
A	Material directly on hard surface	#4
B	Material cemented to plasterboard	#1
C-20	Material with perforated, expanded, or other open facing furred out 20 mm ($^3/_4$")	#5
C-40	Ditto, furred out 40 mm (1$^1/_2$")	#8
D-20	Material furred out 20 mm ($^3/_4$")	#2
E-405	Material spaced 405 mm (16") from hard surface	#7

*ASTM Designation: E-795-83
**Mountings formerly listed by Acoustical and Board Products Manufacturers Association, ABPMA, (formerly the Acoustical and Insulating Materials Association, AIMA).

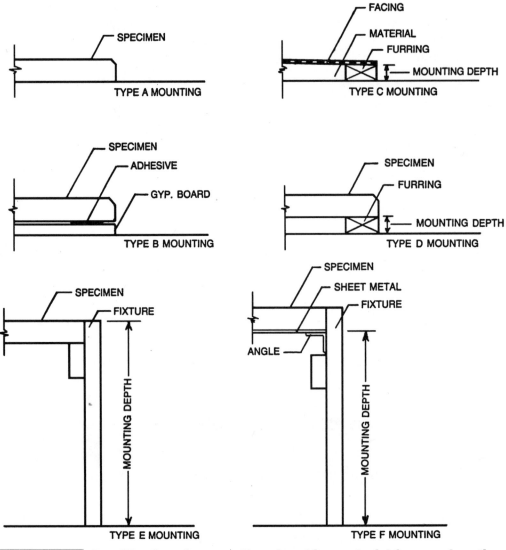

Figure 2-12 **Specifications for mounting absorbing material for reverberation room measurements (see Table 2-2).**

mountings are shown in Fig. 2-12. With the aid of this table and this figure, absorption coefficients from recent or older sources can be compared and understood.

Reverberation Time

Applying the concept of reverberation time to small rooms, such as the 12 designs to follow, has come under considerable scrutiny since the first edition of this book appeared. The argument runs like this: The reverberation time equations have been derived on the assumption that a random sound field exists. Since a true reverberant field seldom exists in a small room, the concept of reverberation time should not be applied.

It is a valid criticism, but practical considerations dictate their continued use, while being aware of their limitations. The normal modes of a small room decay, even though they are too few and too far between. The decay of these modes, excited by voice or music signals, constitutes the sound of the room.

Calculating the reverberation time of a small room gives a basis of apportioning the areas of various absorbing materials to give uniform overall absorption throughout the audible band. This is the procedure followed in the upcoming chapters.

Sabine's early work resulted in the reverberation time equation bearing his name. Eyring and numerous others published different versions of Sabine's equation, some of which are supposed to give more accurate results in small rooms. Young[43] has pointed out that absorption coefficients, such as those in the appendix, are Sabine coefficients which can only be directly applied in the Sabine equation. For this reason, only the simple Sabine equation is used in the chapters to follow.

Foams

Numerous attractive and effective foam panels of various dramatic surface textures are now available under such names as Sonex (manufactured by Illbruck Company), Acoustafoam (manufactured by FM Tubecraft Support Systems, Inc.), and Sound-Sorber (manufactured by Discrete Technology). These companies may be located through the listing in Appendix D.

None of the designs in this book specify any of these foam panels, although they may be used. They are more expensive than glass fiberboards for the same absorption (sabins per dollar). However, they are easier to apply, and are more pleasing to the eye. Compare sound absorption coefficients for yourself and feel perfectly free to make substitutions if the expense seems to be justified.

Corner Absorbers

In recent years several products have become available which have special significance for small room acoustics. These are portable absorbers intended to stand (or be mounted) in the corners of the room. All room modes terminate in the corners of the room. Absorbers placed in the corners therefore act on all modes, not just some of them.

Among these corner absorbers are Tube Traps® (manufactured by Acoustic Sciences Corporation) and Korner Killers® (manufactured by RPG Diffusing Systems). Both companies can be located through the listing in Appendix D.

Although acoustically quite applicable to most of the designs to follow, permanent installations have been favored over the portable units. The latter, admittedly, allow you to "trim" a room by ear until satisfactory conditions prevail, but not everyone is qualified to make such judgments. Try incorporating such interesting new products in an experimental program.

Construction Permit

It is imperative that a construction permit be obtained before work is started. This requires plans and specifications. During construction, inspections can be expected covering structural, electrical, and plumbing installations. Obtaining the permit gives evidence that zoning restrictions are met and gives assurance that fire and other insurance will not be invalidated at a later time.

AUDIOVISUAL BUDGET
RECORDING STUDIO

Features: "Contracarpet" ceiling, standing a room "on end" to get volume, detailed calculation of reverberation time.

Here is the problem presented by the client: to build a small, repeat small, studio and control room suitable for producing sound tracks for audiovisual presentations such as filmstrips, slide sets, and 16 mm motion picture film shorts. It is to be placed inside a large prefabricated building with ample headroom, but floor space at a premium. On top of this, the cost must be held to an absolute minimum. Quality performance; bottom dollar. This message is familiar enough and occurs often enough to suggest a detailed treatment of the solution.

Studio

Speech is the predominant sound to be recorded, to which is added canned music and sound effects from subscription disc or tape libraries in the editing process. This means that the 1500 cubic foot minimum room volume discussed earlier is acceptable for the studio. The floor plan of Fig. 3-1 provides a studio with a floor area of 158.5 square feet and, with a 10 foot ceiling, a volume of 1585 cubic feet. The dimensional ratio of 1:1.14:1.39 distributes axial modes quite well as shown in Fig. 1-4A and Table 1-2A. The two closest modes near 283 Hz are high enough in frequency to be unlikely to cause voice colorations. A studio of these dimensions has a response down to 40 Hz, which is more than adequate for voice.

Control Room

The high ceiling of the prefabricated building allows a control room ceiling of any reasonable height to be specified. This suggested the possibility of standing the control room on end, so to speak, to minimize floor space. The dimensional ratio of Fig. 1-4B and Table 1-2B of 1:1.28:1.54 was selected over the dimensions in Table 1-2A because a longer (or, in this case, higher) room results.

The elevation sketch in Fig. 3-1 shows the ceiling of the control room at 13 feet 9 inches. The 9.0 feet × 11.4 feet × 13.75 feet dimensions of this control room are slightly different from the case of Table 1-2B (10.0 feet × 12.8 feet × 15.4 feet), but are close enough for us to use Fig. 1-4B to get a qualitative view of mode spacings.

The two modes near 124 Hz, only 2.3 Hz apart, alert us to possible voice colorations at that frequency. The three just below 250 Hz are probably high enough in frequency to be less troublesome. Remember, however, that voice recording is done in the studio; this considers only factors which might affect listening conditions in the control room.

Sound Lock

The sketch in Fig. 3-1 includes a tiny sound lock corridor to control the effects of both entering from the noisy exterior, and traffic between the studio and control room. This places two doors in series for any given path which eliminates the need for expensive, special acoustical doors with their awkward clamping hardware and seals. Also, it allows studio access and egress during recording without a blast of noise from monitor loudspeakers or outside. Sound locks should always be a part of any studio suite intended for professional or quasi-professional use.

Stealing space for the sound lock from the control room affects the modal situation in the control room. The pristine rectangular room with the usual three axial modes now has two other modes added, those associated with the M and N dimensions of Fig. 3-1. These dimensions of about 8.0 and 5.5 feet reduce average mode spacing, which is good, but

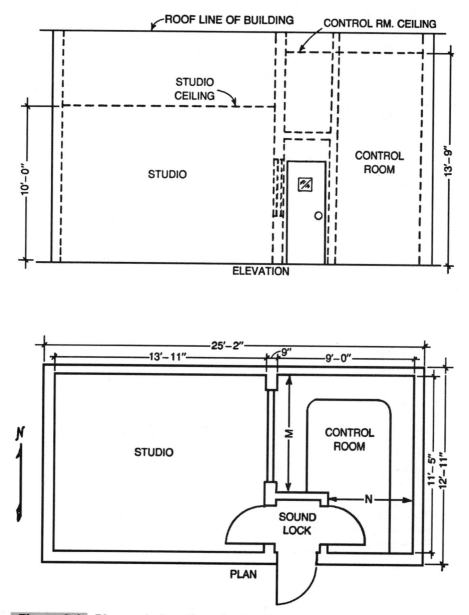

Figure 3-1 Plan and elevation of a budget studio suite for the production of audiovisual materials. Because of limited floor space the control room "stands on end" to obtain the requisite room volume and room proportions.

increase the possibility of coincidences, which is bad. Because these small dimensions resonate at higher frequencies we can expect their effect to be less noticeable. A detailed examination in such cases always alerts you to a potential coloration problem.

Work Table

A control room dedicated to audiovisual activity needs worktable space. For this reason a built-in workbench with some drawer and cabinet space below is suggested. The mixing console for such an activity is normally one of the simple desktop models and the recorders are of the advanced audiophile type which require table space rather than floor space.

Studio Contracarpet

When the client says, "Carpet on the studio floor," the acoustical consultant gulps and bravely says, "Can do!" Carpet plays a dominant role in the acoustics of the studio because the floor area is a substantial part of the total surface area of the room; the problem is that carpet absorbs well at higher frequencies and very poorly at lower frequencies. Carpet on the floor dictates compensating absorbers peaking at low frequencies and this often means tuned Helmholtz units.

In the elevation of Fig. 3-1 the studio and control room walls run all the way to the roof. This is necessary to prevent flanking sound traveling from one room to the other via the "attic." Establishing an acoustical ceiling in the studio can be easily done by suspending it with common metal angles and Tees and wires. However, instead of the usual 24 inch × 48 inch soft fiber lay-in panels, the special *contracarpet* panels detailed in Fig. 3-2 are used.

BBC engineers have used the term *anticarpet* to denote an absorber to compensate for the carpet deficiencies. However the term *contracarpet* seems to be somewhat more descriptive, at least in this case where the contracarpet units are opposite the carpet. The contracarpet units are Helmholtz resonators about 2 inches thick fabricated in the familiar 24 inch × 48 inch size. The side facing the studio is $^3/_{16}$-inch plywood or Masonite perforated so that about 0.1 percent of its area is holes. Holes of $^3/_{16}$-inch diameter spaced 6 inches on centers give a perforation percentage of about the proper magnitude. The back (top) of each unit is of $^3/_4$-inch particleboard (chipboard) which is somewhat denser than plywood. This particleboard constitutes the acoustical ceiling which should be established 10 feet from the floor.

Sections not designated for contracarpet panels (Fig. 3-3) are filled with panels of $^3/_4$-inch particleboard. Thus the acoustical ceiling height varies by 2 inches from place to place. Both types of lay-in panels are set on a continuous bead of nonhardening acoustical sealant on the suspended Tee frames. This makes a virtual hermetic seal between the studio and the "attic" space and reduces the possibility of rattles.

A 2-inch thickness of 3 pounds per cubic foot density glass fiber is jammed into the approximately $1^3/_4$-inch space determined by the 2 × 2s between the particleboard and the perforated facing. (Throughout the book glass fiber of this density is repeatedly specified. Owens-Corning Type 703 is admirably suited. It is available in thicknesses of 1, $1^1/_2$, and

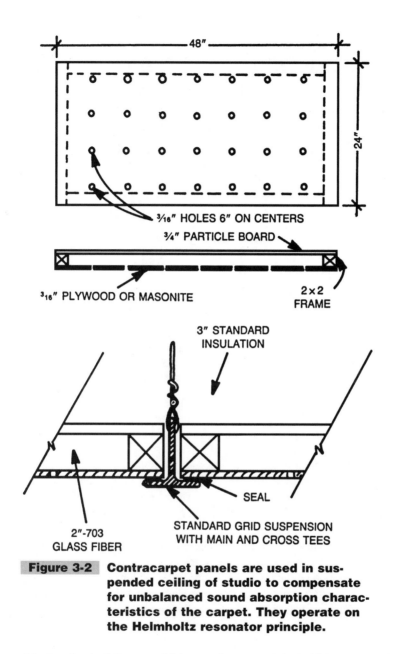

Figure 3-2 Contracarpet panels are used in suspended ceiling of studio to compensate for unbalanced sound absorption characteristics of the carpet. They operate on the Helmholtz resonator principle.

2 inches, but building up a thickness of, say, two 2-inch thicknesses to obtain a 4-inch thickness is acoustically equivalent. Johns-Manville has a product, Series 1000 Spinglass of 3 pounds per cubic foot density which is also acceptable.) This glass fiber broadens the low frequency absorption peak. On top of the contracarpet and blank panels a blanket of common house insulation material of approximately 3-inch thickness is laid. If paper is attached, it should be placed downward. The purpose of the insulation layer is not so much to make the ceiling more impervious to sound as to deaden the space resonances in the "attic."

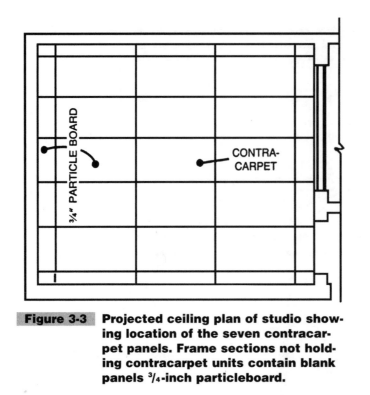

Figure 3-3 Projected ceiling plan of studio showing location of the seven contracar-
pet panels. Frame sections not holding contracarpet units contain blank
panels ³/₄-inch particleboard.

Studio Wideband Wall Units

The third acoustical element in the studio (in addition to the carpet and contracarpet) is a
series of identical wall units constructed as shown in Fig. 3-4. Each of these is nothing
more or less than patches of 4-inch thick glass fiber of 3 pounds per cubic foot density,
each having an acoustically effective surface of 12 square feet. These give essentially perfect sound absorption at 125 Hz and above.

The frame is of ordinary 1-inch lumber. The backing board of ³/₁₆-inch or ¹/₄-inch plywood or Masonite is only to strengthen the frame and to make each unit a self-contained
entity which can easily be mounted or removed. The cloth cover serves both as a cosmetic function and as an aid to contain the irritating glass fibers. This cloth should be treated
with fire retardant chemicals for safety. Loudspeaker grille cloth is ideal, although relatively
expensive. Ordinary burlap or other open-weave cloth of light weight can be employed. This
fabric cover presents an excellent opportunity for color emphasis in the decor of the studio (pink noise penetrates even a purple grille cloth)!

The method of mounting the wideband units to the wall is left to the ingenuity of the
builder, although a simple suggestion is shown at M in Fig. 3-4. If molding M is a strip
running the length of a wall, the units may easily be positioned laterally anywhere
on the wall. Further, a metal hook N on each front edge of the frame allows reversal of
the unit.

In this way complete flexibility is realized: mounting or removing, positioning and
reversing. With the soft side out, reverberation time is decreased. With the hard back

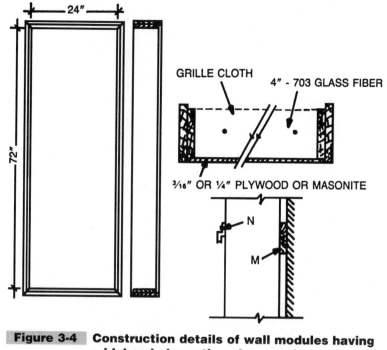

Figure 3-4 **Construction details of wall modules having wideband absorption characteristics. Used in both studio and control room.**

exposed, it is increased, yet retains the advantage of a rectangular protuberance for diffusion of sound and for a measure of control of flutter echoes. There are limits to such adjustments of the acoustical properties of the room, but this degree of flexibility comes with negligible cost.

The suggested locations of the wideband modules on the walls of the studio are shown in Fig. 3-5. Three modules on the west wall oppose the window and the door on the east wall. The pair of modules on the north wall opposes bare areas on the south wall, and vice versa. At first glance, Fig. 3-5 seems to show module opposite module, but remember that these wall elevations are the view one has facing the wall from inside the room and one must "do a 180°" between looking at the north wall and looking at the south wall.

Studio Drywall

If a structural element contributes significantly to sound absorption in the studio, it must be considered as part of the acoustical treatment. The type of wall construction utilized in this studio is illustrated in Fig. 3-10. A layer of gypsum drywall panels is applied to one face of the wall and a double layer to the other face. As far as noise isolation is concerned, either face can be on the studio side. In the ensuing calculations it is assumed that the

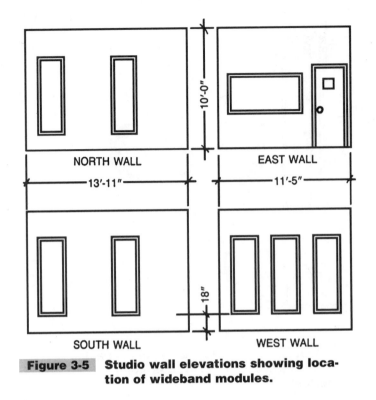

Figure 3-5 Studio wall elevations showing location of wideband modules.

single layer of drywall is toward the studio and the control room although there would be only a minor difference in absorbing effect if it were the other way around. The gypsum panels on both sides of the wall vibrate as diaphragms on the cushion of air contained between them. The sound absorbed is greatest near the resonance frequency of the panel which, in turn, is a function of the depth of air space and mass per unit area of panel.

Absorption coefficients are available for ¹/₂-inch gypsum board on 2 × 4 framing which resonates at about 61 Hz. Using ⁵/₈-inch instead of ¹/₂-inch and a nominal air space of 8 inches instead of 4 inches shifts the resonance frequency down to about 38 Hz. This reduces the absorption coefficients in the 125 Hz-4 kHz range somewhat.

In Table 3-1, however, the available published values for the ¹/₂-inch thickness and 4-inch airspace are used to avoid complicating the procedure. Both the contracarpet and blank panels of the ceiling contribute slightly to low-frequency absorption as diaphragms, over and above the contracarpet Helmholtz resonator effect. This compensates somewhat for the fact that the wall used differs from the one to which the coefficients strictly apply.

Studio Computations

A bit of figuring gives us the required data for the studio: surface area = 823 square feet, volume = 1585 cubic feet. With this we can enter the *sanctum sanctorum* of the *Sabine*

TABLE 3-1 Studio Calculations

SIZE   11'5" × 13'11" × 10'0" CEILING
FLOOR   HEAVY CARPET AND PAD
CEILING   7 – 2' × 4' CONTRACARPET MODULES (SEE FIG. 3-3)
WALLS   7 – 2' × 6' WIDEBAND MODULES (SEE FIG. 3-4 & 3-5)
SURFACE AREA   823 SQUARE FEET.
VOLUME   1,585 CUBIC FEET

MATERIAL	S AREA SQ. FT.	125 HZ A	SA	250 HZ A	SA	500HZ A	SA	1KHZ A	SA	2KHZ A	SA	4KHZ A	SA
CARPET	159.	0.08	12.7	0.24	38.2	0.57	90.6	0.69	109.7	0.71	112.9	0.73	116.1
DRYWALL	665.	0.10	66.5	0.08	53.2	0.05	33.3	0.03	20.0	0.03	20.0	0.03	20.0
CONTRACARPET	56.	0.90	50.4	0.54	30.2	0.30	16.8	0.16	9.0	0.12	6.7	0.10	5.6
WIDEBAND MODULES	84.	0.99	83.2	0.99	83.2	0.99	83.2	0.99	83.2	0.99	83.2	0.99	83.2
TOTAL SABINS, SA			212.8		204.8		223.9		221.9		222.8		224.9
Reverberation time, seconds (sabine)			0.36		0.38		0.35		0.35		0.35		0.35

equation to determine the absorption required to realize our desired reverberation time, which is 0.35 second. The *Sabine equation* is:

$$T_{60} = \frac{0.049\ V}{(S)\ (a)}$$

where

T_{60} = reverberation time, seconds
V = volume of studio, cubic feet
S = surface area of studio, square feet
a = average absorption coefficient

The next step is to determine the number of absorption units in the room required to give us the desired T_{60}, 0.35 second.

$$\text{Total absorption units} = (s)\ (a) = \frac{(0.049)\ (1585)}{0.35}$$

$$= 222 \text{ sabins}$$

Now, what does this mean? Simply that 222 square feet of perfect absorber ($a = 1.00$) in the room yields the desired 0.35 second reverberation time. The original perfect absorber conceived by pioneer acoustician Wallace Sabine was an open window.

All the sound falling on an open window is surely absorbed as far as the room is concerned, but the practical absorbing materials we have to work with are something less than perfect, especially if you add the requirement, "throughout the range of audible frequencies."

The room computation process requires some of what is euphemistically called *engineering estimating*. This is nothing more than guessing, but engineers become better and better guessers as their years of experience pile up. The guessing comes in deciding how much of what land of absorbing materials will give the 222 sabins for each frequency point throughout the band.

Carpet is specified, so there is no guessing about that. The carpet area is entered in Table 3-1. The absorption coefficients for the carpet selected are entered for each frequency. By multiplying the carpet area by each coefficient, the absorption in sabins is found for each frequency and entered in Table 3-1.

By plotting the carpet absorption points in Fig. 3-6, graph A is obtained. As drywall is the other fixed element, its absorption is calculated for each frequency and entered in Table 3-1. By adding carpet and drywall absorption and plotting the resulting sums on Fig. 3-6, graph B is obtained.

The drywall partially makes up for lack of carpet absorption in the low frequencies, but not enough. A few trial calculations show that seven contracarpet ceiling units give us graph C which is reasonably horizontal at roughly 140 sabins. This must be raised to the vicinity of 225 sabins and it is the function of the seven wideband wall modules (essentially perfect absorbers in the 125 Hz-4 kHz range according to the manufacturer's measurements) to do this (graph D of Fig. 3-6). The wiggles of graph D have only a minor effect on reverberation time.

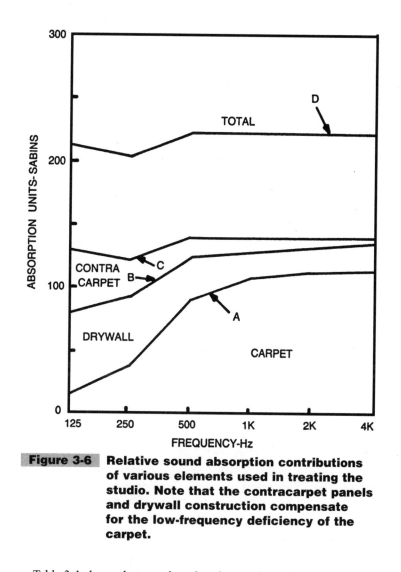

Figure 3-6 **Relative sound absorption contributions of various elements used in treating the studio. Note that the contracarpet panels and drywall construction compensate for the low-frequency deficiency of the carpet.**

Table 3-1 shows that reverberation time varies only from 0.35 to 0.38 second as a result of the fluctuations of graph D, Fig. 3-6. Our precision is not good enough to justify pursuing such calculations further. The calculated studio reverberation time variations with frequency are shown graphically in Fig. 3-7.

Control Room Treatment

The control room is generally admitted to be a work room, especially in audiovisual work. For this reason there is usually a minimum amount of opposition to vinyl tile floors which are especially practical for rolling equipment around the room.

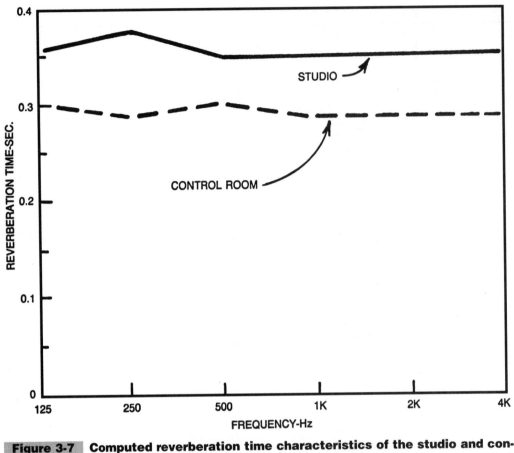

Figure 3-7 Computed reverberation time characteristics of the studio and control room. The evaluation of studio sounds monitored in the control room is aided by having lower reverberation in the control room.

Reverberation time for a control room should be somewhat shorter than that of the studio being monitored. The reverberation associated with studio sounds reproduced on the monitoring loudspeaker are then heard without being masked by control room reverberation. Listening rooms require relatively uniform reverberation time with frequency as do studios.

Control Room Ceiling Treatment

A standard suspended ceiling is specified to oppose the reflective vinyl floor. This ceiling is dropped 16 inches below the acoustical ceiling of drywall which is at a 13 foot 9 inch height (Fig. 3-8).

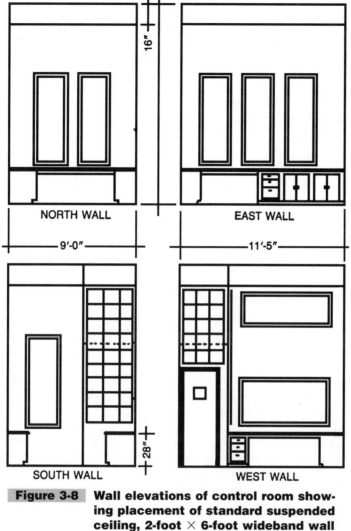

Figure 3-8 Wall elevations of control room showing placement of standard suspended ceiling, 2-foot × 6-foot wideband wall modules and acoustical tile.

The distinction must always be made between the visual and the acoustical dimensions of a room. The space between the suspended lay-in ceiling and the solid ceiling at a 13 foot 9 inch height is acoustically active.

For example, there are specific resonance effects between the lay-in panels and the air space above them which result in good low-frequency absorption. This makes the overall absorption relatively uniform with frequency (Fig. 3-9E).

By using the usual soft fiber lay-in panels with an NRC (*noise reduction coefficient*) rating between 0.75 and 0.85, the suspended ceiling absorption coefficient varies between 0.65 and 0.92 (Table 3-2). By using the 16-inch drop instead of some other distance, the

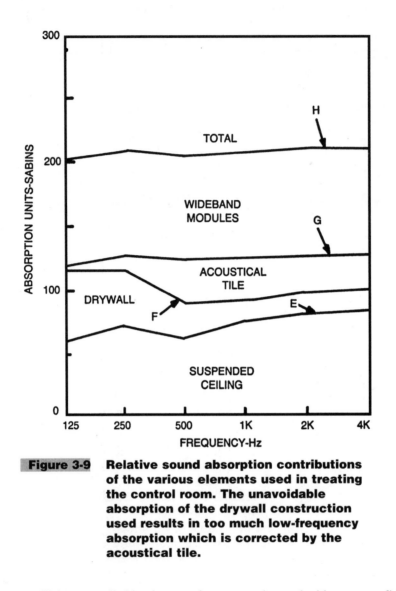

Figure 3-9 Relative sound absorption contributions of the various elements used in treating the control room. The unavoidable absorption of the drywall construction used results in too much low-frequency absorption which is corrected by the acoustical tile.

coefficients supplied by the manufacturer can be used with some confidence. Deviations from this standard 16-inch drop results in no known coefficients to depend upon.

Control Room Drywall

The diaphragmatic absorption of gypsum board surfaces must also be figured into the control room. The floor is vinyl tile covering concrete, so that is not included. Because of the complicating effect of the suspended ceiling we shall neglect the drywall ceiling above it. The walls alone (including doors and window which also act as diaphragms)

TABLE 3-2 Control Room Calculations

SIZE   9'0" × 11'5" × 13'9" CEILING
FLOOR   VINYL TILE
CEILING   STANDARD SUSPENDED CEILING, PANELS NRC 0.75–0.85
WALLS   42 ACOUSTICAL TILES 12" × 12" × 3/4" (FIG. 3-8)
  7 WIDEBAND MODULES 2' × 6' (FIG. 3-8)
SURFACE AREA   742 SQ. FT.
VOLUME   1,243 CU. FT.

MATERIAL	S AREA SQ. FT.	125 HZ		250 HZ		500HZ		1KHZ		2KHZ		4KHZ	
		A	SA	A	SA	A	SA	A	SA	A	SA	A	SA
SUSPENDED CEILING	90.4	0.65	58.8	0.78	70.5	0.67	60.6	0.82	74.1	0.89	80.5	0.92	83.2
DRYWALL	561	0.10	56.1	0.08	44.9	0.05	28.1	0.03	16.8	0.03	16.8	0.03	16.8
ACOUSTICAL TILE	42.	0.09	3.8	0.27	11.3	0.78	32.8	0.84	35.3	0.72	30.2	0.64	26.9
WIDEBAND MODULES	84.	0.99	83.2	0.99	83.2	0.99	83.2	0.99	83.2	0.99	83.2	0.99	83.2
TOTAL SABINS, SA			201.9		209.9		204.7		209.4		210.7		210.1
Reverberation time, seconds (Eynng) (sabine)			0.30		0.29		0.30		0.29		0.29		0.29

total 561 square feet. The sum of drywall and suspended ceiling absorption from Table 3-2 is plotted as graph F in Fig. 3-9. Slight overcompensation now prevails in the low-frequencies.

Control Room Acoustical Tile

Common acoustical tile is probably the most abused and misused product in the annals of sound treatment because people expect too much of it. However, it is inexpensive, easy to apply, and an excellent product if properly used. It is characterized, as is carpet, by good high-frequency absorption, but little at low frequencies.

This is exactly what is needed in the present case. The addition of 42 acoustical tiles 12 inches × 12 inches × $^3/_4$ inch to the control room brings the absorption to an approximately uniform 125 sabin level throughout the frequency range of interest as shown by graph G of Fig. 3-9. It is desirable not to have all of a given type of absorber in a room active in only one mode. Figure 3-8 shows the acoustical tile acting on both the N-S and E-W modes of the room.

Control Room Wideband Modules

The 125 sabin level of graph G in Fig. 3-9 is about 83 sabins below the 208 sabins required for a reverberation time of 0.25 second. Seven 2 foot × 6 foot wideband modules provide this 83 sabins, bringing the total absorption up to graph H of Fig. 3-9 which hovers close to the 208 sabin goal. The control room reverberation tine of Table 3-2 is compared graphically to that of the studio it serves in Fig. 3-7.

A grand total of 14 of the 2 foot × 6 foot wideband modules detailed in Fig. 3-4 are now required, seven for the studio and seven for the control room. Some economy in effort and expense should result from mass production.

Noise Factors

The level of noise outside the studio suite and background noise standards set for inside the studio determine the type of wall construction. A nearby printing press, buzz saw, or another such noisemaker may require exceptional measures.

However, if this is to be a budget studio, wall construction costs must be kept in line. Most small organizations attracted by the budget approach are willing to do some horse-trading. If the printing press operates only part time, the audiovisual people can schedule their recording time accordingly. Perhaps a flashing red light at the buzz saw during a take would suggest to the carpenter that this is a good time for a cup of coffee.

Having to repeat three takes a year because of low-flying helicopters is far cheaper than building a building within a building to get 80 dB transmission loss.

For this budget studio complex, an economical wall that offers good protection (about STC 50) against outside noise is illustrated in Fig. 3-10. The staggered stud principle gives two independent walls attached only at the periphery to the 2 × 4 studs 16 inches on centers on a 2 × 8 plate. A single layer of ⅝-inch gypsum board is nailed on one side and a double layer of ⅝-inch gypsum board on the other. Thermal type insulation of at least a 3-inch thickness minimizes resonances in the space between the walls. The effectiveness of even this staggered construction depends upon tight sealing. To assure this, a bead of nonhardening acoustical sealant should be run around all intersections of walls with ceiling, floor, and other walls. All exterior walls and the wall separating the control room from the studio should be of the construction shown in Fig. 3-10. The north and east walls of the sound lock and the sound lock ceiling may be of normal single stud construction with single layers of gypsum board. The control room ceiling at a 13 foot 9 inch height is standard ⅝-inch gypsum board on frame construction with the suspended ceiling dropped 16 inches below this.

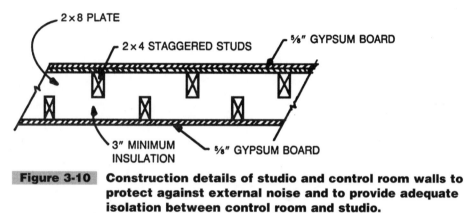

Figure 3-10 **Construction details of studio and control room walls to protect against external noise and to provide adequate isolation between control room and studio.**

4

STUDIO BUILT IN A RESIDENCE

Features: Polys on the ceiling, A/C duct layout for minimum noise and crosstalk.

CONTENTS AT A GLANCE CO HEAD	
Floor Plan	CONTROL ROOM TREATMENT
	CEILING TREATMENT
Studio Treatment	WALL TREATMENT
	REVERBERATION TIME
Ceiling	
WALLS	Air Conditioning
Studio Reverberation Time	Observation Window

Building a studio and control room in the average modern single family dwelling presents major problems: thin walls, low ceilings, and limited floor area. In this case, however, the residence is not average. It has concrete floors, stone and brick walls, and ample headroom. Needless to say, it is located outside the United States. There are a number of lessons to be learned from this case, however, and the solutions to specific problems to be considered are quite applicable to other situations.

Floor Plan

The *as found* floor plan is shown in Fig. 4-1. Walls are either 17-inch thick stone or 10-inch thick brick or glass. The first two warm the acoustician's heart but the last one is mentally placed high on the *get rid of it* list. The dining and living rooms are separated by a 4-foot barrier. The floor of the dining room is 8 inches lower than that of the living room. This is the house that is available. Can it be converted into an effective studio and control room without major alterations?

Figure 4-2 illustrates the changes made. Basically, the living room was visualized as the studio and the dining room as the control room. This required pouring enough concrete into the dining room to bring its floor up flush with the living room floor. The north wall of glass was eliminated and a 10-inch thick brick wall established at the outer center column, enlarging the width of the studio more than 6 feet. A wall of brick to hold the observation window was placed at a 45 degree angle to give the control operator a good view into the studio and to provide certain acoustical advantages in both rooms. This angled wall makes the control room unsymmetrical which would be considered a disadvantage in a professional recording studio. However, in the present case the advantages outweigh the disadvantages.

A sound lock with brick walls was located as shown in Fig. 4-2 and the inner glass wall of the entrance hall was eliminated in the process. The external glass wall of the entrance hall survived as the entrance hall function remained unchanged. Door A in the west wall

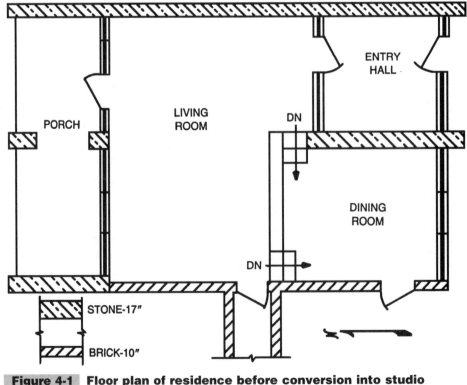

Figure 4-1   **Floor plan of residence before conversion into studio complex.**

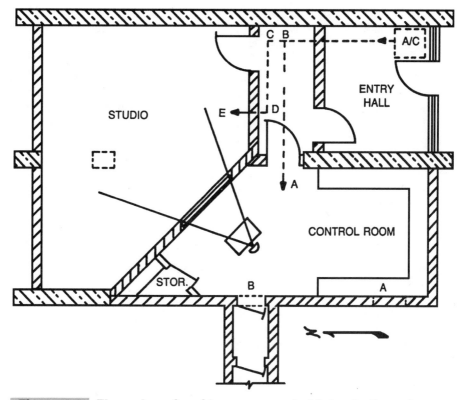

Figure 4-2 **Floor plan of residence converted into studio and control room with sound lock.**

of what is now the control room was bricked up. This routed all traffic between the studio and other parts of the house either through door B in the west wall or by outdoor paths. After objections were voiced by the consultant at having a second door into the control room, it was left to the client to either establish a second sound lock in the west hallway or route all traffic outdoors and through the entrance hall. As this residence is in a tropical country, this latter should create no major problems.

The space in the sharp angled northwest corner of the control room was made into a closet for the storage of tape stock, recorded tape library, etc. Note also that the operator's position is set back at least 5 feet from the glass surface. This gives the operator an acceptable angle with the monitoring loudspeaker(s) and an acoustically better position for critical listening. The pillar in the studio area was slated for removal if not loadbearing. However, it could remain, causing some inconvenience but little adverse acoustical effect. The added studio volume obtained by pushing out the studio north wall is very important acoustically.

Studio Treatment

Stepping into the untreated studio of Fig. 4-2, one is met by a great expanse of brick and stone walls, concrete floor, glass observation window and the wooden underside of the roof. At this time you cannot escape the thought that it would make an excellent reverberation chamber. The first step in making a studio out of it is to determine the correct number of absorption units (sabins) required. Measurements reveal a volume of 4,614 cubic feet and an inside surface area of 1,597 square feet.

What reverberation time should be adopted? As both music and speech are to be recorded in this room, a compromise value of about 0.48 second seemed in order.[2] Cranking these values of volume, and reverberation time into the Sabine equation it is computed that 470 absorption units, or sabins, are required. This is a point of departure and a start in building Table 4-1.

Ceiling

At the high frequencies the carpet is a dominant factor, supplying almost half the required absorption while contributing practically nothing at the low frequencies (Fig. 4-3). This poses the classical problem of introducing other absorbing elements which have the opposite effect. This time semicylindrical panel units are chosen over the Helmholtz resonator approach. Such cylindrical units on the studio ceiling between the concrete roof beams, as shown in Figs. 4-4 and 4-5, will contribute in the following ways:

- They augment a thin roof in protecting against outside noise.
- They act as excellent diffusers of sound in the studio.
- They absorb sound in the studio in a way which tends to compensate for the carpet deficiencies at low frequencies.

These cylindrical elements are basically a thin skin of $3/16$-inch plywood or masonite stretched over bulkheads cut as segments of a circle. The radius and chord of this segment are carefully adjusted so that the arc is 48 inches—the standard width of plywood and masonite. The skin of these cylindrical elements, which vibrates vigorously in response to sound in the room, must not rattle. As protection against rattles a bead of nonhardening acoustical sealant, or better yet, a thin strip of felt, is applied to the edge of each bulkhead before the skin is bent over and nailed in place. The functioning and construction of such cylindrical elements (often called *polycylindrical diffusors* or *polys*) as well as absorption coefficients for units of different sizes are detailed in a companion volume.[3] The space within the semicylindrical units can be stuffed with common thermal type mineral wool or glass fiber.

WALLS

The compensation for carpet deficiencies by the cylindrical elements, as shown in Fig. 4-3, is quite good except for 125 Hz and below. The effects of this on reverberation time are considered more fully later. To approach the required 470 sabins, about 256 square feet of wideband absorber are required. This absorption can be supplied by 4 inches of 703 glass fiber or its equivalent. The suggested wall modules have several advantages. They:

TABLE 4-1 STUDIO CALCULATIONS

SIZE 18'4" × 23'1" (CORNER CUT) × 13'3" AVE. CEILING HT.
FLOOR CARPET
CEILING SEMICLINDRICAL PLYWOOD ELEMENTS
WALLS 16 2' × 8' WIDEBAND MODULES
VOLUME 4,614 CU. FT.

MATERIAL	S AREA SQ. FT.	125 HZ		250 HZ		500 HZ		1 KHZ		2 KHZ		4 KHZ	
		A	SA	A	SA	A	SA	A	SA	A	SA	A	SA
CARPET	349	0.05	17.5	0.15	52.4	0.30	104.7	0.40	139.6	0.50	174.5	0.60	209.4
SEMICYLINDRICAL PLYWOOD	328	0.45	146.6	0.57	187.0	0.40	131.2	0.25	82.0	0.20	65.6	0.20	65.6*
WIDEBAND MODULES	256	0.99	253.4	0.99	253.4	0.99	253.4	0.99	253.4	0.99	253.4	0.99	253.4
TOTAL SABINS, SA			418.5		492.8		489.3		475.0		493.5		528.4
Reverberation time, seconds (sabine)			0.54		0.46		0.46		0.48		0.46		0.43

*Mankovsky

55

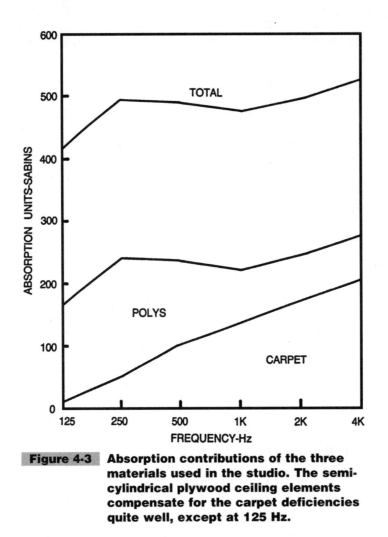

Figure 4-3 Absorption contributions of the three materials used in the studio. The semi-cylindrical plywood ceiling elements compensate for the carpet deficiencies quite well, except at 125 Hz.

■ Contribute to the diffusion of sound in the room, thus making microphone placement less critical.

■ Allow for greater trimming of acoustics if measurements indicate the necessity for this.

■ Result in economy by being used in both studio and control room.

A total of sixteen 2 foot × 8 foot units provide the required 256 square feet. They are positioned on the walls as shown in Fig. 4-6.

Studio Reverberation Time

Table 4-1 brings together the specific absorption contribution of the carpet, the cylindrical ceiling elements, and the 16 wall modules at each frequency. The resulting reverberation

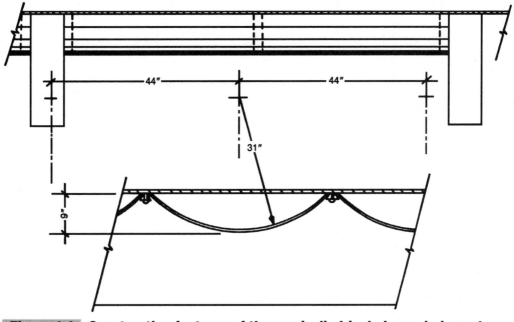

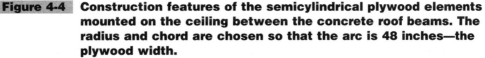

Figure 4-4 **Construction features of the semicylindrical plywood elements mounted on the ceiling between the concrete roof beams. The radius and chord are chosen so that the arc is 48 inches—the plywood width.**

time is plotted in Fig. 4-7. There are some deviations from the goal of 0.48 second. The slight drooping at high frequencies is no problem. It is actually preferred by many.

The bass rise shall be examined a bit more closely. BBC engineers have looked into this with characteristic thoroughness.[10] They found that the degree of impairment of speech quality by such bass rise was affected by the voice of the person speaking, the type of microphone used, and the distance between the speaker and the microphone (which determines the relative effect of room reverberation). As a tentative conclusion for the average situation, they suggested that a rise of reverberation time at 125 Hz over the 500 Hz value of no more than roughly 20 percent should be allowed for voice work. The rise in Fig. 4-7 is very close to this amount. They suggested that no more than about 90 percent rise of the 63 Hz reverberation time over that at 500 Hz be allowed. Table 4-1 stops at 125 Hz but we know that absorption of both ceiling elements and wall modules falls off below 125 Hz. It seems unlikely that the reverberation time at 63 Hz, however, would be greater than the allowed 0.9 second.

CONTROL ROOM TREATMENT

Deciding in favor of a hard floor (concrete, vinyl tile, parquet wood, etc.) reduces the treatment of the control room to a single type of absorber, 4 inches of 703. Table 4-2 combines the 124 square feet of 703 assigned to the ceiling, the 184 square feet in wall modules on the east and west walls, and 48 square feet in wall modules on the south wall. It treats the 356 square feet total together. This total absorption, of course, was calculated from the

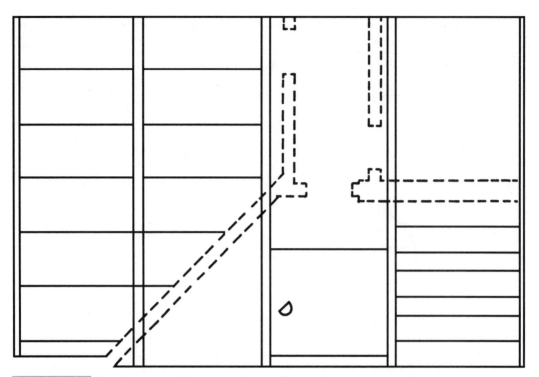

Figure 4-5 **Projected ceiling plan for both studio and control room showing placement of semicylindrical plywood elements in the studio and patches of 4-inch glass fiber in the control room.**

Sabine equation by inserting the desired reverberation time of 0.41 second, and the volume of 2,964 cubic feet, turning the crank and coming out with a total of 352 sabins.

CEILING TREATMENT

First, it is important that the ceiling over the operator be absorbent. The operator's position is indicated in Fig. 4-5 by the same symbol used in Fig. 4-2. Remember that this ceiling slopes as shown in Fig. 4-6 which is favorable for the floor-ceiling mode. The bare ceiling area must be distributed for the best diffusion effect and the pattern of Fig. 4-5 was selected.

The 4 inches of 703 is held to the ceiling in the manner detailed in Fig. 4-8. There is really considerable freedom in how this is done. If 703 of 4-inch thickness (instead of two 2-inch layers) is employed, the glass fiberboard is rigid enough to allow straightforward cementing to the ceiling. This method does not provide for the cosmetic grille cloth cover nor does it offer protection from small glass fibers falling, but acoustically it is excellent. It may be desirable to stretch zig-zag wires between the edges of the frame of Fig. 4-8 under the cloth cover to hold the 703 in place and to keep the cloth cover from sagging.

WALL TREATMENT

The 703 glass fiber is applied to control room walls in the same modular form as shown in Fig. 3-4. In the present case both 8-foot and 6-foot modules are used, both of similar

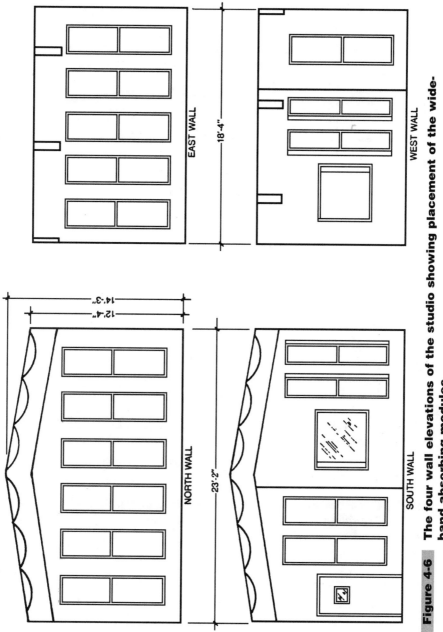

Figure 4-6 The four wall elevations of the studio showing placement of the wide-band absorbing modules.

EAST WALL

18'-4"

WEST WALL

14'-3"

12'-4"

NORTH WALL

23'-2"

SOUTH WALL

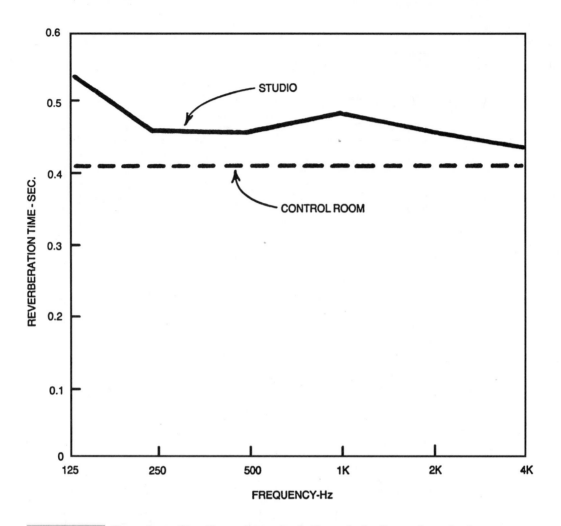

Figure 4-7   **Reverberation time characteristics of studio and control room. The bass rise in computed studio reverberation is within BBC limits.**

construction. The 8-foot module really needs a divider in the middle for strength and to help the unit keep its shape. Unlike the 6-foot module of Fig. 3-4, a similar central divider is suggested in this case for a more uniform appearance. The wall modules are positioned as shown in Fig. 4-9.

REVERBERATION TIME

With only 4 inches of 703 entering into our calculations for the control room, only a uniform reverberation time between 125 Hz and 4 kHz can be expected. In Fig. 4-7 this uniform reverberation time for the control room is shown by the broken line. It is somewhat lower than that of the studio it serves, which is as it should be.

TABLE 4-2 CONTROL ROOM CALCULATIONS

SIZE
FLOOR — 11'0" × 16'2" × 13'4" AVE. CEILING HT.
CEILING — VINYL TILE OR CONCRETE OR PARQUET
WALLS — PATCHES OF 4" 703 GLASS FIBER
VOLUME — 102' × 8' AND 62' × 6' WIDEBAND MODULES
2,964 CU. FT.

MATERIAL	S AREA SQ. FT.	125 HZ A	125 HZ SA	250 HZ A	250 HZ SA	500 HZ A	500 HZ SA	1 KHZ A	1 KHZ SA	2 KHZ A	2 KHZ SA	4 KHZ A	4 KHZ SA
CEILING: 703 GLASS FIBER	124												
E AND W WALLS WIDEBAND MODS	184												
S WALL WIDEBAND MODULES	48												
TOTAL AREA	356	0.99	352.4	0.99	352.4	0.99	352.4	0.99	352.4	0.99	352.4	0.99	528.4
TOTAL SABINS, SA			3524		352.4		352.4		352.4		352.4		352.4
Reverberation time, seconds (sabine)			0.41		0.41		0.41		0.41		0.41		0.41

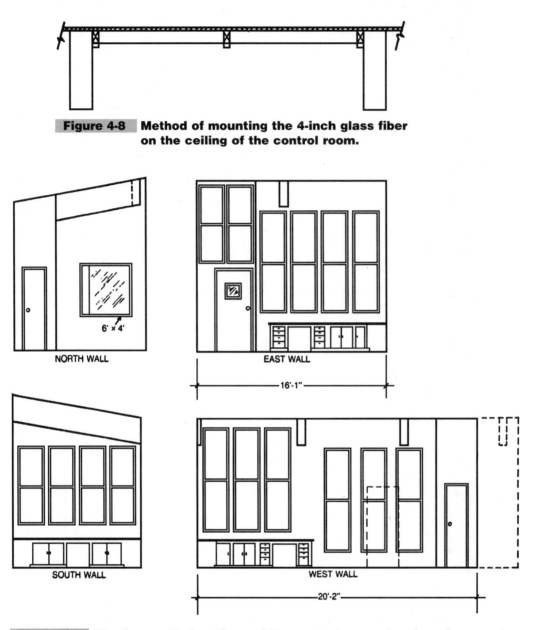

Figure 4-8 Method of mounting the 4-inch glass fiber on the ceiling of the control room.

NORTH WALL

6' × 4'

EAST WALL

16'-1"

SOUTH WALL

WEST WALL

20'-2"

Figure 4-9 The four wall elevations of the control room showing placement of wideband wall modules.

Air Conditioning

Figure 4-2 shows a possible location for the air conditioning equipment behind the entry door. A suspended ceiling in the entry hall would hide the ducts in this room. Metal ducts for both supply and exhaust lined with acoustically absorbent board are suggested. The paths of the supply ducts are shown by broken lines in Fig. 4-2 as an example. This supply duct for the studio follows the path b-c-d-e. The supply duct to the control room follows the path b-a. These duct routings would accomplish the following:

- Give one 90 degree bend at b between the control room grille and the HVAC unit plus about 20 feet of lined duct length.
- Give two 90 degree bends at c and d between the studio and the HVAC unit plus about 20 feet of lined duct length.
- Give 90 degree bends at b, c, and d between the control room and studio plus about 20 feet of lined ducting.

This ducting plan should reduce HVAC machinery and fan noise to an acceptable level and minimize the *speaking tube* effect between the control room and studio. The air velocity at the grilles should be kept below 500 feet per minute, a limit easily met. The exhaust duct routing should follow a similar plan to that of the supply ducts.

Observation Window

The construction of this most important part of the studio complex must be carried out carefully, following the general plan of Fig. 2-11 except that with 10-inch thick brick walls, the frame need not be divided as with the staggered stud wall. The frame should preferably be made of sturdy 2-inch thick lumber and mounted as the brick wall is being laid. The frame should be supported in the center by bracing during bricklaying so that the weight of bricks and mortar does not distort the window frame. Beads of acoustical sealant must be run between frame and mortar and plaster on both sides as hairline cracks commonly develop as the mortar and plaster dry.

A SMALL STUDIO FOR
INSTRUCTION AND CAMPUS RADIO

*Features: Window plugs, reversible wall modules, large poly diffuser/
absorber, discrepancies between published coefficients and
experience.*

CONTENTS AT A GLANCE

Many institutions of higher education have communication departments and most of these departments teach courses in electronic media (radio and television). Students in radio broadcasting need hands-on experience and this requires at least a small recording facility to serve as a practical laboratory. It is common for such students to produce programs to be broadcast over the campus system. The studio and control room described in this chapter are for precisely this purpose.

Studio Plan

A classroom was made available for space for the studio and control room. Dimensions of 18 feet 10 inches × 21 feet 8 inches with a 10 foot 3 inch ceiling height are not what you would call munificent, but they represent a volume of almost 4200 cubic feet, which is fairly generous. The ceiling height of 10 feet 3 inches gives relief from 8-foot heights often encountered in budget renovation jobs. The floor plan of Fig. 5-1 was settled upon after a bit of horse-trading. Many students were to be accommodated in the studio at one time as observers and performers, fewer in the control room. A studio volume of 2525 cubic feet, a control room volume of 1138 cubic feet, and a sound lock were carved out of the class-room. This means that the control room volume was below the recommended minimum of 1500 cubic feet. This sacrifice made it possible to have a larger studio.

The angled wall separating the control room and the studio reduces the chances for flut-ter echo in both rooms, tends toward spreading out of model resonances, and gracefully provides for reduction of the volume of the sound lock.

As for penetration of outside noise, concrete walls, ceiling, and floor were comforting, but almost the entire west wall was taken up by four windows overlooking a very busy thoroughfare with many trucks growling up a steep hill. These windows were plugged by

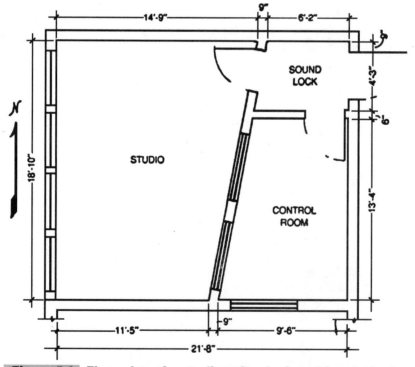

Figure 5-1 Floor plan of a studio suite designed for student instruction in radio broadcasting in a college communication department. A former classroom was converted for this purpose.

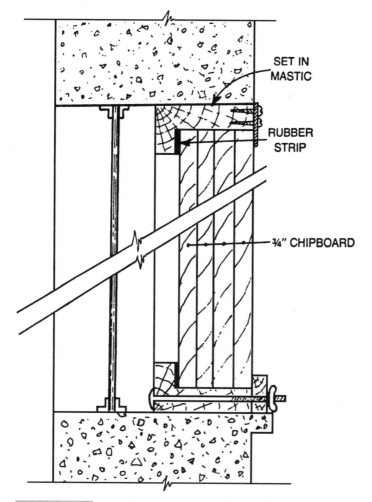

Figure 5-2 Nondestructive and inexpensive window plug designed to protect the studio from exterior noise.

four thicknesses of $3/4$-inch chipboard (particleboard) as shown in Fig. 5-2. This $3/4$-inch chipboard has a surface density of about 3 pounds per square foot. Four thicknesses bring the surface density to about 12 pounds per square foot. Considering only the mass and neglecting for the moment any resonance effort, such a well-sealed window plug should give a transmission loss of about 38 dB at 500 Hz, less for lower and more for higher frequencies. The frame holding the chipboard panels was sealed tightly to the concrete window opening. The four chipboard panels were then sealed by a soft rubber strip as the panels were pressed into place. The optional carriage bolt arrangement makes possible the nondestructive removal of the panels if necessary.

Studio Ceiling Treatment

Common suspended ceilings are rarely seen in studios. Before the end of this chapter is reached, some of the reasons for this state of affairs may be apparent. In the present case, such a ceiling was selected because it was attractive, cheap, and promised some easily obtained low-frequency absorption. Because of eye appeal, Johns-Manville Acousti-Shell Textured Vault 3-dimensional 24 inch × 24 inch ceiling panels were selected. Laying out the suspension grid on paper for such an odd shape emphasized the need for flat panels around the edges as shown, in Fig. 5-3. The coordinated Acousti-Shell Textured Flats logically fill this role.

Semicylindrical Unit

For the south wall a large diffuser/absorber of semicylindrical shape was selected. The construction of this unit is detailed in Fig. 5-4. The skin is of ¹/₄-inch plywood over which a very thin veneer was cemented for appearance. The frame is of 2 inch × 2 inch lumber and the space behind the skin is divided into nine sections by dividers of 1-inch lumber and curved bulkheads of 2-inch lumber over which the skin is stretched. This also gives the cylindrical segment its shape.

Bulkheads and dividers are sealed where they meet the plastered wall so that each of the nine segments is essentially airtight. A self-adhering foam rubber strip was applied to the edges of the 2-inch thick shaped bulkheads before the plywood skin was mounted. This made for a tight seal and made the structure rattle-free.

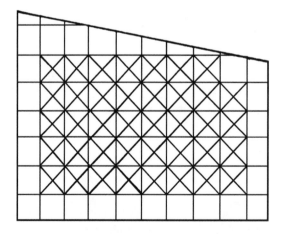

☒ JOHNS-MANVILLE ACOUSTI-SHELL
   TEXTURED VAULT, 24″ × 24″ × ⅛″

☐ JOHNS-MANVILLE ACOUSTI-SHELL
   TEXTURED FLAT, 24″ × 24″ × ⅛″

Figure 5-3 Layout plan for suspended ceiling in studio.

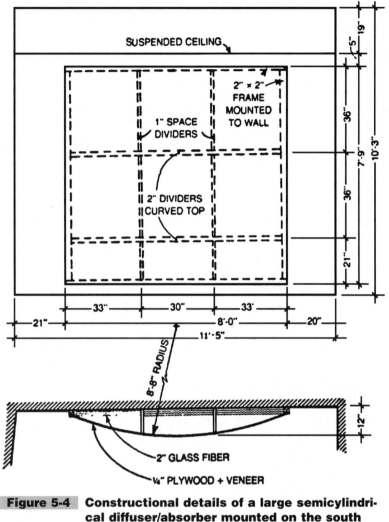

Figure 5-4 **Constructional details of a large semicylindrical diffuser/absorber mounted on the south wall of the studio.**

Reversible Wall Panels

The west wall required some absorbing and diffusing elements to go over the plugged windows. Reversible panels of nominal 4 foot × 6 foot size (Fig. 5-5) were selected. The inside dimensions of the frame were adjusted to accept 24 Johns-Manville ³/₄-inch Temper-Tone acoustical tiles of 12-inch × 12-inch size without cutting. These comprise the acoustically *soft* or absorbent side. The *hard*, reflective side is made of ³/₄-inch plywood. The center of this reflecting surface is higher than the edges, to minimize flat reflections back to nearby microphones. The slope of this reflective panel determines, of course, that such a diffusing effect is primarily at the higher audible frequencies.

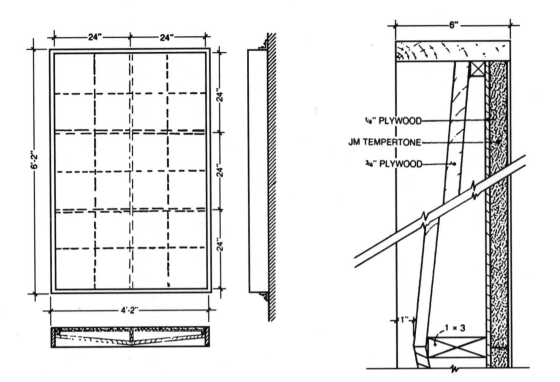

Figure 5-5 Construction details of reversible units on west all over the plugged windows. The reversible feature is intended more for allowing some final adjustments in room acoustics during measurements than for changing between program types.

Although these are called *reversible,* it was recognized in advance that once hung they would probably never be changed. However, the dual-sided approach kept open some options during the testing phase. Either simple angle brackets can be used in mounting such wall elements or, if reversals are probable, two pins in the bottom of each unit can fit into holes in metal brackets affixed to the wall to carry the weight. A simple latch arrangement can also be installed to hold the top of the unit against the wall.

Figure 5-6 shows all four studio wall elevations and the treatment for each. The north wall has 36 Johns-Manville Temper Tone 360 acoustical tiles of ³/₄-inch thickness mounted in two patches. The south wall is dominated by the 8-foot semicylindrical element previously described. The east wall is almost completely taken up by two observation windows and the door. The west wall has three of the reversible wall elements of Fig. 5-5. The location of the plugged windows is indicated by broken lines.

The suspended ceiling line is indicated on each wall elevation 19 inches below the structural ceiling. This distance allows the accommodation of air conditioning ducting. This means that some uncertainty is introduced in the absorption coefficients which are given for a standard Mounting #7 (new designation, E-405) distance of 16 inches.

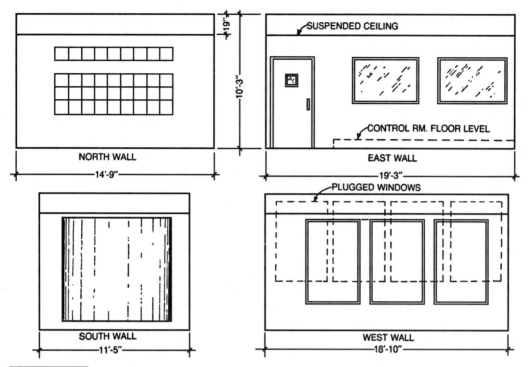

Figure 5-6 **Placement plan for acoustical elements on four walls of studio.**

Studio Calculations

For those who wish to follow through on the calculations of studio reverberation time, Table 5-1 is included. It is really "gilding the lily" to separate out absorption of door and observation window glass in view of the uncertainties in all coefficients of absorption, but no harm is done by including them as long as limitations of our overall precision is kept in mind. Computations are included for the three units on the west wall with both the reflective (hard) and absorptive (soft) sides facing the room. Calculated reverberation times for these two conditions, taken from Table 5-1, are shown graphically in Fig. 5-7. By turning the units from soft to hard, an increase in reverberation time of about 30 percent is obtained for frequencies above 500 Hz. Looking at things the other way around, by changing from hard to soft a reduction in reverberation time of about 23 percent results for the higher frequencies.

Control Room

The control room, as mentioned previously, is substandard in volume, a compromise designed to accommodate a greater number of students in the studio. Figure 5-8 illustrates the very practical equipment arrangement utilized. Room for two standard racks for

TABLE 5-1   STUDIO CALCULATIONS

SIZE 18' 10" × 14' 9" × 11 5" (ONE WALL SPLAYED)
CEILING JOHNS-MANVILLE ACOUSTI-SHELL TV&TF 24" × 24" × 1/8" SUSPENDED 19'
FLOOR VINYL TILE
WALLS PLASTERED CONCRETE TREATED AS BELOW
VOLUME 2525 CU FT

MATERIAL	S AREA SQ. FT.	125 HZ A	125 HZ SA	250 HZ A	250 HZ SA	500 HZ A	500 HZ SA	1 KHZ A	1 KHZ SA	2 KHZ A	2 KHZ SA	4 KHZ A	4 KHZ SA
J-M ACOUSTI-SHELL TEXTBRED VAULT	156	0.64	99.8	0.66	103.0	0.67	104.5	0.75	117.0	0.72	112.3	0.70	109.2
J-M ACOUSTI-SHELL TEXTBRED FLAT	90	0.70	63.0	0.69	62.1	0.66	59.4	0.80	72.0	0.84	75.6	0.83	74.7
FLOOR, VINYL TILE	246	0.02	4.9	0.03	7.4	0.03	7.4	0.03	7.4	0.03	7.4	0.02	4.9
GLASS	30	0.05	1.5	0.03	0.9	0.02	0.6	0.02	0.6	0.03	0.9	0.02	0.6
PLASTER	458	0.02	9.2	0.03	13.7	0.04	18.3	0.05	22.9	0.04	18.3	0.03	13.7
DOOR	20	0.24	4.8	0.19	3.8	0.14	2.8	0.08	1.6	0.13	2.6	0.10	2.0
CYLINDRICAL ELEMENT, SOUTH WALL	62	0.50	31.0	0.35	21.7	0.22	13.6	0.14	8.7	0.11	6.8	0.10	6.2
J-M TEMPER TONE TILE, NORTH WALL	36	0.09	3.2	0.25	9.0	0.70	25.2	0.85	30.6	0.83	29.9	0.89	32.0
			217.4		221.6		231.8		260.8		253.8		243.3
WEST WALL PANELS, HARD	72	0.28	20.2	0.19	13.7	0.14	10.1	0.11	7.9	0.08	5.8	0.05	3.6
TOTAL SABINS, HARD, SA			237.6		235.3		241.9		268.7		259.6		246.9
REVERB. TIME, SA			0.52		0.53		0.51		0.46		0.48		0.50
WEST WALL PANELS, SOFT	72	0.2	14.4	0.5	36.0	1.0	72.0	1.0	72.0	1.0	72.0	1.0	72.0
TOTAL SABINS, SOFT, SA			231.8		257.6		303.8		332.2		325.8		315.0
REVERB TIME, SEC.			0.53		0.48		0.41		0.37		0.38		0.39

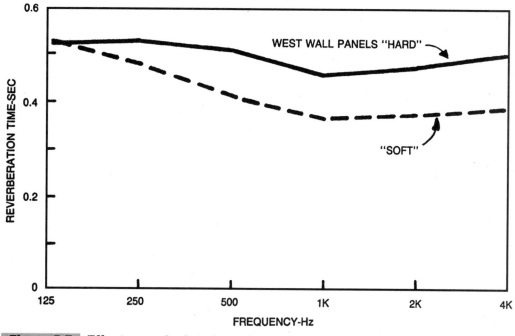

Figure 5-7 **Effect on calculated studio reverberation time of reversing the three panels on west wall.**

19-inch panels of ancillary equipment was allowed at the north wall. Along the south and west walls are work surfaces with built-in drawers and storage cabinets below. The window in the south wall looks into the television studio. Although the television studio has its own control room, this window, along with interconnecting tie lines, makes it possible to use both studios for special productions. Normally this south window is covered by drawn drapes.

The floor of the control room was raised about 12 inches and a 4-inch × 4-inch trough runs around the east, south, and west walls for interconnecting cables. In spots not covered by cabinets, access to the trough is by hinged lids flush with the floor.

Measurements

Listening tests in the studio revealed that with the hard surfaces of the three west wall panels exposed, the sound was somewhat too bright; that is, the high frequency components of music and speech were too prominent. For a studio of 2525 cubic feet the optimum reverberation time for music is about 0.5 second and for voice about 0.3 second. This studio, to be used for both, should have a compromise reverberation time somewhere between the two. Flipping the west wall units first one way for a music program and then the other for speech is just too much trouble; many programs involve both. The object, then, is to determine by actual measurements which way these panels give the best compromise effect and then leave them that way.

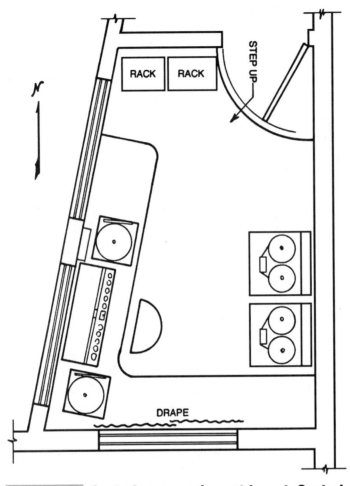

RACK RACK

STEP UP

DRAPE

Figure 5-8 Control room equipment layout. Control room space is considerably less than ideal in order that many students can be accommodated in the studio at one time.

Reverberation measurements were performed using interrupted octave bands of pink noise to evaluate the accuracy of the computations and to determine the proper exposure of the west wall panels. In Fig. 5-9 the measured and calculated values of reverberation time are compared for the west wall panels hard condition. Between 500 and 1,000 Hz the agreement is perfect, but calculated values are too high at low frequencies and too low at high frequencies. This means that there is greater absorption at low frequencies and less absorption at high frequencies than the coefficients of Table 5-1 indicate. These measured results surely explain why the sound was too bright with hard panels!

With the west wall panels exposing the Johns-Manville Temper Tone acoustical tile (the soft condition, see Fig. 5-5), the calculated values are again too high at low frequencies and too low at high frequencies as shown in Fig. 5-10. In the mid-frequency region there is excellent agreement at 500 Hz and 1 kHz.

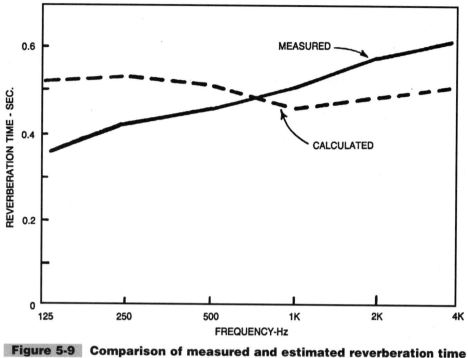

Figure 5-9 Comparison of measured and estimated reverberation time of studio in exercise to illustrate basic problems involving sound absorption coefficients in computations.

Figure 5-11 compares measured reverberation time between the hard and soft conditions of the west wall panels. The measured comparison of Fig. 5-11 shows the same type of separation for the higher frequencies as the calculated comparison of Fig. 5-7. Otherwise the agreement is not too good.

Glancing down the various materials of Table 5-1 and the absorption coefficients used for each, you may ask, "Which ones are in error?" One relatively unfamiliar component was the suspended ceiling. In Table 5-1 good absorption is attributed to it both at low and high frequencies. Is this substantial portion of the total absorption actually realized in practice across the band? Measurements were made both with the suspended AcoustiShell panels in place and with them removed from the room with the results shown in Fig. 5-12. Aha! There is the expected absorption in the low frequencies, but essentially none at 2 and 4 kHz!

Is the measured low-frequency absorption really the expected amount? A simple computation can settle that question. The measured reverberation time at 125 Hz with the suspended ceiling in place was 0.33 second. With the ceiling removed, and no other change in the room, it was 0.65 second. By feeding these values back into Sabine's equation, corresponding values of 375 and 190 sabins of absorption for the two conditions are obtained. The difference of 185 sabins can only be attributed to the suspended ceiling. In our calculations of Table 5-1, 99.8 + 63.0 = 162.8 sabins were assigned to the ceiling. This means the suspended ceiling yields a modest 14 percent more absorption at 125 Hz than the manufacturer's coefficients would indicate.

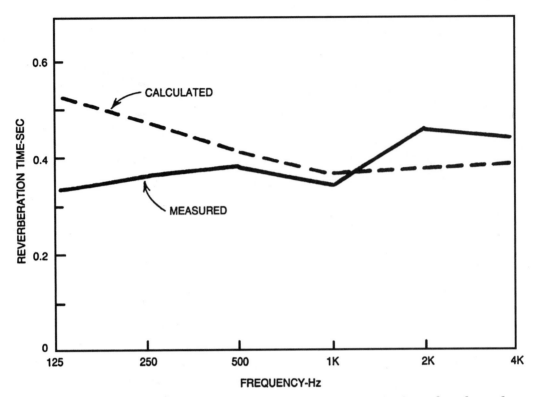

Figure 5-10 **Comparison of measured and calculated reverberation time of studio with west wall panels soft side out. Calculated values of reverberation time are the same as those in Fig. 5-7.**

   The Acousti-Shell ceiling is actually 19 inches from the structural ceiling rather than the standard 16 inches for which the coefficients were obtained. This may account for the difference. Yet in Figs. 5-9 and 5-10 the calculated low- frequency reverberation times are too high, not too low. Where does all this absorption at 125 Hz come from? The semicylindrical element on the south wall is suspect. The assumed absorption coefficient of 0.5 may be too low. If this unit were resonant at 125 Hz, the highest possible coefficient it would be 1.0. If this were the case, the calculated reverberation time for the soft west wall panel condition would be brought down to about 0.47 second. This is in the right direction but not enough to account for the difference.

   Where else can you look for some unexpected low-frequency absorption? The graphs of Figs. 5-9, 5-10, and 5-12 are for the west wall panels in the soft position with the Johns-Manville Temper Tone acoustical tile facing the room. The coefficients used in Table 5-1 for the Temper Tone tile on the north wall are those provided by the manufacturer for the tile cemented to a solid wall. In Fig. 5-5 there is a substantial air space behind the tile on the panels. This would surely enhance both low- and high-frequency absorption of the west wall units with the Temper Tone tile exposed.

   The coefficients used in Table 5-1 for the west wall units, soft side out, are wild estimates designed to allow for the effect of the air space backing as well as the diaphragm effect of the ¼-inch plywood backing. The absorption coefficient assumed for these tiles,

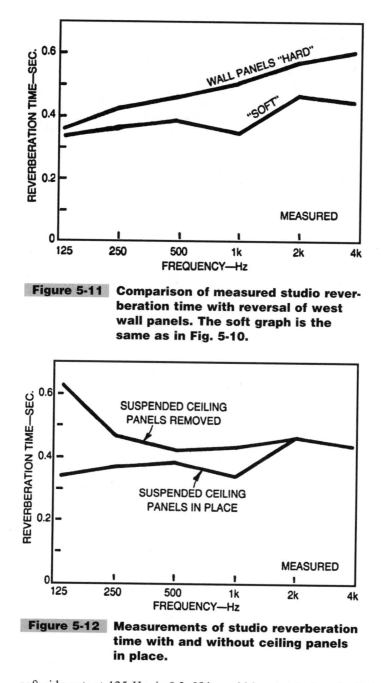

Figure 5-11 Comparison of measured studio reverberation time with reversal of west
wall panels. The soft graph is the
same as in Fig. 5-10.

Figure 5-12 Measurements of studio reverberation
time with and without ceiling panels
in place.

soft side out, at 125 Hz, is 0.2. If it could be stretched to the limit of 1.0, most of the
remainder of the lost absorption at low frequencies would be accounted for. Perhaps 100
percent absorption at 125 Hz is a rather optimistic assumption for both the semicylindrical unit on the south wall and the three panels on the west wall, but something close to 100
percent for these and some slack in other 125 Hz coefficients helps correlate measure-

ments and calculations. It also reveals the flimsy nature of some of these coefficients. So much for disparity between low-frequency measurements and calculations.

In both Figs. 5-9 and 5-10 the measurements reveal that less absorption is realized in the high frequencies than the coefficients of Table 5-1 give. The measurement of Fig. 5-12, with and without the suspended ceiling, reveal that for some reason the AcoustiShell elements seem to give no appreciable absorption at 2 kHz and 4 kHz. In Table 5-1, 109.2 + 74.7 = 183.9 sabins are attributed to the suspended ceiling at 4 kHz. If these sabins are eliminated, only 131 sabins remain for the soft wall panel condition which gives a reverberation time of 0.89 second, far higher than the 0.44 second measured. Total absorption of 249 sabins is required to account for the 0.44 second measured reverberation time. It would appear that something like 66 sabins are being obtained either from the ceiling or some of the other sound absorbing elements listed in Table 5-1.

This has been a tedious excursion into the field of practical absorption coefficients. As James Moir, prominent British acoustician, has said, "Anything that is obvious in acoustics is nearly always wrong."[11] Perhaps the tedium is worthwhile if we learn only one thing: that absorption coefficients are the insecure basis of reverberation time calculations and that the computed results are no more dependable than the coefficients used.

Absorption coefficients supplied by manufacturers for their proprietary materials may or may not be realized in practice, depending on how closely the practical mounting and surroundings approximate those of the measuring conditions.

For the nonproprietary acoustical elements, such as the semicylindrical south wall unit, finding even approximate coefficients is a great problem and we are even more vulnerable to error. Computations, carefully done, serve only as a rough guide. Measurements and subsequent tuning adjustments are essential to accurate acoustical treatment of studios, control rooms, and other critical listening spaces.

SMALL AD AGENCY STUDIO FOR
AVS AND RADIO JINGLES

Feature: Use of midband absorbers.

An advertising agency had a small, makeshift recording facility which was both cramped and poorly laid out. Expansion of agency business required moving up one floor in their large commercial building to provide the space needed. In the process the recording facilities were also to be enlarged and restructured. As is so often the case, those doing the actual recording work found themselves on the wrong end of the totem pole with their space shrinking daily as front office ideas grew during the planning stage. A space at one end of the floor between two concrete walls was eventually designated for studio use. The walls effectively blocked expansion north and south.

The primary use of this facility is in the production of radio advertising announcements. A secondary (but growing) activity is production of audiovisuals, principally slide sets and filmstrips. Two rooms were envisioned, one to be devoted principally to the recording and audiovisual functions, the other to be a combination control room for recording and a general work room in which mixing, editing, and dubbing would also be done. Several people, each working on a different project or at least different aspects of the same project, were bound to get in each other's way from time to time but the "gigantic step forward for mankind" which the new facility offered over the old one made such conflicts seem trivial.

Floor Plan

The floor plan emerging from the smoke and fire of space allocation, incorporating the best functional ideas of production personnel and the acoustical consultant, is shown in Fig. 6-1. It incorporates some basic problems, such as volumes below the 1500 cubic foot minimum (but not much below). Because sound lock space had to be taken from the studio, a cavity is created in the recording studio near the observation window. The temptation to sit in this cozy indentation at the built-in table near the window is great, especially for those nurtured on the radio tradition of the announce booth. The better position for the narrator during recording is back in the main part of the room and not in this indentation. The indentation would not have existed if sound lock space could have been allotted outside the studio area.

The control work room has a built-in work surface along the south and west sides of the room. This bench carries a mixing console as well as numerous advanced audiophiletype magnetic recorders.

Room Proportions

As discussed in Chapter 1, axial mode distribution is something of a problem even when we are free to specify the three dimensions of a room. In this case, two of the three dimensions of both rooms were fixed by circumstances and tight constraints were placed on the third. All that can be done in such circumstances is to study the distribution of fundamental resonance frequencies and harmonics of the space in an attempt to evaluate the threat of colorations on paper before construction is started.

Figure 6-2A is a plot of modal frequencies for the audiovisual/recording studio. The solid lines are associated with the basic 18-foot 4-inch length, 10-foot 2-inch width, and 8-foot 11-inch ceiling height (the length of the lines of Fig. 6-2 holds no significance). The broken lines are associated with the 5-foot 8-inch alcove in the N-S mode and the 12-foot 3-inch step in the E-W mode.

Some of these secondary dimensions within the room (broken lines) occur in rather large gaps between major resonances (solid lines), which is favorable, while others are almost coincident with major dimension modal frequencies, which can be unfavorable. The triple coincidence at about 277 Hz is probably no threat because few colorations are found to be problems

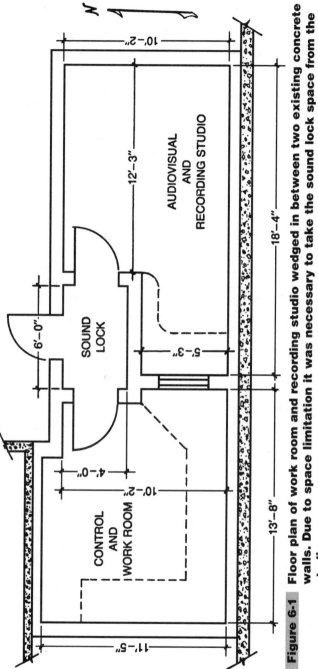

N

10'-2"

AUDIOVISUAL
AND
RECORDING STUDIO

12'-3"

18'-4"

SOUND
LOCK

6'-0"

5'-3"

4'-0"

10'-2"

13'-8"

CONTROL
AND
WORK ROOM

11'-5"

Figure 6-1 Floor plan of work room and recording studio wedged in between two existing concrete walls. Due to space limitation it was necessary to take the sound lock space from the studio.

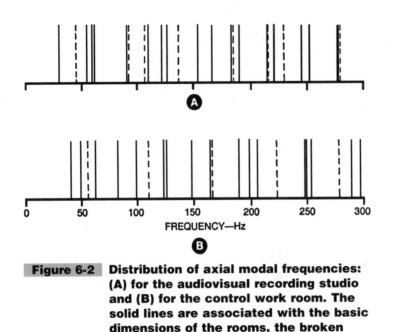

Figure 6-2 Distribution of axial modal frequencies: (A) for the audiovisual recording studio and (B) for the control work room. The solid lines are associated with the basic dimensions of the rooms, the broken lines with secondary dimensions of alcove and step.

above 200 Hz. The three or four double pileups or near pileups below 200 Hz may or may not be troublesome. These will require the application of a keen ear for evaluation.

The solid lines of Fig. 6-2B represent the modal resonance picture for the control work room major dimensions of a 13-foot 8-inch length, 11-foot 5-inch width and 8-foot 11-inch ceiling height. The broken lines are associated with the 10-foot 2-inch N-S secondary step in the room width. With the exception of one at 167 Hz, all the secondary resonances land nicely between the major dimension resonances. In our keen ear analysis of this room, particular attention should be given to the possibility of colorations due to the double coincidences near 125 Hz and 166 Hz.

Wall Construction

It was immediately recognized that other diverse and noisy activities in the building could easily be carried to and radiated in the sound sensitive spaces by the concrete structure of the building. For example, elevator equipment mounted securely to the structure sends impulses into the reinforced concrete walls and pillars and these can be radiated into the studios by concrete surfaces acting as diaphragms. Isolating against such noises took on a high priority. Floating concrete floors were ruled out by budget limitations, but something could be done about walls and ceilings.

Walls paralleling existing concrete surfaces are set back, creating a 3-inch air space filled with glass fiber insulation. Metal studs 2½ inches thick form the framework and a double layer of ⅝-inch gypsum board make up the mass of the wall surface facing the studio. Other walls, including the one in which the observation window is set, are constructed as double metal stud walls separated by 3 inches. The space is filled with glass fiber insulation. Double gypsum board on both faces yield an overall wall 10½ inches thick with a rating in the vicinity of STC-50. The ceiling is suspended from the structural ceiling with a vibration isolation hanger on each wire. A black iron angle frame holds the double ⅝-inch gypsum board ceiling. All gypsum board edges are staggered and all joints caulked with nonhardening acoustical sealant.

This plan provides a reasonable degree of isolation from building sounds on all surfaces but the floor. It was decided that if the floor did become a problem, a wooden floating floor could be added at a later time at minimum cost.

Such structureborne sounds can be a serious problem. For example, I visited a fabulous new government broadcasting house in a certain foreign land. The architect had claimed 90 dB transmission loss as protection against nearby jet landing pattern noise by the studio-within-a-studio technique. Stepping into one of the beautifully treated and decorated studios, however, a distinct hammering noiser from another part of the building was clearly heard to the embarrassment of the engineer-host.

Audiovisual Recording Studio Treatment

The acoustical treatment of the audiovisual recording studio involves four basic elements: the carpet, wideband wall units (2 feet × 4 feet), midband wall units (2 feet × 2 feet), and the low frequency ceiling units (4 feet × 5 feet). The absorption of the gypsum board walls and ceiling has been neglected in the discussion to follow, but will be treated later in the chapter.

Figure 6-3 shows the placement of the wall units, Fig. 6-4A the placement of the ceiling units and Table 6-1 tabulates the computations for this room. Entering into Sabine's formula the room volume of 1390 cubic feet and the reverberation time goal of 0.3 second gives a required absorption of 227 sabins. The problem now becomes one of juggling the areas of the four types of absorbers to give close to 0.3 second reverberation time across the band.

To obtain a uniform reverberation time throughout the audible spectrum requires a constant number of absorption units (sabins) with frequency. In Table 6-1 the total sabins at each frequency is fairly close to the calculated 227 required, varying from 195 to 211.3. To see more clearly how absorption of each type of material varies with frequency, the data of Table 6-1 is graphically presented in Fig. 6-5. The greatly unbalanced carpet absorption (Fig. 6-5A) is quite well compensated by the equally, but opposite, absorption of the low frequency ceiling units. However, there is a sag at midrange frequencies of the low frequency plus carpet curve (Fig. 6-5B) and the midrange units, tuned to the 500 Hz-1 kHz region, are designed to straighten out the carpet + LF + midrange curve (Fig. 6-5C). Once this is done, enough wideband absorber is introduced to raise the total to approximately the 227 sabin level (Fig. 6-5D). This gives close to 0.3 second reverberation time across the audible range shown in Fig. 6-6.

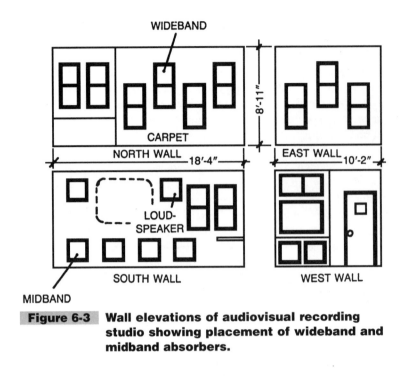

Figure 6-3 **Wall elevations of audiovisual recording studio showing placement of wideband and midband absorbers.**

Low Frequency Units

The low frequency units are most properly placed opposite the carpet for which they compensate, in the usual contracarpet position. The area required for these LF units means that 100 square feet, or 64 percent of the total ceiling area be covered with these 4 foot × 5 foot boxes, but leaving enough space for illumination fixtures. The facings of the low frequency units are quite reflective at the higher frequencies, but vertical flutter echoes are controlled by the carpet absorption.

The construction of the ceiling low frequency units is detailed in Fig. 6-7. The frame and center divider are made of 1 × 8 lumber strengthened by a backing of ½-inch chipboard on plywood. A facing of ³⁄₁₆-inch tempered hardboard perforated with ³⁄₁₆-inch holes spaced 2 ⁹⁄₁₆ inches on centers covers the entire frame. In intimate contact with the perforated cover inside the box is a 4-inch thick layer of Owens-Corning Type 703 Fiberglas of 3 pounds per cubic foot density.

If the glass fiber material is loosely fitted, something is needed to hold it against the perforated facing. The 1 × 4 spacers with fine wire tacked to the edges in a zig-zag form will do this in a very positive way, but if gravity and friction can be depended upon to hold the glass fiber snugly against the back of the perforated cover, so much the easier and cheaper. The air space plays an active part in the performance of this absorber. The boxes can be mounted to the ceiling in any convenient way. Painting these units and the exposed parts of the ceiling flat black will render them visually unobtrusive, especially if the illumination fixtures direct the light downward. Track lights are ideal for this.

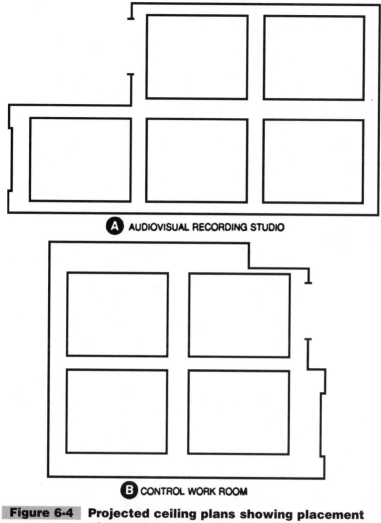

A AUDIOVISUAL RECORDING STUDIO

B CONTROL WORK ROOM

Figure 6-4   Projected ceiling plans showing placement of the 4-foot × 5-foot low frequency units on the ceiling: (A) audiovisual recording studio and (B) control work room.

TABLE 6-1 AUDIOVISUAL RECORDING STUDIO CALCULATIONS

SIZE 10'2" × 18'4" × 8'11" CEILING HT.
FLOOR CARPET
CEILING 5 LOW FREQUENCY ABSORBERS
WALLS 12 WIDEBAND, 6 MIDBAND ABSORBERS
VOLUME 1390 CUBIC FEET

MATERIAL	S AREA SQ. FT	125 HZ A	125 HZ SA	250 HZ A	250 HZ SA	500 HZ A	500 HZ SA	1 KHZ A	1 KHZ SA	2 KHZ A	2 KHZ SA	4 KHZ A	4 KHZ SA
CARPET	157	0.05	7.9	0.15	23.6	0.30	47.1	0.40	62.8	0.50	78.5	0.60	94.2
LOW FREQ. ABSORBERS 5 – 4' × 5'	100	1.0	100.0	0.68	68.0	0.39	39.0	0.17	17.0	0.13	13.0	0.10	10.0
MIDBAND ABSORBERS 6 – 2' × 2'	24	0.35	8.4	0.63	15.1	0.88	21.1	0.84	20.2	0.66	15.8	0.35	8.4
WIDEBAND ABSORBERS 12 – 2' × 4'	96	0.99	95.0	0.99	95.0	0.99	95.0	0.99	95.0	0.99	95.0	0.99	95.0
Total sabins, Sa			211.3		201.7		202.2		195.0		202.3		207.6
Reverb. Time, second			0.32		0.34		0.34		0.35		0.34		0.33

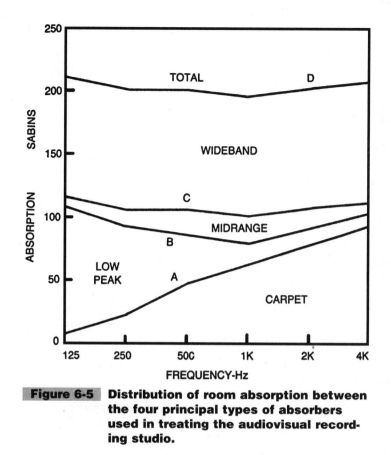

Figure 6-5 **Distribution of room absorption between the four principal types of absorbers used in treating the audiovisual recording studio.**

Hand drilling almost 500 holes in each of the nine ceiling box covers can be a staggering job. The obvious way to minimize this is to stack all covers, drilling all with one set of holes. Each cover can be split at the center divider if desired so that 18 pieces of 30 inch × 48 inch hardboard could be stacked for drilling.

Midband Units

The midband units have a relatively minor, but important, role to play in the overall treatment of the audiovisual recording studio as illustrated in Fig. 6-5. They are mounted on the wall under the window table and along the lower edge of the south wall in Fig. 6-3. Their simple construction is detailed in Fig. 6-8.

The covers are Johns-Manville Transite panels which come perforated with 550 ³/₁₆-inch holes per square foot. These are autoclaved asbestos cement boards ³/₁₆-inch thick and their 24 inch × 24 inch size determines the size of the supporting frame. This frame is made of

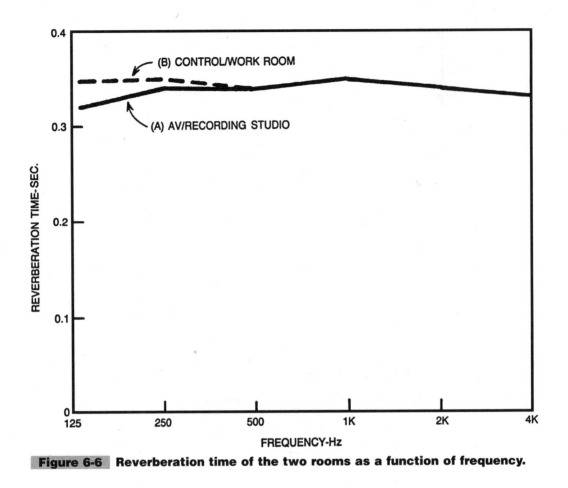

Figure 6-6 Reverberation time of the two rooms as a function of frequency.

1 × 3 lumber with 1 × 2 spacers inside. The 2-inch dimension should be met to accommodate the 2-inch thickness of the Owens-Corning Type 703 Fiberglas.

Wideband Units

Acoustically speaking, the 2 foot × 4 foot wideband units are nothing more or less than 4 inches of 703 Fiberglas of 3 pounds per cubic foot density. The rest is mechanical mounting and cosmetic cover. Figure 6-9 shows how the 1 × 6 frame, 1 × 4 divider and spacers, and the ½-inch chipboard backboard fit together.

The manufacturer of the glass fiber stops short of attributing 100 percent absorption to 4 inches of 703, listing 0.99 as the coefficient from 125 Hz to 4 kHz.

Below 125 Hz the absorption does fall off, of course, but in this low frequency region the diaphragmatic absorption of the five gypsum board surfaces tends to compensate.

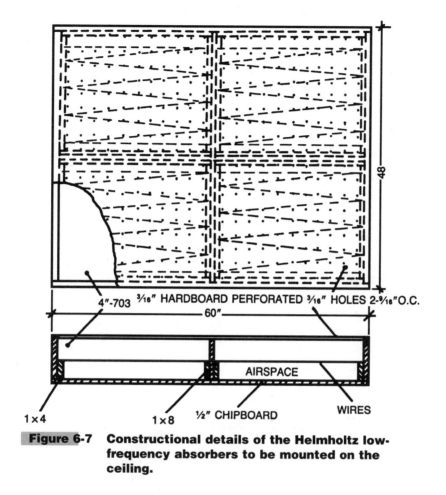

4"-703   ³⁄₁₆" HARDBOARD PERFORATED ³⁄₁₆" HOLES 2-³⁄₁₆"O.C.

AIRSPACE

1×4   1×8   ½" CHIPBOARD   WIRES

Figure 6-7 Constructional details of the Helmholtz low-frequency absorbers to be mounted on the ceiling.

The double ⅝-inch gypsum board surfaces resonate well below 125 Hz. Calculations indicate that the walls paralleling the existing concrete walls on the north and south sides with 5-inch air space have an absorption peak at about 32 Hz. The other walls with their 8-inch air space peak near 26 Hz. These cavities are filled with insulation which increases the breadth of the absorption region markedly, but the resonance frequency little. No values of measured absorption coefficients are available for double ⅝-inch gypsum board walls with these cavity depths. However, by taking the values available for ½-inch gypsum board on 2 × 4s 16 inches on centers and shifting them to take into account the different resonance frequencies involved, an absorption coefficient of 0.05 to 0.08 is estimated for 125 Hz.

With a gypsum board area of about 580 square feet involved in the audiovisual recording studio, something like 30 to 40 additional sabins may be at work at 125 Hz because of wall absorption. This would reduce the reverberation time at 125 Hz to something like 0.23 second and be less effective than this at 250 Hz and above. This possible 23 percent reduction of reverberation time at 125 Hz requires measurements to tie it down specifically, but

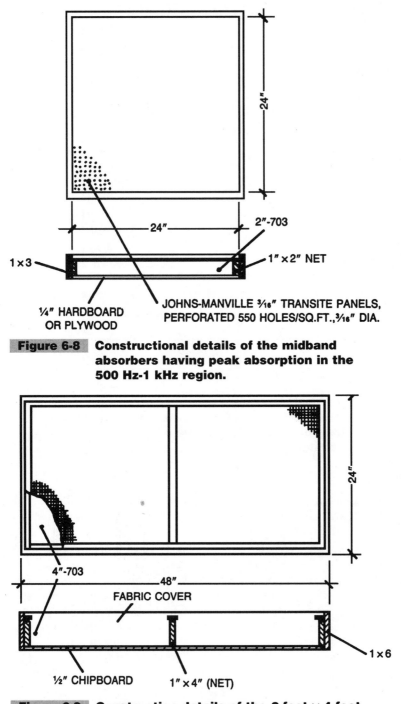

24"

24"

2"-703

1×3

1" × 2" NET

¼" HARDBOARD
OR PLYWOOD

JOHNS-MANVILLE ³⁄₁₆" TRANSITE PANELS,
PERFORATED 550 HOLES/SQ.FT.,³⁄₁₆" DIA.

Figure 6-8 **Constructional details of the midband absorbers having peak absorption in the 500 Hz-1 kHz region.**

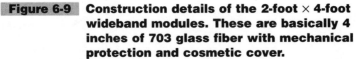

24"

4"-703

48"

FABRIC COVER

½" CHIPBOARD

1" × 4" (NET)

1×6

Figure 6-9 **Construction details of the 2-foot × 4-foot wideband modules. These are basically 4 inches of 703 glass fiber with mechanical protection and cosmetic cover.**

this discussion points out that such wallboard absorption might reduce the required area of ceiling low frequency absorbers.

Control Work Room Treatment

The control work room is treated with the same four elements as the audiovisual recording studio: carpet, wideband and midband wall units, and ceiling mounted low frequency absorbers. Figure 6-10 shows the suggested placement of each unit on the walls and Fig. 6-4B the placement of the four low frequency units on the ceiling. Table 6-2 lists vital statistics of this room as well as the absorption units (sabins) expected of each of the four elements at each of the six frequencies.

There is little need to plot the contribution of each of the four elements. Although the values differ somewhat, the general apportionment principle revealed in Fig. 6-5 for the audiovisual recording studio applies to this room as well. The calculated reverberation times of Table 6-2 are plotted in Fig. 6-6 for ready comparison with those for the other room.

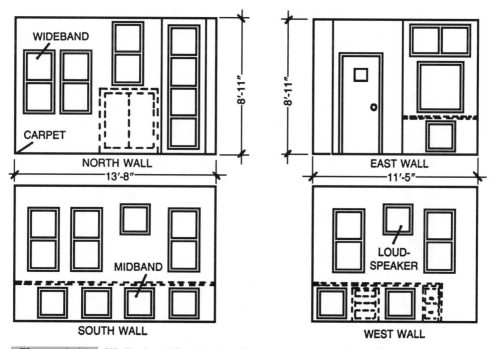

Figure 6-10 Wall elevations of control work room showing placement of wideband and midband absorbing units.

TABLE 6-2 CONTROL WORK ROOM CALCULATIONS

SIZE 11'5" × 13'8" × 8'11" CEILING HT.
FLOOR CARPET
CEILING 4 LOW FREQUENCY ABSORBERS
WALLS 11 WIDEBAND, 6 MIDBAND ABSORBERS
VOLUME 1304 CUBIC FEET

MATERIAL	S AREA SQ. FT	125 HZ		250 HZ		500 HZ		1 KHZ		2 KHZ		4 KHZ	
		A	SA	A	SA	A	SA	A	SA	A	SA	A	SA
CARPET	147	0.05	7.4	0.15	22.1	0.30	44.1	0.40	58.8	0.50	73.5	0.60	88.2
LOW FREQ. ABSORBERS 4 – 4' × 5'	80	1.0	80.0	0.68	54.4	0.39	31.2	0.17	13.6	0.13	10.4	0.10	8.0
MIDBAND ABSORBERS 7 – 2' × 2'	28	0.35	9.8	0.63	17.6	0.88	24.6	0.84	23.5	0.66	18.5	0.35	9.8
WIDEBAND ABSORBERS 11 – 2' × 4'	88	0.99	87.1	0.99	87.1	0.99	87.1	0.99	87.1	0.99	87.1	0.99	87.1
Total sabins, Sa			184.3		181.2		187.0		183.0		189.5		193.1
Reverb. Time, second			0.35		0.35		0.34		0.35		0.34		0.33

MULTITRACK IN A
TWO-CAR GARAGE

Feature: Understanding the room resonance problem.

CONTENTS AT A GLANCE

There has been a proliferation of small studios in basements, barns, garages, and other locations in and around private residences. Some of these are built by the members of new bands who figure that such a facility will help them develop their musical techniques and enable them to record demonstrations and other records at leisure without high studio charges. Some are built by advanced audiophiles who have become jaded to further improvements in the living room hi-fi, but are challenged by recording techniques. Such a private studio is a logical next step after experimenting with multichannel recording with a four-track consumer type tape recorder. This type of experimentation soon runs headlong into the frustrations of household noise, limitations in the number of tracks, and problems created by haphazard, temporary lashups.

Some may look to a private recording studio as a stepping stone to getting into professional recording. As far as gaining experience is concerned, excellent, but if renting the studio to musical groups is contemplated, beware. Most communities look with disfavor on commercial activities in residential areas. Construction of the studio to be described in this chapter definitely requires a building permit and the usage planned for the facility is sure to come up. Now that this point has been made, the studio has limited commercial possibilities. The greater the experience one has, the greater the emphasis on that word limited.

Floor Plan

The two-car garage to be converted into a multitrack studio is described in Fig. 7-1. It is almost square and is covered with a simple A-roof. The open ventilation louvers in the north wall, a 15-foot 3-inch overhead door opening in the east wall, and a small rear door in the south wall emphasize that the garage is just a wide open shell with unfinished walls inside.

The first step is to make the garage into a tight structure and to make provisions for a monitor control room. Figure 7-2 shows one way of distributing the 464 square feet of total area between the studio and control room. This gives a studio floor area of 352 square feet and a control room area of only 85 square feet. The studio size is over twice the minimum prescribed volume of 1500 cubic feet but the control room is only about half the minimum.

This may be sufficient incentive to drive the recording technician to using high quality headphones instead of monitor loudspeakers. The fact is that with a square garage, any location for the control room other than a corner results in serious degrading of studio space. The floor plan of Fig. 7-2 favors the studio. Perhaps that demo record may have greater impact with a reasonable studio and operator wearing headphones than a tiny studio with its poor separation and more accurate monitoring room.

The use of high quality headphones,[12] if not a first choice, is at least a viable alternative for listening critically in acoustically difficult situations. They are being improved much faster than monitoring loudspeakers.

The louver ventilators in the north wall are abandoned and the opening framed in and covered to conform to the other external walls. The overhead door probably should be retained for external appearance, although the bulky hardware may be removed and stored for possible future use. A new frame closes off the 15-foot 3-inch door opening. A door (3 feet wide to accommodate instruments) is cut in the east wall for access to the studio. This should be a 1¾-inch solid core door and well weatherstripped. The sound lock corridor in this case is the great outdoors. The existing doorway in the south wall serves the control room, but the hollow core door is replaced by a 1¾-inch solid core door and also weatherstripped.

Wall and Ceiling Construction

The internal wall and ceiling surfaces are covered with ⅝-inch gypsum board as shown in Fig. 7-3. Great care should be exercised to assure tightness as this layer is the chief assurance against complaints from the neighbors. This requires filling of all cracks and taping

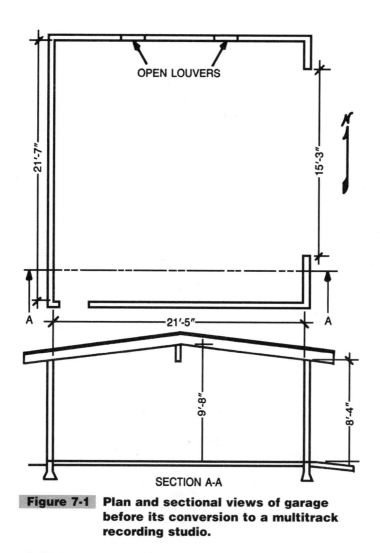

OPEN LOUVERS

21'-7"

15'-3"

N

A   21'-5"   A

9'-8"

8'-4"

SECTION A-A

Figure 7-1 **Plan and sectional views of garage before its conversion to a multitrack recording studio.**

of all joints as well as a liberal use of nonhardening acoustical sealant at all intersections of surfaces. This drywall layer goes in the control room as well as the studio. The diagonal wall between the studio and the control room has a $5/8$-inch gypsum board on each side as shown in Fig. 7-3, Section C-C.

The conversion of this garage to a studio is fraught with compromises. The result will be midway between the living room and a first class studio. The wall between the studio and control room is such that studio sounds will sometimes be heard in the control room without benefit of amplifiers and loudspeakers. Neighbors can probably hear a lively number being played in the studio, but hopefully at a low enough level that they will not call the police.

Unless the two single doors are made impervious to sound, there is little advantage in strengthening the walls. A good, tight wall such as shown in Section C-C of Fig. 7-3 has an STC rating of 30 to 35 and the stucco plaster of Section A-A is only slightly better. Great care must be exercised in weatherstripping doors to get STC 30. The $3\frac{1}{2}$ inches of

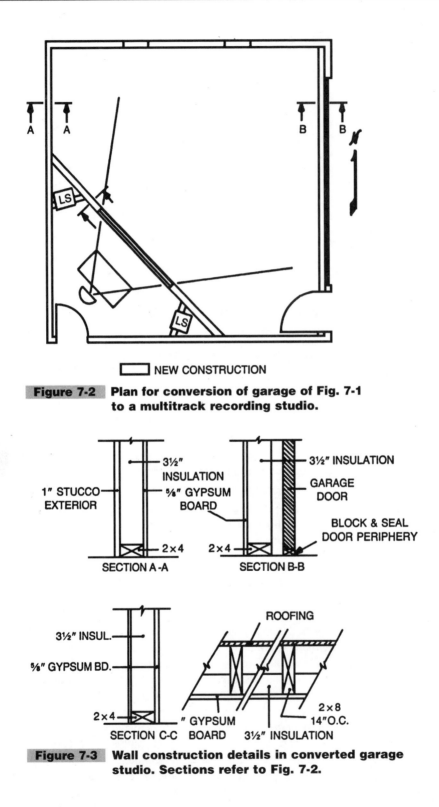

NEW CONSTRUCTION

Figure 7-2 Plan for conversion of garage of Fig. 7-1 to a multitrack recording studio.

3½" INSULATION

1" STUCCO EXTERIOR

⅝" GYPSUM BOARD

2×4

SECTION A-A

3½" INSULATION

GARAGE DOOR

BLOCK & SEAL DOOR PERIPHERY

2×4

SECTION B-B

3½" INSUL.

⅝" GYPSUM BD.

2×4

SECTION C-C

" GYPSUM BOARD

ROOFING

3½" INSULATION

2×8 14"O.C.

Figure 7-3 Wall construction details in converted garage studio. Sections refer to Fig. 7-2.

insulation fill indicated in all sections of Fig. 7-3 contributes very little (about 2 dB) to the transmission loss. It does help some in discouraging cavity resonances. Because of its minor effect, it could be omitted if the budget is very tight.

Studio Treatment

Multitrack recording requires acoustical separation of the sounds of instruments or groups in instruments which are recorded on separate tracks. One way of achieving such separation is to physically separate the sources and place a microphone close to each source. Space is limited in this studio, but this logical and desired approach is more effective in an acoustically dead space than in a live one. There are other ways separation can be achieved, such as the use of microphone directivity, or baffles.

ACOUSTICAL GOALS

Musician reaction places a limit on the deadness of such a studio because they must hear themselves and other musicians to play effectively. The studio of Fig. 7-4 has been made as dead as practical to allow the achievement of reasonable track separation, even though the space is small for this type of recording.

FLOORS AND CEILING

Heavy carpet and pad are applied to the entire concrete floor except the drum booth area. This opposes the sloping, reflective ceiling surfaces which are bare gypsum board. Reflections from this ceiling could contribute to leakage between tracks. If this proves to be a limiting factor, absorbent material could be applied to critical areas of the ceiling.

Walls

Much of the wall is faced with 4 inches of 703 semirigid glass fiber boards. These glass fiber panels are inserted between vertical 2 × 4s which are mounted against the gypsum board wall covering and run from floor to ceiling. Figure 7-4 shows the 2 × 4s spaced 16 inches center to center, lined up with the studs of the exterior wall to which they are nailed. After the glass fiber is installed between these inner studs, a fabric cover is stretched over all, tacked in place, and finished strips nailed on the edge of each 2 × 4 to complete the floor to ceiling job. The south half of the east wall and the south wall are left reflective to provide an area in the southeast corner of the studio, near the door, which would have somewhat brighter acoustics than the other areas near absorptive walls. Such localized acoustics can be of great help in instrument placement.

On the north wall are two swinging panels 4 feet wide running from an inch above the carpet to a height of 6 feet to 8 feet. These panels are framed of 2 × 4s with ³⁄₄-inch particleboard or plywood backs for strength holding 4 inches of 703 glass fiber covered with

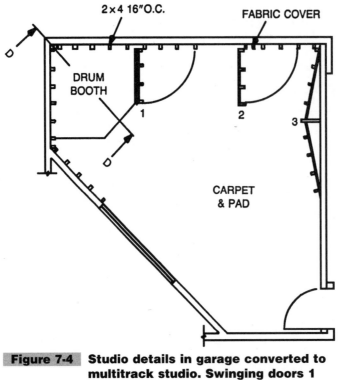

Figure 7-4 Studio details in garage converted to multitrack studio. Swinging doors 1 and 2 serve as baffles between instruments. Element 3 on east wall helps to absorb very low frequencies.

fabric like the walls. The back of panel 1 in Fig. 7-4 in open position presents the drummer with his only reflective surface apart from the floor. The space between swinging panels 1 and 2 can be occupied by one instrumentalist (or group) while the space between panel 2 and the east wall can be occupied by a second. Others will have to be positioned in the open area.

On the east wall the same 2 × 4 framing filled with 4 inches of 703 is followed except that here a space behind is provided to augment absorption in the very low frequencies (element 3). At these frequencies the sound penetrates the 4 inches of glass fiber, causing the ¾-inch plywood or particleboard to vibrate as a diaphragm, absorbing sound in the process. It may be found that the instrumentalist in space 1 is too close to the high level drum sound. It would be quite acceptable to move the low frequency element to the north wall and the instrument alcoves 1 and 2 to the east wall if this would meet separation needs better. Such a move would increase the distance between instrumentalists (and their microphones) and the drum kit. Barriers 1 and 2 would provide separation only between instruments, not between instruments and drums unless more complicated double-hinged panels were installed.

Drum Booth

The corner area for the drum booth is indicated in Fig. 7-4. The concrete floor of the booth is left bare under the drums to give the desired effect. Another reflective surface for the benefit of the drummer is the surface of swinging panel 1. Apart from these, all surfaces around the booth are highly absorptive to contain the drum sounds and thus improve separation from the sounds of other instruments.

Without adequate separation the advantages of multitrack recording disappear. There is general agreement on two points:

- That sounds from the drum kit are hard to contain.
- A good drum sound is basic to any group.

This is justification for doing something extra for the drummer, even in a budget facility such as this. It would be nice if we could do the same for the vocalist.

Leaving the bare ceiling over the drum booth would defeat the whole purpose of the booth. Drum sounds must be absorbed rather than allowed to float over the entire area. A canopy of 2 × 4 framing is dropped down from the ceiling to a point 6 feet above the concrete floor as shown in Fig. 7-5. The shape of this canopy follows the general drum booth front edge shown in Fig. 7-4.

The face of the canopy toward the studio is covered with a double layer of ⅝-inch gypsum board, but the inside of the entire 2 × 4 framing on 16-inch centers is left open to receive 4 inches of 703 glass fiberboard.

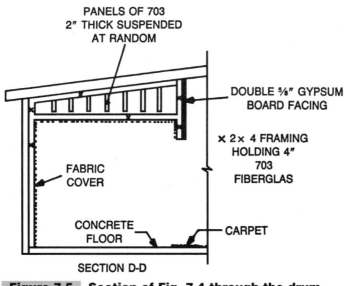

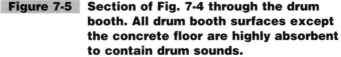

Figure 7-5 Section of Fig. 7-4 through the drum booth. All drum booth surfaces except the concrete floor are highly absorbent to contain drum sounds.

A fabric cover is then applied and held in place with finished strips nailed to the 2 × 4s. This 4 inches of 703 treatment faces the drummer on all sides including walls, underside of canopy, and inside of canopy lip. The construction leaves an *attic* between the canopy ceiling of 4 inches of 703 and the 4 inches of 703 affixed to the uppermost ceiling gypsum board. Because drums have a hefty low frequency content, the attic can be made into an effective absorber for very low frequencies, frequencies lower than the 4 inches of 703 can handle. The term *basstrap* has been applied to such absorbers. Such a catchy term, which seems only to confuse the populace, is not used in this book, at least until *trebletrap* is accepted to describe the acoustical effect of carpet. Low frequency absorbers, however, do come in varying degrees of *lowness*. The 4 inches of 703 is essentially a perfect absorber down to 125 Hz and its absorbing effectiveness decreases as frequency is lowered. For this drum booth attic, a trick of the builders of early anechoic, *free-field chambers* or *dead rooms* is used. This is the hanging of spaced absorbing panels to extend the useful low frequency range of the room. The spacings of the panels of 2-inch thickness 703 may be random rather than following meticulous rules used in those early days.

Computations

Reverberation time, of itself, does not play too important a role in multitrack recording because our primary goal is adequate track separation. Distinctive sound, not naturalness, is the goal of rock recording. Following through on the computations of Table 7-1, however, gives a good feel for comparing treatment of this type of studio with the more traditional speech and music studios. The drywall, carpet, and 4 inches of 703 areas are estimated. The extra bass absorption effect of the drum booth attic and the structure on the east wall are neglected as their effect over and above that of the 4 inches of 703 is largely below 125 Hz, the lowest frequency of Table 7-1.

The calculated reverberation times of the studio range from 0.30 to 0.22 second as compared to 0.35–0.60 second if the same studio were treated for recording speech and traditional music (Fig. 7-6). This comparison emphasizes the general deadness of multitrack studios in the quest for track separation.

Studio deadness, of course, is not the only step toward adequate track separation. The use of baffles, microphone directivity, close placement of microphones, and other factors have their important effects.

Control Room Treatment

Figuring the axial mode frequencies of a triangular room is far beyond the scope of this book, but the average ceiling height of about 9 feet would yield a fundamental of around 60 Hz; the others would not be too far from this. Cutting off the sharp corners near the window provides a shelf for the loudspeakers. To remove the bad effects of the cavity in which the loudspeakers sit, a heavy plywood baffle should be fitted around the face of the loudspeaker (Fig. 7-7).

TABLE 7-1 Studio Calculations

SIZE: 21'5" × 21'7" WITH CORNER CUT
FLOOR: CARPET AND PAD (EXCEPT FOR DRUM BOOTH)
CEILING: ⅝" GYPSUM BOARD
WALLS: ⅝" GYPSUM BOARD, PARTIALLY COVERED WITH 4" 703
VOLUME: 3,170 CUBIC FEET

MATERIAL	S AREA SQ. FT	125 HZ		250 HZ		500 HZ		1 KHZ		2 KHZ		4 KHZ	
		A	SA	A	SA	A	SA	A	SA	A	SA	A	SA
Drywall	1,000	0.1	100.	0.08	80.	0.05	50.	0.03	30.	0.03	30.	0.03	30.
Carpet 4"703	310	0.05	15.5	0.15	46.5	0.30	93.	0.40	124.	0.50	155.	0.60	186.
	500	0.99	495.	0.99	495.	0.99	495.	0.99	495.	0.99	495.	0.99	495.
Total sabins, Sa			510.5		621.5		638.0		649.0		680.0		721.0
Reverberation time, sec.			0.30		0.25		0.24		0.24		0.23		0.22

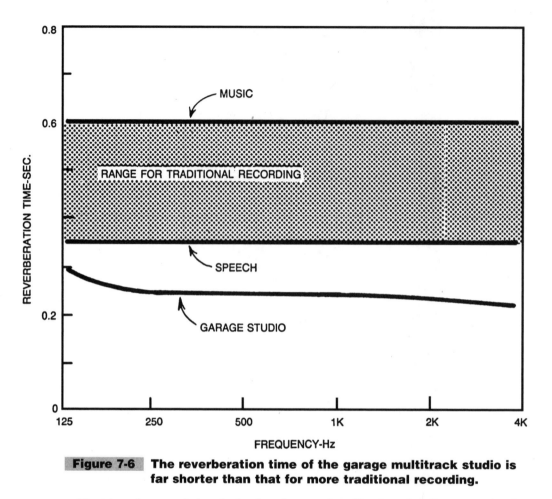

Figure 7-6 **The reverberation time of the garage multitrack studio is far shorter than that for more traditional recording.**

The triangular space below the loudspeaker may be utilized as a low frequency absorber to help counteract the carpet effect. To coin a euphonious phrase, this is a slit resonator utilizing the slots between the slats and the cavity behind. It is made simply of 1 × 4s spaced about ¼ inch with 2 inches of 703 pressed against the rear of the slats. The cavity itself serves only to contain the springy air.

On the walls behind the operator about thirty 12 inch × 12 inch × ¾ inch acoustical tiles are cemented to each wall in a 6 × 5 array. This totals 60 square feet. Carpet covers the floor and the ceiling is drywall. The two resonators plus some 350 square feet of gypsum board compensate for the low frequency deficiencies of the tile and carpet, but not completely. This brings the reverberation time of the control room to about 0.34 second, rising to about 0.46 second below 250 Hz. With all the compromises involved there seems to be little justification for further acoustical adjustment.

OUTSIDE NOISE TO INSIDE

No matter how high the average and peak levels of music are in the studio, there are those soft, sweet, and sentimental passages for contrast. The barking of a neighbor's dog heard

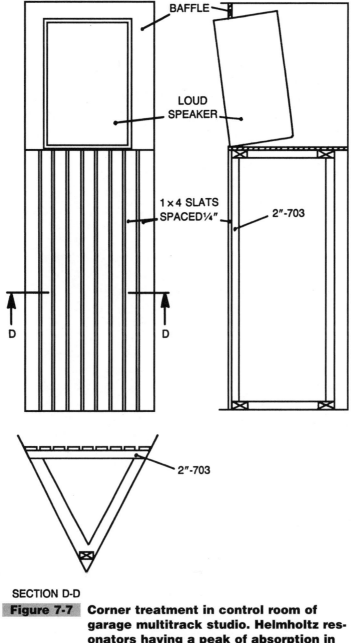

BAFFLE

LOUD
SPEAKER

1 × 4 SLATS
SPACED ¼"

2"-703

D   D

2"-703

SECTION D-D

Figure 7-7 Corner treatment in control room of
garage multitrack studio. Helmholtz resonators having a peak of absorption in
the low frequencies are built into the
corners. These are slat type resonators.

on the vocal track during such a passage is guaranteed to raise the emotional level even higher than the vocalist could hope for. Therefore, concrete walls 6 feet thick are longed for at such times, but they are just too expensive.

The simple walls of Fig. 7-3 are all this budget could stand. Statistics are on the side of the amateur or low budget recording job. How often do soft passages occur? How often do screaming motorcycles or other interfering noises occur? The permutations and combinations are such that redoing a very occasional ruined take is usually the answer for this type of studio. However, it is quite a different story if it costs $5,000 in studio time and for talent to redo the passage.

INSIDE NOISE TO OUTSIDE

Very square neighbors have been known to identify that beautiful stuff being laid down in multitrack within the studio as noise. The difference in point of view can bring the police. In some communities the noise ordinance is as broad and general as this one:

> *It shall be unlawful for any person to make, cause or permit to be made, any loud or unusual noise which directly causes an unreasonable interference with the use, enjoyment and/or possession of any real property owned or occupied by any other person.*

In an increasing number of communities a certain maximum allowable noise level is set for the boundary of the property on which the studio rests. In brief, the sounds escaping from the studio may give far more trouble than exterior sounds spoiling takes. Walls offer the same transmission loss both ways and the walls of Fig. 7-3 are certainly minimum with respect to noise going either way.

MULTITRACK RECORDING

Multitrack recording techniques require studios quite different from the traditional kind. Numerous books have been written on the multitrack subject [13-15] as well as many articles in the technical press.[16-26] I refer you to them for information on the host of pertinent points which cannot be covered in this book.

BUILDING A STUDIO FROM SCRATCH FOR RADIO PROGRAM PRODUCTION

Feature: Adjustable acoustics, service areas, splayed walls.

CONTENTS AT A GLANCE

Many small studio projects must be warped around to fit into space in an existing building. This means compromises, compromises, and more compromises. When this particular client said that the studio building was to be built from the ground up, the news was received with delight. That was the good news. Then came the bad news: The new building was to be jammed between existing buildings on two sides and a brick wall on the third side. But even this means a certain amount of shielding from exterior noise by these masonry structures.

The space available was 18×6 meters (the site is in a foreign country) or about 59 feet 0 inches $\times$ 19 feet 8 inches. This was taken as the outside dimensions of the building. Besides a studio and control room, space was required for tape and film storage, toilet facilities, and an office.

The floor plan of Fig. 8-1 was developed with some give and take. Walls of spaced double brick tiers were specified to surround the studio and control room as protection against outside noise. The studio size and shape was the first task. What maximum studio length could best fit a maximum inside width of 18 feet with a practical, economical ceiling height? Several factors were considered: one variable (length); one constant (width); and one variable with definite constraints (ceiling height).

When the opportunity to splay two of the four walls presents itself, as it did in the present case, it is wise to do so. This eliminates (or at least reduces) the chance for flutter echoes between parallel surfaces.

Placement of acoustical materials can also reduce flutter echoes, but if it can be done with geometry independent of the acoustical treatment, a greater degree of flexibility accrues in the placement of such materials.

The east wall containing the control room window was inclined at about 5 degrees from its rectangular position as shown in Fig. 8-2. The south wall was given two triangular protuberances with sides inclined 8.6 degrees. Any angle between 5 degrees and 10 degrees is usually sufficient for control of flutter echo. This inclining of wall surfaces takes care of the N-S and E-W flutter modes. The vertical mode flutter could be cared for by constructing wrinkles of some sort in the ceiling (or the floor!) but other methods will be used.

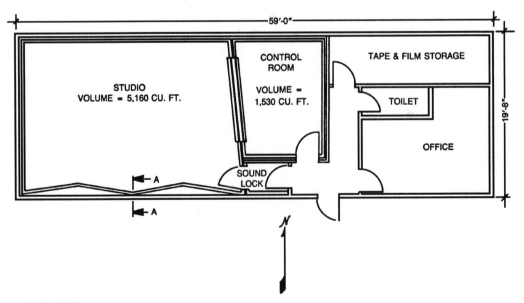

Figure 8-1   **Floor plan of studio complex which includes service areas as well. Splayed east and south walls help control flutter echo.**

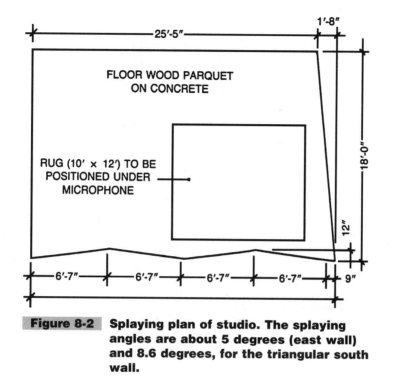

FLOOR WOOD PARQUET
ON CONCRETE

RUG (10′ × 12′) TO BE
POSITIONED UNDER
MICROPHONE

Figure 8-2 Splaying plan of studio. The splaying
angles are about 5 degrees (east wall)
and 8.6 degrees, for the triangular south
wall.

Distribution of Modal Resonances

Again, considering only the axial modes of the studio and disregarding the less influential tangential and oblique modes, let us select studio length and ceiling height which will give reasonable distribution of room resonances. With a splayed east wall and a south wall broken up by four splayed surfaces, the average length is taken to be 26 feet 3 inches and the average width 17 feet 6 inches. With a height of 11 feet 3 inches, the distribution of modal resonances is as illustrated in Fig. 8-3A.

One very great advantage of a larger studio is that the average spacing of room resonances is reduced. The average spacing of resonances in this studio (5160 cubic feet) is 11 Hz. The average spacing of resonances in the 1530 cubic feet control room (Fig. 8-3B) is about 16 Hz. This closer spacing, if not too close or coincident, tends toward fewer colorations, thus better quality.

The small numeral 2 above certain room resonance lines in Fig. 8-3 indicates a coincidence of two modes at that frequency. You can disregard the two coincidences in the studio at higher frequencies and the one at 65 Hz because of the BBC experience that audible colorations are rare in those frequency regions.[6] This leaves the coincidence at 129 Hz and the close pair near 150 Hz as possible threats.

In the control room (Fig. 8-4) coincidences at 249 Hz are high enough to not cause concern. The close pair at 166/169 Hz is aggravated by wide spacings on each side. As sound decays in the room, these two could beat together at a 3 Hz rate.

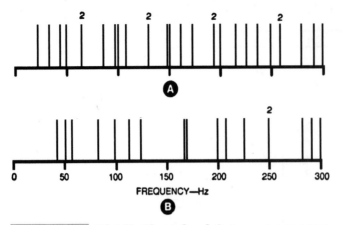

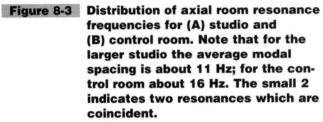

FREQUENCY—Hz

Figure 8-3   Distribution of axial room resonance frequencies for (A) studio and (B) control room. Note that for the larger studio the average modal spacing is about 11 Hz; for the control room about 16 Hz. The small 2 indicates two resonances which are coincident.

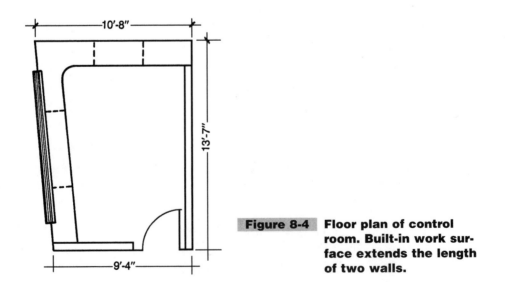

Figure 8-4   Floor plan of control room. Built-in work surface extends the length of two walls.

Other factors must be favorable before coincidences result in observable colorations, hence they may or may not be troublesome. It is impractical to push further an analysis such as this; suffice it to say that there is the advantage of a warning of potential problems at 129 Hz and 150 Hz in the the studio and 166/169 Hz in the control room.

Noise Considerations

How heavy should the walls be made? In general, the heavier the walls the better protection offered against external noise. On the other hand, the heavier the walls the greater the cost. In the present case, brick, a favored and economical local building material, seemed both logical and economical for walls. The question is, single tier wall or double tier?

The answer is to be found in two categories: (a) the level of environmental noise at the studio site, and (b) the level of noise allowable inside the studio. The walls (and other parts of the enclosure, such as doors) are called upon to reduce (a) to (b).

The hoped for lowness of studio noise may be expressed by selection of a standardized noise contour. In this example the NC-15 contour was selected which is the lower curve in Fig. 8-5. This is a reasonably stringent requirement. Studio noise levels somewhat above this (e.g., NC-20) would not seriously impair most types of recording.

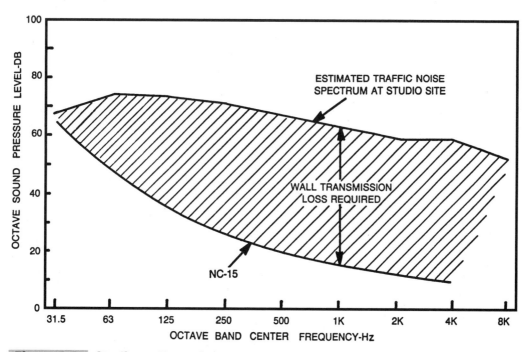

Figure 8-5   Studio wall requirements are based on the environmental noise level at the site and the noise contour selected for inside the studio. The shaded area between these two curves represents transmission loss which the walls must provide.

A lower background noise level is required for mono recording than for stereo. As this studio is engaged only in mono recording at present, with plans to covert to stereo in the future, it seemed wise to select the more conservative NC-15 criterion. Because of the increase in sensitivity of the human ear as frequency is increased, the NC-15 contour has its characteristic downward slope.

Evaluating the environmental noise at the studio site in any specific, direct way is a rather complicated procedure. This noise is usually anything but steady, having peaks and low points between peaks. There is usually a variation with time.

This studio is located behind an office building, about 100 feet from a city street carrying typically heavy downtown traffic. Because actual around-the-clock measurements were not feasible, similar measurements made by others and reported in the literature were used as the basis of estimation.[27] This procedure gives the upper curve in Fig. 8-5. The shaded area between the two curves represents the transmission loss which the walls must give to reduce the exterior noise to the NC-15 level in the studio.

The differences between the two curves of Fig. 8-5 at selected frequencies gives the transmission loss required which is replotted as the heavy line in Fig. 8-6. Adequate data on the transmission loss of brick walls is difficult to find, but a paper published in Europe gives measurements on single and double tiered brick walls which are adapted in Fig. 8-6.

The double tiered brick wall has one face cement plastered, the single tiered wall has no plaster. If the transmission loss offered by these brick walls is greater than the transmission loss required, well and good. If below, the walls are falling short of the requirement.

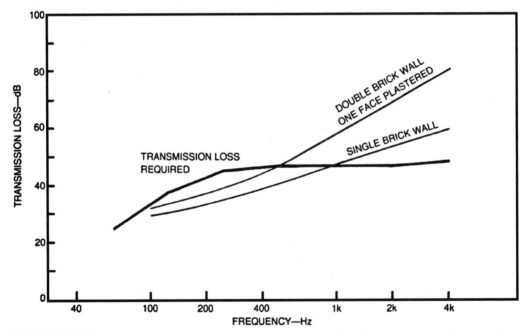

Figure 8-6 The transmission loss required curve is obtained from the number of dB separation between the two curves of Fig. 8-5. The transmission loss offered by different types of wall construction may then be compared to that loss required.

In Fig. 8-6 the single brick wall falls far short of the required loss from about 100 to 1,000 Hz. The double brick wall is considerably better, falling short a maximum of only 6 dB between 100 and 400 Hz. The site noise level and the selection of the NC-15 contour are not precise enough to make much of a fuss about this 6 dB.

Further, plastering both the exterior and interior faces of the double brick wall increases its transmission loss to the point where the 6 dB difference is partially made up. The conclusion, then, is that a double brick wall, plastered on exterior face and studio face, should bring the studio background noise level to about the NC-15 contour, a satisfactory level for the recording work contemplated.

Wall Construction

Figure 8-7 is Section A-A of Fig. 8-1. It shows one method of supporting the ceiling joists on the inner tier of bricks. Solid ties between inner and outer brick tiers, intentional or otherwise, seriously degrade the transmission loss of the wall. Local codes may require such ties, but they should be used as sparingly as possible or not at all.

In building a studio from the ground up there is the opportunity of including flutter echo insurance with no significant increase in cost. The inner brick wall is built into triangular shapes on the south wall and the splayed east wall helps both the studio and the control room.

Sound Lock

The normal studio layout would open both studio and control room doors into the sound lock. In the present case high control room traffic resulted in the decision to open the control room into the entrance lobby (Fig. 8-1). It must be recognized that this somewhat negates the double east wall of the control room as the single control room door will give only some 30 dB isolation, compared to something like 50 dB for the wall. Office noise may be a greater threat than exterior noise.

Studio Floor

The floors of the studio are specified to be covered with wood parquet. This is a beautiful floor covering which is ruled out by high cost in many areas, but not in this one. This is a highly reflective floor as far as sound waves are concerned. It is opposed by a highly absorbent ceiling to be described later. A microphone placed near any reflective surface receives both a direct component of sound from the source and another reflected from the surface. The latter arrives later and creates comb filter distortion in the electrical signal obtained from the microphone.[29] For a close source, the 10 foot × 12 foot rug under the microphone and source will minimize this form of distortion. For a distant

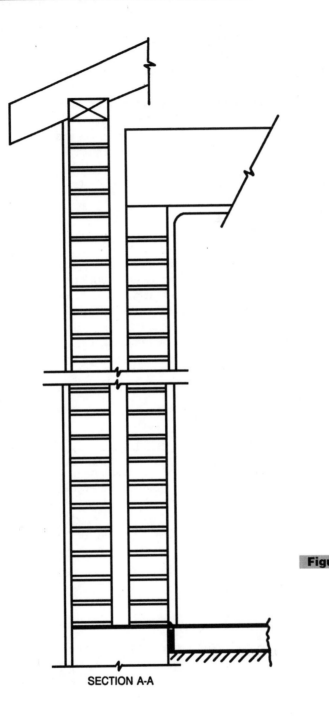

SECTION A-A

Figure 8-7 Studio wall construction of double-tier brick, plastered inside and out. Section A-A specified in Fig. 8-1.

pickup the microphone should be placed very close to the floor so that the two path lengths are approximately equal, or on a stand so that the bounce takes place on the other end of the rug.

Studio Walls

Three types of wall treatment (Fig. 8-8) are used in the studio:

- wideband (WB) absorbers
- acoustical tile
- low frequency (LF) absorbers to compensate for low frequency deficiencies in WB absorbers and acoustical tile, especially the acoustical tile

WIDEBAND MODULES

The wideband modules, described in Fig. 8-9, are similar to others in previous chapters. For example, these modules are similar to those of basically 4 inches of 703 type of glass fiber with a suitable mounting structure. Instead of a fabric face cover, a board is used. An ideal cover is Owens-Corning glass cloth covered Fiberglas boards 1 inch thick.

In some foreign countries such a board may not be readily available, in which case a soft wood fiberboard ½-inch thick could be substituted with modest degradation of absorption. Similarly, glass wool or glass fiber of other types could be substituted for the 703 Fiberglas if density approaches 3 pounds per cubic foot.

A degree of variation of studio acoustics can be introduced by mounting the eight wideband modules as shown in Fig. 8-10. Such a mounting allows reversal of each module by the simple expedient of flipping the top latch, lifting the module off the pin, reversing the module, replacing it on the pin and reengaging the top latch. In this way either soft or hard faces of some or all of the modules could be exposed.

LOW FREQUENCY ABSORBERS

Figure 8-8 shows four 4 foot × 8 foot and one 8 foot × 8 foot low frequency absorbers on the studio walls. These can be built with either perforated facings or slat-slit facings, the latter being chosen in this particular design.

The frames of 2 × 8 lumber are first constructed and then the cross pieces of 2 × 8 are added as shown in Fig. 8-11A. At this point the frames are mounted on the wall with suitable angles and expansion screws (Fig. 8-11B). The crack between the frame and wall is then sealed on the inside with running beads of nonhardening acoustical sealant type of mastic so that each smaller section is essentially airtight as far as the wall junction is concerned.

The glass fiber material will probably be stiff enough so that, cut slightly large, the pieces will be held by friction against the backs of the slats without support. If support is needed, zig-zag wires tacked to the 2 × 8s ensure close contact of glass fiber and slats. To avoid stroboscopic optical illusions as eyes are swept horizontally across the vertical slats, the face of the glass fiber should be covered with lightweight black cloth. Instead of

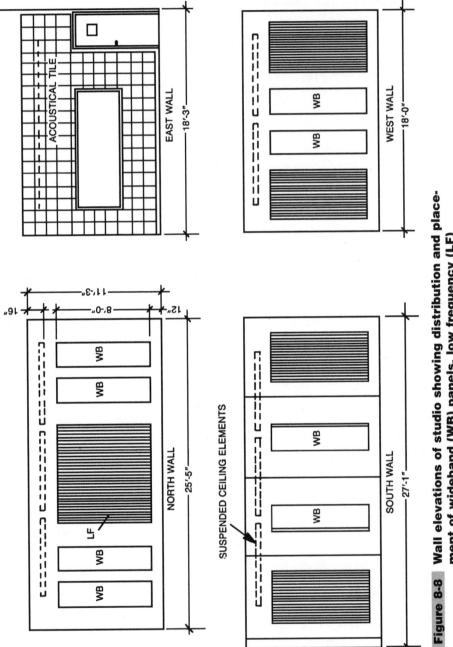

Figure 8-8 Wall elevations of studio showing distribution and placement of wideband (WB) panels, low frequency (LF) absorbers, and acoustical tile.

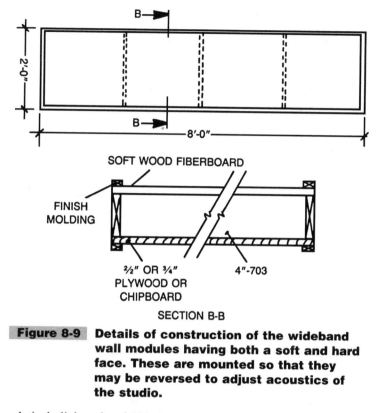

Figure 8-9 Details of construction of the wideband wall modules having both a soft and hard face. These are mounted so that they may be reversed to adjust acoustics of the studio.

relatively light colored 703, Owens-Corning and perhaps other manufacturers, also make a black duct liner which has been used in this service. This eliminates the need for the black cloth. The slats should be stained and varnished before nailing in place. The spacings should be alternately ⅛ inch and ¼ inch. The use of temporary shims while nailing assures uniform slits from top to bottom.

ACOUSTICAL TILE

The entire east wall (with the exception of the observation window and door) is to be covered with ½-inch acoustical tile (Fig. 8-8). These tiles may have perforations or slits or neither if of high quality. They are cemented in place in the usual way. The lower tiles are subjected to considerable abrasion and storing a few dozen matching tiles for later repairs is a good idea.

Studio Ceiling

The studio ceiling has plaster 1 inch thick like the walls. To reduce the reflectivity of the ceiling, six elements are suspended by wires so that the lower edge of the 2×6 frames is 16 inches below the plaster surface. These elements hold absorbing material and light fixtures.

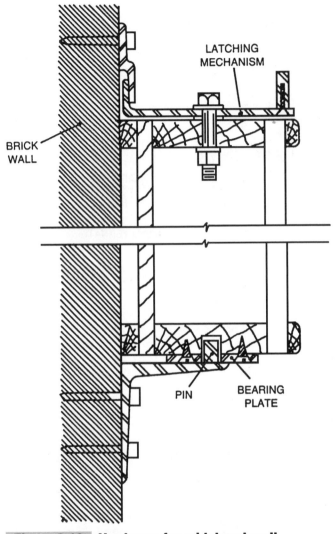

Figure 8-10 Hardware for wideband wall modules to allow reversal.

Figure 8-12A shows how the 6 foot 6 inch × 6 foot 6 inch frame of 2 × 6 lumber is subdivided into nine smaller sections. The center one holds a specially fabricated sheet metal rectangular box containing four 18-inch fluorescent tubes (remember that the starter reactors should be mounted outside the studio to avoid buzzing noises).

The inside lower edges of this metal box should be painted flat black. The deep recessing of the tubes and the black paint are to remove the tubes and their bright reflections from the view of those in the control room to avoid glare in the eyes and in the observation window.

Section B-B of Fig. 8-12B reveals the basic constructional features of each element. Metal tees and angles, such as used in conventional lay-in suspended ceilings, form the 24 inch ×

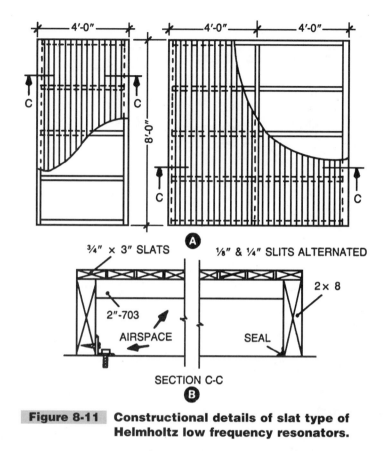

¾" × 3" SLATS   Ⓐ   ⅛" & ¼" SLITS ALTERNATED

SECTION C-C
Ⓑ

Figure 8-11   **Constructional details of slat type of Helmholtz low frequency resonators.**

24 inch sections, eight of which have plastic grilles with honeycomb or square openings resting in them. The 4 inches of 703 is held up by these grilles. A lightweight fabric should be placed between the grille and the 703 it supports to prevent small bits of glass fiber from sifting down the artists' necks. Of the 48 24 inch × 24 inch sections, only 38 are used to hold glass fiber in the design to follow. This means that there will be one or two vacant sections in each element. The vacant sections should be distributed randomly within the element and across the studio.

The six ceiling elements are suspended by wires from the ceiling joists according to the projected ceiling plan of Fig. 8-13.

Control Room Treatment

The control room floor is also wood parquet. Opposing it is a conventional lay-in suspended ceiling dropped down 16 inches from the plastered surface. The walls are treated as shown in Fig. 8-14 with the same type of low frequency absorber used in the studio. They are constructed as shown in Fig. 8-11 and use the same acoustical tile.

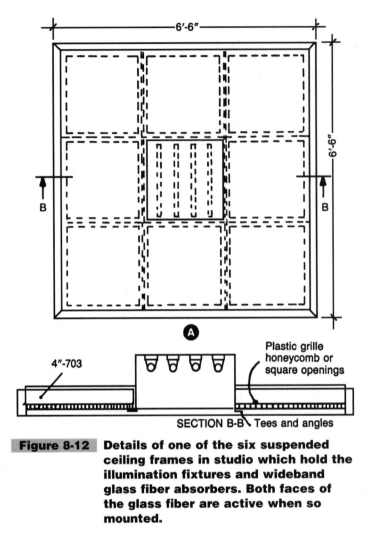

Figure 8-12 Details of one of the six suspended ceiling frames in studio which hold the illumination fixtures and wideband glass fiber absorbers. Both faces of the glass fiber are active when so mounted.

Studio Computations

The generally accepted optimum reverberation time for a studio of 5160 cubic feet volume is about 0.67 second for music and 0.4 second for speech. The studio is to be used for both, hence some compromise reverberation time must be used. This could be warped upward or downward, depending upon whether music or speech were to be favored. The treatment of this studio has been presented as a *fait accompli* in previous paragraphs, but retracing the calculation steps should be a profitable exercise as this will provide the basis for the reader who wants to adapt this information to his own studio situation.

Table 8-1 gives reverberation time calculations for two conditions. Condition A is when the soft fiberboard sides of all eight wideband modules are facing the studio. Condition B is when the hard faces of all eight wideband modules are exposed. The results of these calculations are presented in graphical form in Fig. 8-15.

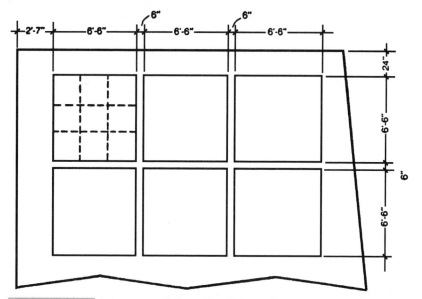

Figure 8-13 Placement details of the six suspended frames in the studio.

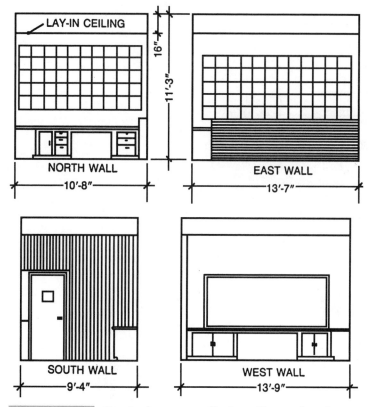

LAY-IN CEILING

NORTH WALL
10'-8"

EAST WALL
13'-7"

SOUTH WALL
9'-4"

WEST WALL
13'-9"

Figure 8-14 Control room wall elevations showing distribution of slat resonators and acoustical tile. A conventional suspended ceiling is used in the control room.

TABLE 8-1 STUDIO REVERBERATION TIME CALCULATIONS

SIZE 25'5" × 18'3" × 27'1" × 18'0 (TWO WALLS SPLAYED)
FLOOR WOOD PARQUET
CEILING PLASTERED WITH 6 SUSPENDED ELEMENTS
WALLS 8 WIDEBAND, 5 LOW FREQUENCY MODULES, ACOUSTICAL TILE
VOLUME 5,160 CUBIC FEET

MATERIAL	S AREA SQ. FT	125 HZ A	SA	250 HZ A	SA	500 HZ A	SA	1 KHZ A	SA	2 KHZ A	SA	4 KHZ A	SA
Floor; parquet	238	0.04	9.5	0.04	9.5	0.07	16.7	0.06	14.3	0.06	14.3	0.07	16.7
Floor: rug 10' × 12'	120	0.05	6.0	0.10	12.0	0.15	18.0	0.30	36.0	0.50	60.0	0.55	66.0
LF units	160	0.98	156.8	0.72	115.2	0.33	52.8	0.21	33.6	0.16	25.6	0.14	22.4
Acous. Tile 1/2"	137	0.10	13.7	0.25	34.3	0.65	89.1	0.73	100.0	0.73	100.0	0.68	93.8
Ceiling elements:	152	0.99	150.5	0.99	150.5	0.99	150.5	0.99	150.5	0.99	150.5	0.99	150.5
Lower surface													
Upper surface	152	0.20	30.4	0.20	30.4	0.20	30.4	0.20	30.4	0.20	30.4	0.20	30.4
Total sabins, Sa without wall modules			366.9		351.9		357.5		364.8		380.8		379.8
Condition A													
Wall modules, soft side out 128		0.88	112.6	0.99	126.7	0.99	126.7	0.99	126.7	0.99	126.7	0.96	122.9
Total sabins, Sa			479.5		478.6		484.2		491.5		507.5		502.7
Reverb. Time, Sa			0.52		0.53		0.52		0.51		0.50		0.50
Condition B													
Wall modules, hard side out 128		0.28	35.8	0.22	26.2	0.17	21.8	0.09	11.5	0.10	12.8	0.11	14.1
Total sabins, Sa			402.7		380.1		379.3		376.3		393.6		393.9
Reverb. Time, Sec.			0.63		0.66		0.67		0.67		0.64		0.64

The reverberation time can be increased about 30 percent by flipping the eight wideband wall modules from soft to hard side out. Although this must be determined by many listening tests, the best average condition would appear to be Condition A with wall modules soft. This would be essentially a speech condition shifted somewhat in the music direction.

The idea of reversing wall modules for every recording session is too idealistic and just too much work. The type of recording carried out in any studio invariably falls into one, two, or a few categories and usually a single reverberatory condition meets the needs of most of the jobs. It is nice, however, to know that if especially bright conditions are desired for a certain type of music recording, the facilities are available to attain these conditions. The diffusion of sound in the studio is certainly better with soft wall modules than with hard modules. Good diffusion is especially needed for recording of speech.

Even greater flexibility in adjusting studio acoustics is available if the 24 inch × 24 inch pads of 703 in the six ceiling elements are involved. The design represented by Table 8-1 and Fig. 8-15 includes 152 square feet, or only 38 of the 48 possible absorbent sections in the ceiling frames. This would allow adding 10 sections or 40 sabins, or removing 10 sections with only minor sound diffusion effects if a distributed pattern of sections remaining is maintained.

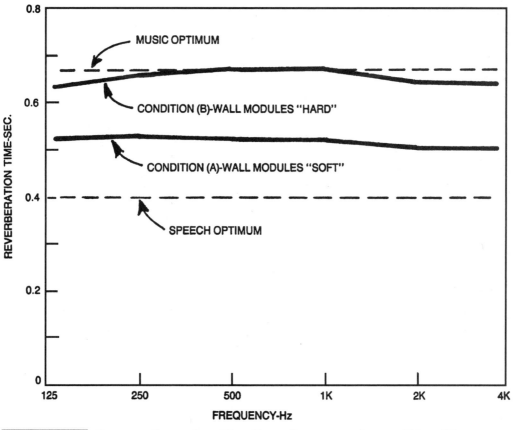

Figure 8-15 Degree of reverberation time change made possible with reversible wideband wall modules.

Table 8-1 takes advantage of the fact that the ceiling element pads of 703 are capable of absorbing sound on both the lower and upper faces. As the upper face is somewhat shielded by the framework, an estimated 20 percent absorption is applied to the top surface. Acoustical measurements in the studio could very well demonstrate that the upper surface absorbs more than this.

Control Room Reverberation

The reverberation time goal for the control room is about 0.3 second which is suitably shorter than the studio it serves. The treatment shown in Fig. 8-14 approaches this as the computations of Table 8-2 reveal. Note that the low frequency absorption of the cabinets is a significant contribution. Although specific proprietary materials are listed for both the lay-in ceiling boards and the acoustical tile, many other products of the same type would serve just as well if their different absorption coefficients are known and areas properly adjusted.

TABLE 8-2 CONTROL ROOM REVERBERATION TIME CALCULATIONS

SIZE 13'9" × 13'7" × 10'8" × 9'4" (ONE WALL SPLAYED)
FLOOR WOOD PARQUET
CEILING LAY-IN CELOTEX SAFETONE CELOTONE, NAT FISSURED, 3/4" MOUNTING#7
WALLS SLAT TYPE KF ABSORBERS AND SIMPSON PYROTECT
VOLUME 1,530 CUBIC FEET

MATERIAL	S AREA SQ. FT	125 HZ		250 HZ		500 HZ		1 KHZ		2 KHZ		4 KHZ	
		A	SA	A	SA	A	SA	A	SA	A	SA	A	SA
Floor parquet	108	0.04	4.3	0.04	4.3	0.07	7.6	0.06	6.5	0.06	6.5	0.07	26
Cabinets-1/2" plywood	90	0.30	27.0	0.23	20.7	0.18	16.2	0.14	12.6	0.11	9.9	0.10	9.0
Lay-in ceiling	136	0.50	68.0	0.54	73.4	0.55	74.8	0.77	104.7	0.87	118.3	0.88	119.7
LF slat absorbers	128	0.98	125.4	0.72	92.2	0.33	42.2	0.21	26.9	0.16	20.5	0.14	17.9
Acoustical tile	110	0.08	6.5	0.27	29.7	0.81	89.1	0.91	100.1	0.68	50.8	0.48	52.8
Total sabins, Sa			231.3		220.3		229.9		250.8		206.0		207.0
Reverb, time, sec.			0.32		0.34		0.33		0.30		0.36		0.36

9

STUDIOS FOR A COMMERCIAL
RADIO STATION

Feature: Measurements for trimming acoustics, perils of concrete block walls.

CONTENTS AT A GLANCE

The client in this example was represented by a very knowledgeable engineer who had already laid out a floor plan suitable for their needs. He wanted it checked out for room resonance distribution, a complete acoustical treatment plan made for each room, and then measurements done to verify the design.

The floor plan already decided upon, shown in Fig. 9-1, includes master control, production control, and a talk booth. The word *booth* infers that it is small and so it is. This one is 805 cubic feet, only half the minimum volume recommended. The other two rooms are somewhat larger than the minimum 1500 cubic feet 2000 and 1943 cubic feet.

In an operation such as this there is generally a great reluctance to be generous in the area devoted to recording, editing/listening, and live broadcasting. After all, abstract and intangible acoustics are locked in a battle with down-to-earth factors such as space for offices for the station manager, production manager, and engineer, as well as space for library, traffic and continuity, accounting, cafeteria, news and farm department, announcers' work area, and last and most important, ample room for *SALES*.

To keep the studio area under control there are pessimists around who would even question straining for acoustical quality when the audio signal is routinely highly processed to get the effect of greater transmitter power. The signal night not be so good, but signal coverage and resulting sales apparently are best served by such a procedure. Another argument tending to scuttle support for acoustical quality of broadcast studios is that little goes out live anymore, except voice, and the announcer always close-talks the microphone. It is true that room effects are most noticeable for greater source to microphone distances but the imprint of the room is always there, no matter how close the microphone to the source.

Construction

The studio area of Fig. 9-1 is one corner of a 55 foot × 60 foot single story structure. The walls of the studio are of 6-inch concrete block, an air space of 2 inches and an inner wall

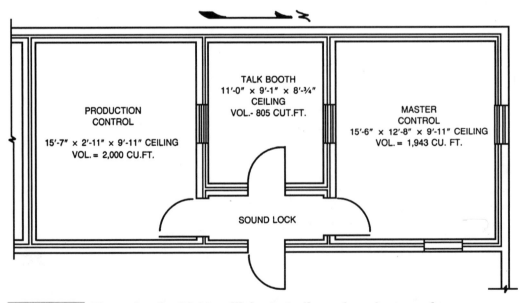

Figure 9-1 Floor plan for highly efficient studio and work space for a commercial radio station. The talk booth is substandard in size.

of 4-inch concrete block as shown in Fig. 9-2. The hollow spaces in the concrete blocks of both tiers were filled with concrete, well-rodded to eliminate air pockets. The external walls have added thermal protection because of the northern location.

The exterior face of the external walls is covered with stucco or siding. The 2-inch air space is filled with thermal type of glass fiber. The inner surface of the 4-inch block tier of external walls is covered with ⅝-inch gypsum board furred out on 2 × 2s with the space behind also filled with a thermal type of glass fiber.

The interior walls of the studio area are the same basic construction as the external walls as far as the 6-inch and 4-inch concrete block tiers and the 2-inch air space are concerned. The similarity ends there as the air spaces of the interior walls have no glass fiber in them nor is there a gypsum board layer. The inner faces of all walls were to be painted.

An exterior window was insisted upon in the north wall of master control. Another window in the east wall allows other staff workers to see what is going on in master control without entering. The two windows in the talk booth are lined up so that personnel in the three rooms can see their colleagues in the other rooms. This allows the use of hand signals. In fact, someone in production control can look through both the talk booth and master control to the next range of mountains! The production value of this feature is not too evident unless it is to be able to say during a weather broadcast, "It's snowing hard here." From the acoustical standpoint lining up windows is very bad, especially if the glass plates are parallel.

Acoustical Treatment

In this chapter constructional details of wideband wall modules, suspended and other ceiling elements, etc., are passed over in favor of some new and, hopefully, more interesting features. The treatment of master control (Fig. 9-3), production control (Fig. 9-4), and the talk booth (Fig. 9-5) are quite similar to those of other earlier chapters with one simplifying factor: There is no carpet in these rooms. This means no low frequency resonators are required to compensate for the carpet. Presumably, all we need are areas of 4 inches of 703 suitably disposed around each studio to ensure against flutter echo and to give the best sound diffusion. These areas are provided in three forms:

- wideband wall modules
- suspended ceiling elements in master and production control rooms
- an acoustically similar ceiling frame in the talk booth

General Measurements

The three studios treated as shown in Figs. 9-3, 9-4, and 9-5 were subjected to acoustical measurements. Only the reverberation results are discussed in this chapter. These measurements as well as the the design were all accomplished by mail. The consultant

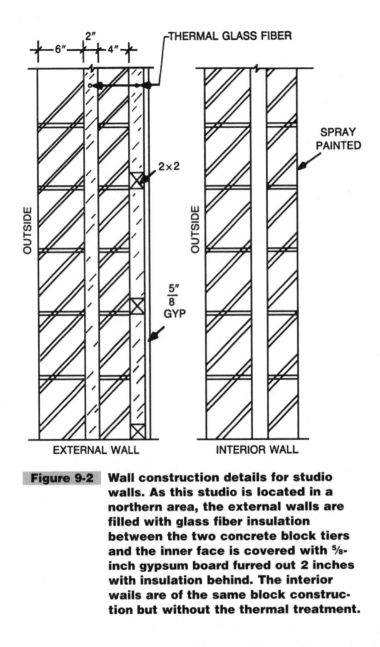

Figure 9-2   Wall construction details for studio walls. As this studio is located in a northern area, the external walls are filled with glass fiber insulation between the two concrete block tiers and the inner face is covered with ⅝-inch gypsum board furred out 2 inches with insulation behind. The interior wails are of the same block construction but without the thermal treatment.

did not visit the studio site. Working at a distance, the consultant sent a magnetic tape with test signals on it to the engineer who played them according to detailed instructions in each room, picking up each room's response to these signals on a suitable microphone placed as directed and recording the response on another tape recorder. Upon receiving the response tapes in the mail the consultant proceeded to analyze them. In very general terms, reverberation time, swept sine test, and the rooms' response to impulses were recorded and analyzed.

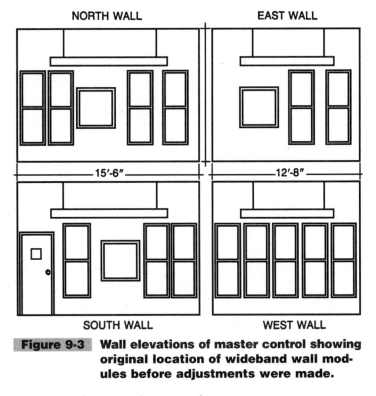

NORTH WALL   EAST WALL

15'-6"   12'-8"

SOUTH WALL   WEST WALL

Figure 9-3 **Wall elevations of master control showing original location of wideband wall modules before adjustments were made.**

Reverberation Time

The goal for reverberation time was taken as a nominal 0.3 second, relatively uniform 125 Hz to 4 kHz. The average measured reverberation time of the three studios is shown in Fig. 9-6. All three are substantially below 0.3 second. Well, you win some and you lose some! But the detective in each of us says, "Why?" If this business is not based on rational physical principles as we were led to believe by our physics teacher, perhaps the gnome in the cave up in the mountain might cease his peeping and muttering long enough to design studios for us. Not that the reverberation characteristics of Fig. 9-6 are uniformly low. A studio having a reverberation time of 0.2 second is quite usable, but its sound would tend to be, stated subjectively, dry, dead, and outdoorish in character. The difference between the 0.3 second goal and 0.2 second reality, however, is a distinct challenge and a problem worth solving.

Theory vs. Practice

The consultant and the radio engineer exchanged views and information by letter, telephone, and one visit of the engineer with the consultant. A number of examples of questionable communication accuracy were disclosed:

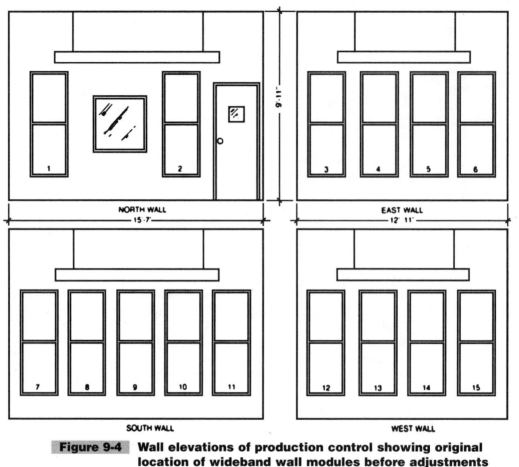

Figure 9-4 **Wall elevations of production control showing original location of wideband wall modules before adjustments were made.**

- Knowledge of the furred out gypsum board on the inside of exterior walls came after the basic design was completed
- The block surfaces were to be painted. The idea of the original painting specification was to close the tiny surface pores of the coarse concrete blocks which absorb the sound, but the kind of paint to use was not stated clearly by the consultant. The paint used evidently was nonbridging in nature, serving only to stabilize the surface somewhat and to give color, but not to seal off the interstices.
- The third factor was the uncertainty as to how effective the top surface of the 4 inches of 703 in the suspended ceiling would be in absorbing sound. In other words, how well is it shielded from sounds of various frequencies filling the room? The manufacturer's coefficients are not given for such double-sided exposure for this type of product because only one side is normally exposed. In the original design the consultant assumed that the top surface of the 703 would absorb about half as well as the lower surface. The evidence indicates that both sides are fully effective.

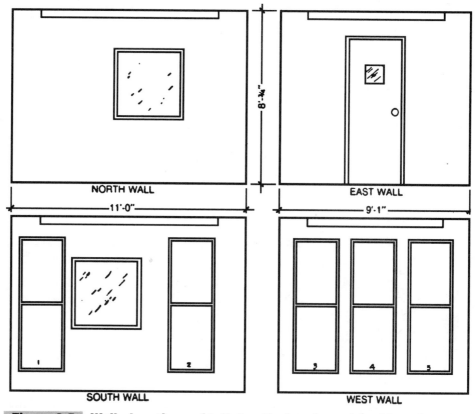

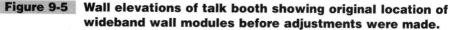

Figure 9-5 Wall elevations of talk booth showing original location of wideband wall modules before adjustments were made.

The calculations leading to the original design gave essentially 0.3 second reverberation time 125 Hz-4 khz for each of the three studios of Fig. 9-1. But these original calculations did not include the uncertainties introduced later which are listed in the previous paragraph.

The problem now is how to account for the measured reverberation time graphs shown in Fig. 9-6. Our confidence in the whole procedure hinges on our ability to account for the actual reverberation time prevailing in these rooms. Accordingly, new calculations were made in which the absorption coefficients differ in the following ways from those of the original calculations which predicted about 0.3 second reverberation time in each room:

1 The area of the 703 used in the suspended ceiling frame was doubled rather than increased by 50 percent. This assumes that the top face is just as effective as the lower face in absorbing sound.

2 The absorption coefficients for concrete block walls were taken from an obsolete source[30] as seen in Table 9-2. The original design, assuming that the paint filled the surface pores, used the coefficients in the right column. Learning that the walls were spray painted, it was judged that the coefficients of the left column applied more closely.

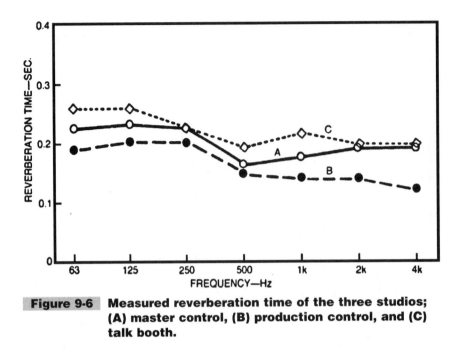

Figure 9-6 **Measured reverberation time of the three studios; (A) master control, (B) production control, and (C) talk booth.**

3 The areas of concrete block were reduced by the exterior wall area covered with furred out gypsum board. Master control has two such walls and the other rooms each have one.

To make a long story short, the recalculated reverberation time for master control, taking into account the factors in (1) to (3) above, are listed at the foot of Table 9-1. These calculated points agree reasonably well with the measured values as shown in Fig. 9-7.

Reverberation time was recalculated for production control using the identical coefficients and the results are displayed in Fig. 9-8. Here the calculated values are very close except at 500 Hz and above. The recalculated values of reverberation time for the talk booth, using the same coefficients, are plotted in Fig. 9-9. The measured and recalculated values of reverberation time, average for six frequencies, are compared in Table 9-3.

These are the types of problems encountered in predicting studio acoustical performance. Figures 9-7, 9-8, and 9-9 show far superior agreement between measured and calculated values, however, than a comparison of the three graphs of Fig. 9-6 with the 0.3 second goal of the original design. The agreement would have been even better had the unpainted block coefficients really fitted the type of concrete blocks actually used.

There are a number of lessons here. First, when one of the absorption elements used in the rooms is as unpredictable as concrete block (unpainted or painted with nonbridging paint), calculations are little more than rough guesses. However, concrete block walls, painted in a way that fills the interstices, become predictable; at the same time, they also lose most of their absorption. The absorption of unpainted concrete blocks varies widely because of variations in the density of the materials of which they are made. In the present case, the block walls painted with nonbridging instead of bridging paint introduced significant absorption which resulted in reverberation times near 0.2 second rather than the predicted 0.3 second.

TABLE 9-1 MASTER CONTROL CALCULATIONS

SIZE 15'8" × 12'8", 9'11" CEILING
FLOOR VINYL TILE ON CONCRETE
CEILING DOUBLE 5/8" GYPSUM BOARD ON CONCRETE. SUSPENDED 1 × G FRAME HOLDING 4" OF 703 GLASS FIBER
WALLS CONCRETE BLOCK EXTERIOR WALLS COVERED WITH 5/8" GYPSUM BOARD FURRED OUT ON 2 X 2S, SPACE FILLED WITH THERMAL GLASS FIBER (FIG. 9-2). INTERIOR CONCRETE BOOK, COARSE, SPRAY PAINTED (FIG. 9-2). WIDEBAND MODULES, 2' X 6', 4" 703 (FIG. 9-3)
VOLUME 1,943 CU. FT.

MATERIAL	S AREA SQ. FT.	125 HZ A	SA	250 HZ A	SA	500 HZ A	SA	1 KHZ A	SA	2 KHZ A	SA	4 KHZ A	SA
Ceiling 4" 703 69 sq. ft., both sides active	138	0.99	136.6	0.99	136.6	0.99	136.6	0.99	136.6	0.99	136.6	0.99	136.6
N&W walls 5/8" gyp board, 2" fill	270	0.15	40.5	0.14	37.8	0.12	32.4	0.10	27.0	0.08	16.2	0.05	13.5
E&S walls coarse concrete block	244	0.36	87.8	0.44	107.4	0.31	75.6	0.29	70.8	0.39	95.2	0.25	61.0
Ceiling: dbl 5/8" gypsum board	196	0.08	15.7	0.05	9.8	0.03	5.9	0.03	5.9	0.03	5.9	0.03	5.9
Wideband wall modules (14)	168	0.99	166.3	0.99	166.3	0.99	166.3	0.99	166.3	0.99	166.3	0.99	166.3
Total sabins, Sa			446.9		457.9		416.8		406.6		420.2		383.3
Reverb. Times, sec.			0.21		0.21		0.23		0.23		0.23		0.25

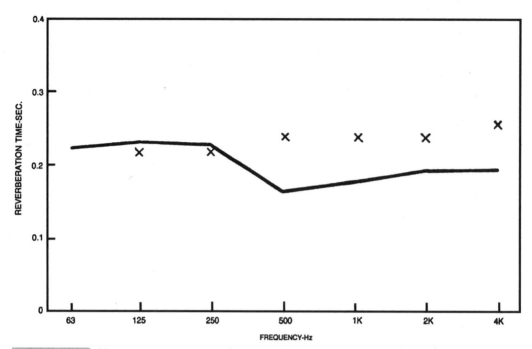

Figure 9-7 **Measured reverberation time in master control repeated from Fig. 9-6 (solid line). The calculated points (x) are in reasonably close agreement.**

TABLE 9-2 CONCRETE BLOCK ABSORBTION COEFFICIENTS		
FREQUENCY HZ	**COARSE, UNPAINTED**	**PAINTED (BRIDGING PAINT)**
125	0.36	0.10
250	0.44	0.05
500	0.31	0.06
1k	0.29	0.07
2k	0.39	0.09
4k	0.25	0.08

Source: Ref. 35

Another lesson is that there is no substitute for measurements to reveal gross problems. If all materials used in a room behave according to the published absorption coefficients, calculated reverberation time can be much closer than Table 9-3 indicates, but if the uncertainty of concrete block absorption is present, accuracies in the −11 percent to +39 percent range must be expected.

A third lesson is that the use of wideband absorption modules allows easy trimming.

TABLE 9-3 COMPARISON OF MEASURED AND CALCULATED REVERBERATION TIMES			
	REVERBERATION TIME, SECONDS		
	MEASURED	RECALCULATED	
Master Control	0.198	0.237	+20%
Production Control	0.160	0.223	+39%
Talk Booth	0.222	0.198	−11%

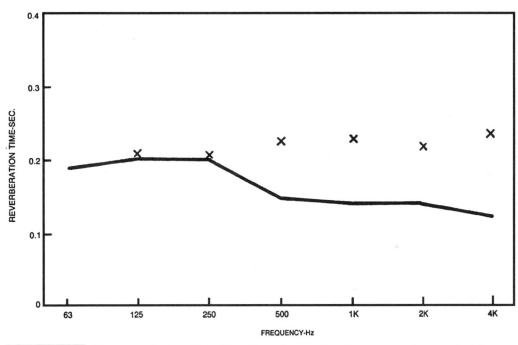

Figure 9-8   **Measured reverberation time in production control repeated from Fig. 9-6 (solid line). The calculated points (x) show somewhat poorer agreement than Fig. 9-7.**

Master Control Trimming

To increase the reverberation time of master control from the measured values of graph A of Fig. 9-6 to approximately 0.3 second across the band, considerable reduction in sound absorption is required. The first impulse is to remove wideband modules. There is a problem here. With the concrete block absorption unchanged, almost all the modules must be removed. Bare walls would be a problem. To increase reverberation time to 0.3 second there is only one thing to do: Paint the concrete block walls with a bridging paint and start all over by recalculating the room. This shows that about a dozen wideband modules are required, which must then be positioned carefully to control flutter echoes. The measurements have

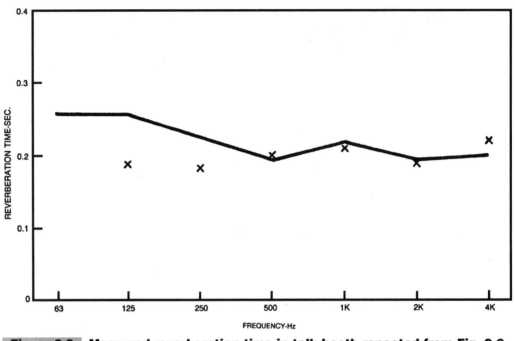

Figure 9-9   **Measured reverberation time in talk booth repeated from Fig. 9-6 (solid line). The calculated points (x) show reasonably good agreement.**

given a base from which to work. With this added insight a couple of wideband modules can be removed to trim the room.

Production Control Trimming

Painting the concrete block wall with a bridging paint is also recommended for production control. Once this is done, the original design of Fig. 9-4 is quite reasonable. In this case, removing a single wideband module from the room may trim sufficiently.

Talk Booth Trimming

Painting the north, east, and south concrete block walls of the talk booth also brings things close to the original design. This raises the reverberation time to about 0.3 second.

Summary

What a tremendous difference a little thing like the type of paint used on the concrete block walls makes! This is a primary lesson to be learned from this chapter. Unless one has the resources to set up to test a concrete block wall and to measure the absorption coefficients of the specific blocks to be used, it is better to paint the stuff with a heavy paint and depend on known materials. If your heart is set on using the absorption of a particular local concrete block with its natural surface, another approach would be to approach the treatment of the room in increments, measuring reverberation time step by step as materials are added until the desired characteristics are achieved. Most of us do not have the means, the patience, or the love for certain concrete blocks to justify this.

ONE CONTROL ROOM
FOR TWO STUDIOS

Features: Small modules for treatment, use of cork, semicylindrical elements, room air conditioners.

CONTENTS AT A GLANCE

Whether or not using one control room to serve two studios is satisfactory depends upon the intensity of the recording schedule. If recording sessions are well spaced and the schedule is flexible, having only one control room might work very well. Certainly, the advantages of requiring only one console, one set of recording equipment, and using only one operator are self-evident.

The problem arises when it is necessary to use both studios simultaneously. Even if duplication of equipment and the use of two operators were accepted as the price for a fuller recording schedule, the conflict of monitor loudspeakers becomes apparent. Which recording job gets the loudspeaker(s)? Which operator uses headphones?

For the activity that does not require such overlapped scheduling, however, the single control room for two studios can work out very well. And, as we have seen in Chapter 7, headphone monitoring is becoming an ever more viable alternative for mono work as headphone quality is undergoing rapid and dramatic improvement.[12]

The incentive for such improvement, it must be admitted, comes from the hi-fi market, not from recording engineers who still prefer loudspeakers. How much of this is inertia from the recent past when loudspeaker quality far outstripped headphone quality is difficult to say. The point here is that this is a new day and the quality difference is much less. Headphones might very well be used by one operator as he records speech from one studio while another operator records music from another using monitor loudspeaker(s). However, the loudspeaker level would have to be kept down.

Studio Suite Layout

The example to be considered is another of those cases in which the studio suite had to be fitted into space available in an existing building: After the usual battle for space, the prostudio faction came up with the end of the third and top floor of a concrete building overlooking a quiet patio farthest from a busy thoroughfare. The overlooking part was immediately cancelled as all windows in the studio area were bricked up and plastered on both sides.

After the usual preliminary consultation period, the floor plan of Fig. 10-1 was agreed upon. The client was insistent upon one control room for two studios and was willing to go to headphones for a second operator if the recording load increased that much in the future.

The sound lock is a strangely shaped space, but for a good reason. The existing hall required an offset sound lock to reach the control room without robbing the music studio of too much area. The walls of the sound lock, if straight, would cut off corners in both studios with approximately 45-degree walls. Looking into these corners from either studio emphasized that such a straight wall across the corner tends toward a concave effect when considered along with the adjoining walls.

Acousticians always get worried when confronted with concave or quasi-concave surfaces, but nothing makes them happier than convex surfaces. So, why not make convex sound diffusing surfaces out of these sound lock walls and let the concave sides be in the sound lock where they can do no harm? Walls of brick are quite amenable to shaping in this fashion. In fact, leaving the rough brick tecture aids diffusion just that much more and provides an attractive visual feature in each room.

The volumes of the control room (1550 cubic feet) and the speech studio (1432 cubic feet) are very close to the 1500 cubic foot minimum, yet are adequate for their intended purposes. The music studio at 3410 cubic feet provides reasonably adequate space for the largest music group contemplated.

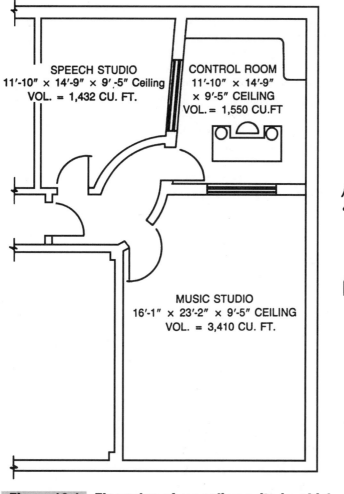

Figure 10-1   **Floor plan of recording suite in which both a speech studio and music studio are served by a single control room. This arrangement works best if the work load is modest and the schedule flexible. A second operator can handle one studio (e.g., the speech studio) by monitoring with high quality headphones.**

Acoustical Treatment

The acoustical treatment of the three rooms is strongly based on a modular plan. There are three types of modules of 24-inch × 24-inch outside dimensions. Each has a different absorption characteristic:

■ A wideband (WB) module which absorbs equally well over the frequency range 125 Hz-4 kHz.

■ A low peak (LP) module having a peak absorption at about 125 Hz.

■ A midpeak (MP) module which peaks at about 800 Hz.

The absorption of these three modules is constrasted in Fig. 10-2. These are three basic building blocks capable of compensating for the absorption of carpets, drapes, acoustical tile, or other materials having unbalanced absorption and performing other tasks in the acoustical treatment of a room. As there is no reason to be limited to these three, versatile as they are, other materials such as cork and convex panels will be used.

Music Studio Treatment

The wall treatment of the music studio is shown in Fig. 10-3, the floor and ceiling treatment in Fig. 10-4.

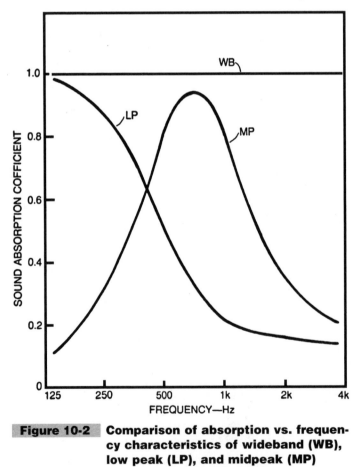

Figure 10-2 Comparison of absorption vs. frequency characteristics of wideband (WB), low peak (LP), and midpeak (MP) types of sound absorbers.

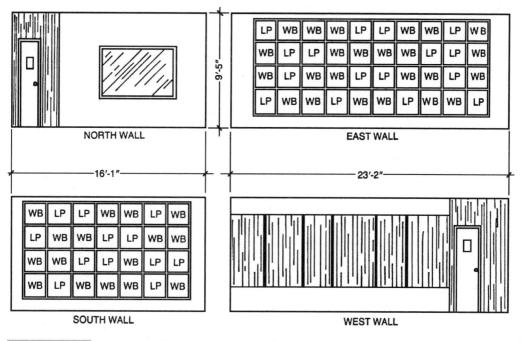

Figure 10-3 Wall elevations of music studio. The door is set in a semicylindrical brick sound lock wall. Plywood polycylindrical diffuser/absorbers cover the west wall and wideband (WB) and low peak (LP) modules cover the east and south walls.

The north wall leaves little space for acoustical elements because of the observation window and the convex brick wall in the northwest corner. The east and south was are practically covered with 24-inch × 24-inch modules of two types, low peak and wideband. The west wall is dominated by three large and three smaller convex panels.

The ceiling (Fig. 10-4) has 96 cork tiles 12 inches × 12 inches × 1 inch cemented to it in an irregular pattern built up of groups of four cork tiles. It was not possible to attain the desired reverberation time if the floor was covered with carpet wall to wall as requested, hence a compromise of an 8-foot × 10-foot rug is specified. Each absorptive element will be discussed individually in connection with the music studio, although the same elements are used in other rooms as well.

CEILING TREATMENT

The cork tiles of 12 inches × 12 inches × 1 inch were selected because they were readily available in this particular foreign country and are considered very attractive by some people. Absorption coefficients for cork may be difficult to locate. As it was a popular acoustical material 50 years ago before the acoustical industry sprouted, coefficients can be found in old textbooks.

Inspecting these coefficients (Table 10-1), low absorption at low frequencies and better absorption at higher frequencies are noted. The absorption varies much the same as the rug

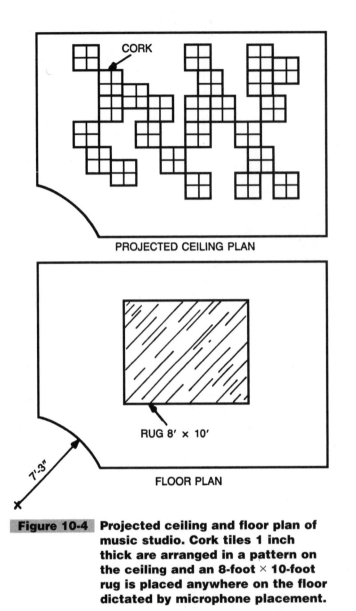

Figure 10-4 Projected ceiling and floor plan of music studio. Cork tiles 1 inch thick are arranged in a pattern on the ceiling and an 8-foot × 10-foot rug is placed anywhere on the floor dictated by microphone placement.

on the floor and about the same order of magnitude. If cork is to be replaced by acoustical tile, it should be the thinnest and cheapest tale available because ¾-inch modern acoustical tile is a far more efficient absorber than 1-inch cork.

Actually, it would be better to have a compensating type of absorber on the ceiling opposite the rug, but there is the potential problem of not knowing just where the rug will be in the studio and mounting compensating (LP) modules on the wall is easier than on the ceiling.

TABLE 10-1 Music Studio Calculations

SIZE: 16'1" × 23'2", CEILING × 9'5"
FLOOR: WOOD PARQUET, 8' × 10' RUG
CEILING: 96 CORK TILES 12" × 12" × 1" RANDOM PATTERN
WALLS: 40 WIDEBAND, 28 LOW PEEK MODULES, CONVEX PANELS
VOLUME: 3,410 CU FT.

MATERIAL	S AREA SQ. FT.	125 HZ		250 HZ		500 HZ		1 KHZ		2 KHZ		4 KHZ	
		A	SA	A	SA	A	SA	A	SA	A	SA	A	SA
Floor: 8' × 10' rug	80	0.05	4.0	0.10	8.0	0.15	12.0	0.30	24.0	0.50	40.0	0.55	44.0
Floor: Wood parquet	282	0.04	11.3	0.04	11.3	0.07	19.7	0.06	16.9	0.06	16.9	0.07	19.7
Cork tiles: 12" × 12" × 1"	96	0.05	4.8	0.10	9.6	0.20	19.2	0.55	52.8	0.60	57.6	0.55	52.8
Walls: low peak 28 3.62 sq ft.	101	0.98	99.0	0.88	88.9	0.52	52.5	0.21	21.2	0.16	16.2	0.14	14.1
Walls: wideband (40)	145	0.99	143.6	0.99	143.6	0.99	143.6	0.99	143.6	0.99	143.6	0.99	143.6
Convex panels: 3 large, empty	66	0.41	27.1	0.40	26.4	0.33	218	0.25	16.5	0.20	13.2	0.22	14.5
Convex panels: 3 small, empty	44	0.32	14.1	0.35	15.4	0.30	13.2	0.25	11.0	0.20	8.8	0.23	10.1
Total sabins, Sa			303.9		303.2		282.0		266.0		296.3		298.8
Reverb. Time, sec.			0.55		0.55		0.59		0.58		0.56		0.56

WALL TREATMENT

The modules on the east and south walls are constructed according to the plan of Fig. 10-5. The wideband (WB) module is simply a frame holding 4 inches of 703 (or two layers of 2 inches) or similar glass fiberboards of about 3 pounds per cubic foot density.

This glass fiber is held in place by zig-zag wires at the appropriate level creating an air space of approximately 1¼ inches. This air space helps to extend good absorption to lower frequencies, or, looking at it another way, it helps us come closer to realizing in practice the 0.99 coefficient at 125 Hz and 250 Hz supplied by the manufacturer.

The face of the 703 is held in place by a lightweight expanded metal lath appropriately spray painted before installation. It may be advisable to cover the 703 with a stretched lightweight fabric before the expanded metal is applied. The backs of both of these modules are uncritical; ¼-inch hardboard or plywood is adequate.

The LP module having an absorption peak at about 125 Hz is housed in a frame identical to that of the wideband absorber. Only 2 inches of 703 is required in this case and the facing is ³⁄₁₆-inch hardboard or plywood drilled with ³⁄₁₆-inch diameter holes spaced 1⁹⁄₁₆ inches center to center. By stacking the panels a number can be drilled simultaneously, reducing the tedium somewhat. A loosely-woven fabric cover over this perforated cover for esthetic reasons is optional.

One of the weightier problems of treating these studios is how to mount the 24-inch × 24-inch modules on walls and ceilings. One of the advantages of the modular approach is to facilitate trimming room acoustics if measurements indicate the need.

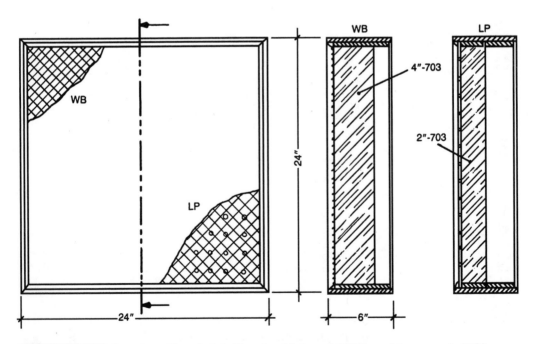

Figure 10-5 Constructional details of wideband (WB) and low peak (LP) absorbing modules which utilize identical boxes.

Therefore, ideally, the best type of mounting is the one which allows easy removal or interchange of modules. This is best accomplished by building a frame into which the individual modules may be inserted or removed at will. A frame of this type is described in Fig. 10-6.

Another less desirable approach is to affix the box itself firmly to the wall and to consider the contents interchangeable, but not the external box. If the contents were removed, a fabric face would preserve visual continuity, but its acoustical effect would be essentially removed.

On the other hand, a WB absorber could be changed to an LP by changing glass fiber depth and cover. Making a midpeak (MP) absorber, to be described later, out of a WB or

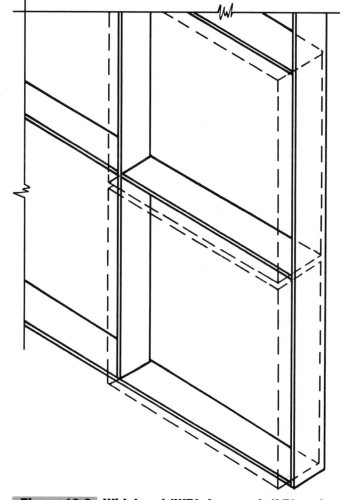

Figure 10-6 Wideband (WB), low peak (LP) and midpeak (MP) absorbing modules can be conveniently slid in and out of a simple wall frame of 1 × 4 lumber.

LP would require using the bottom of the box, leaving some empty space behind the grille cloth cover, but it would serve its acoustical function satisfactorily. It is also possible to stack the boxes like cord wood on a low support, making the stack stable with well-placed cleats or other ties.

Each WB and LP box weighs about 20 pounds. With the adhesives available today the boxes could be cemented directly to the wall. With the acoustical guidance in this and other chapters, it is not too much to expect the mounting problem to be solved, along with a host of other problems, by the ingenuity in residence.

WEST WALL POLYS

The polycylindrical or convex acoustical elements on the west wall of Fig. 10-3 have two things going for them: bass absorption and excellent diffusing characteristics.[3] They require only modest skill in construction.

The details of Fig. 10-7 show a skin of $\frac{3}{16}$-inch hardboard or plywood stretched over bulkheads previously cut on the arc of a circle with a bandsaw. The larger unit requires a sheet about 48-inches wide and the smaller one about 32 inches. No acoustical filling is required inside.

When music fills the room the vibration of the covering skin is easily felt by placing the fingers on it. It is important that rattles be prevented. This can best be done by placing a thin foam rubber strip with a self-adhesive backing along the edge of each bulkhead before nailing the skin. The bulkheads divide the space within each convex element into three unequal, essentially airtight, compartments. All three large and three small convex elements rest on

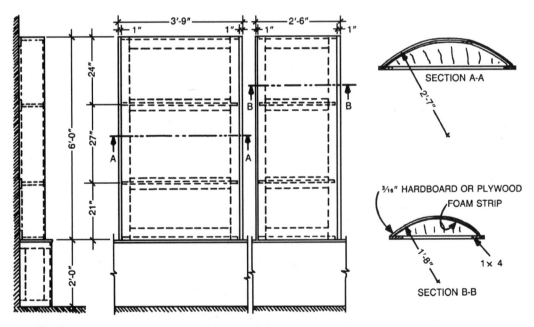

Figure 10-7 Constructional details of polycylindrical diffusers/absorbers on west wall of music studio. They are fastened to the wall and rest on a simple base enclosed with plywood.

a low, enclosed shelf. As this is covered with ¹/₂-inch plywood, it also is a fair bass absorber of the flat panel type. Because this 18-foot shelf contributes only 2.5 percent of the absorption of the room at the most, its effect is minor and is not included in Table 10-1.

With the treatment described the reverberation time of the music studio hovers around the desired 0.55 second level as shown numerically in Table 10-1 and graphically in Fig. 10-8.

Speech Studio Treatment

A reverberation time of approximately 0.3 second is the goal for the speech studio. Carpet was specified in this room. In fact, an indoor-outdoor type of carpet was specifically requested for economy. This is included wall-to-wall. Opposing the carpet, 27 low peak (LP) modules are mounted to the ceiling, arranged as shown in Fig. 10-9. The matter of attachment to the ceiling, admittedly somewhat more complicated than to the wall, is left to local ingenuity.

The treatment of the walls of the speech studio is shown in Fig. 10-10. The east and south walls are dominated by the convex masonry wall shared with the sound lock. The

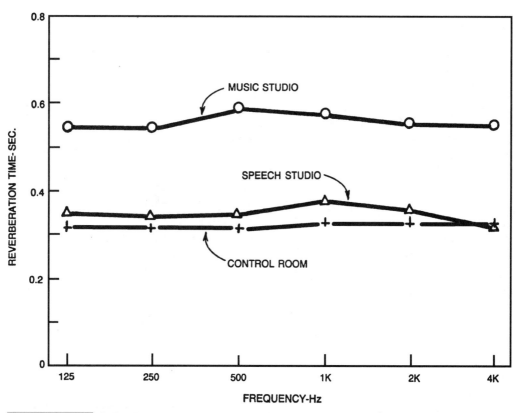

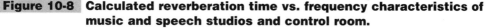
Figure 10-8 Calculated reverberation time vs. frequency characteristics of music and speech studios and control room.

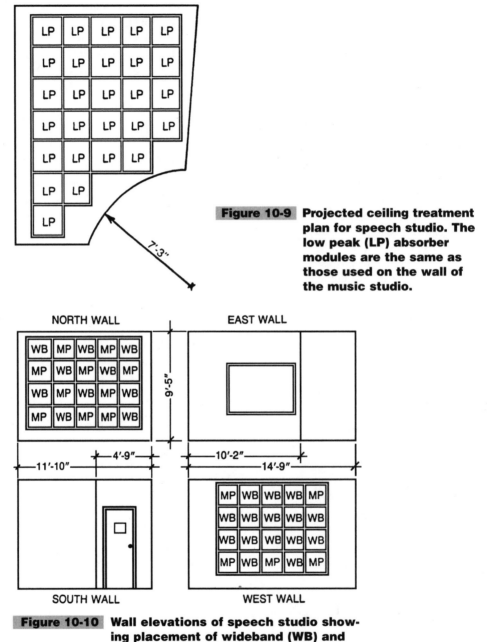

Figure 10-9 Projected ceiling treatment plan for speech studio. The low peak (LP) absorber modules are the same as those used on the wall of the music studio.

Figure 10-10 Wall elevations of speech studio showing placement of wideband (WB) and midpeak (MP) absorber modules.

north and west walls each have 20 modules, most of them (27) are the wideband (WB) modules as used in the music studio. A third type of 24-inch × 24-inch module, the midpeak (MP), is used only in this studio.

Combining the absorption of the indoor-outdoor carpet with the low peak (LP) ceiling modules, a deficiency of absorption around 800 Hz results. One method of boosting this sagging midband absorption is to introduce 13 modules having peak absorption in this frequency region as shown in Fig. 10-2. The net result is a calculated 0.35 second reverberation time for the speech studio which is a quite uniform 125 Hz to 4 kHz as shown in Fig. 10-8. The computations and other details are found in Table 10-2.

The construction of the midband (MB) modules is detailed in Fig. 10-11. It is a very shallow module with a 1-inch cavity filled with 703 or other type of glass fiber material covered with a perforated panel. Like the low peak (LP) module, it is a tuned Helmholtz type of resonator. The resonance frequency is determined by such things as the percentage perforation (the percentage of the hole area to the entire cover area), the thickness of the panel, and the depth of the enclosed cavity.

The glass fiber broadens the absorption peak. The cover is common stock pegboard. Different types of pegboard have different hole configurations and hence different perforation percentages. The diamond hole configuration is square when the panel is rotated 45 degrees. A common diamond type of pegboard comes with $\frac{5}{32}$-inch holes spaced $\frac{3}{4}$ inches center to center on the square.

The perforation percentage of this type is about 3.4 percent. A perforation percentage between 3 and 6 percent is required. Figure 10-11 includes a sketch showing how the perforation percentage for the entire cover of any type of pegboard can be readily calculated on a unit basis knowing only hole diameter and spacing.

Control Room Treatment

The goal is a reverberation time in the control room somewhat shorter than either of the studios it serves. Making it shorter than the 0.35 second of the speech studio, however, runs head-on into another request. The control room acoustics has to be suitable for the occasional recording of interviews, etc., in that room. A satisfactory compromise value of reverberation time for the control room is 0.32 second as shown in Fig. 10-8.

The treatment of the control room with its hard, reflective floor is a very straightforward task as only wideband (WB) modules are required. With a volume of 1550 cubic feet and surface area of 818 square feet it is easy to plug these figures and a reverberation time of 0.28 into Sabine's equation and come up with a 0.32 second reverberation time.

As shown in Fig. 10-12, 24 WB modules are mounted on the ceiling. Figure 10-13 shows 15 WB units on the north and 21 on the east walls of the control room. The enclosed cabinets under the work table, which are of $\frac{1}{2}$-inch plywood enclosing an air space, must be considered panel absorbers. However, the 61 square feet of such a surface yields only about 17 sabins; hence, the reverberation time is pulled down only a slight amount at the low end of the spectrum.

As all walls are masonry, their absorptive effect is negligible. Squeezing out a few sabins from the wood parquet floor and glass surface the total 231 total is approached. If the walls were of drywall, the low frequency absorption could be significant. Absorption of walls

TABLE 10-2 Speech Studio Calculations

SIZE 11'10" × 14'9", EAST WALL SPLAYED
FLOOR INDOOR-OUTDOOR CARPET WALL TO WALL
CEILING 27 LOW PEAK MODULES
WALLS 27 WIDEBAND AND 13 MIDBAND MODULES
VOLUME 1,432 CU. FT.

MATERIAL	S AREA SQ. FT.	125 HZ A	125 HZ SA	250 HZ A	250 HZ SA	500 HZ A	500 HZ SA	1 KHZ A	1 KHZ SA	2 KHZ A	2 KHZ SA	4 KHZ A	4 KHZ SA
Floor: indoor/outdoor carpet	152	0.01	1.5	0.05	7.6	0.10	15.2	0.20	30.4	0.45	68.4	0.65	98.8
Ceiling: 27 low peak 3.62 sq. ft.	96	0.98	96.0	0.88	86.2	0.52	51.0	0.21	20.6	0.16	15.7	0.14	13.7
Walls: 27 wideband	95	0.99	97.0	0.99	97.0	0.99	97.0	0.99	97.0	0.99	97.0	0.99	97.0
Walls: 13 midband	47	0.09	4.2	0.30	14.1	0.80	37.6	0.80	37.6	0.35	16.5	0.20	9.4
Total sabins, Sa			196.7		204.9		200.8		185.6		197.6		218.9
Reverb Time, sec.			0.35		0.34		0.35		0.38		0.36		0.32

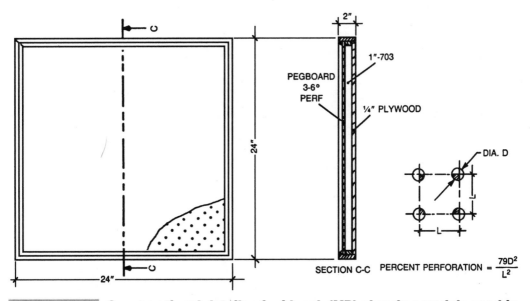

2"

1"-703

PEGBOARD
3-6°
PERF

¼" PLYWOOD

24"

DIA. D

L

L

24"

SECTION C-C

PERCENT PERFORATION = $\dfrac{79D^2}{L^2}$

Figure 10-11 Constructional details of midpeak (MP) absorber module used in speech studio.

WB	WB	WB	WB
WB	WB	WB	WB
WB	WB	WB	WB
WB	WB	WB	WB
WB	WB	WB	WB
WB	WB	WB	WB

PROJECTED CEILING PLAN
CONTROL ROOM

Figure 10-12 Projected ceiling treatment plan of control room showing positions of 24 wideband (WB) modules.

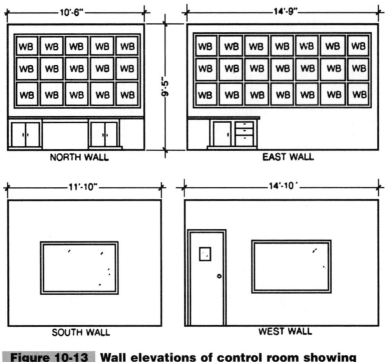

Figure 10-13   Wall elevations of control room showing
locations of wideband (WB) modules on
north and east walls.

and cabinets provide the only unbalanced absorptive effect in a room such as this, and it is usually of nominal magnitude.

Air Conditioning

Tearing holes in studio walls comes under the heading of bad news to acoustical consultants. The client in the present case considered a central air conditioning system too expensive and elected to use wall-mounted room units and live with the resulting inconvenience, discomfort, and increase in background noise.

The sectional view of Fig. 10-14 shows how an acoustical shield can be built over the face of the air conditioning unit to be used when quiet conditions are required. Two layers of ¾-inch particleboard are used because its density is about 30 percent greater than plywood. Such a double panel offers 35 dB attenuation at 500 Hz on a mass law basis. The trick is to seal this lid around the edges in a way which matches or exceeds this. Figure 10-14 shows an arrangement of a double 2 × 4 or 2 × 3 frame which can accomplish this if carefully constructed. The secret is well-caulked joints and use of heavy felt for a snug wiping fit. The lid can be hinged if desired and some device to clamp it in place is needed. This is not a very convenient solution to the air conditioning problem as the lid must be continually closed for a recording and opened to cool things off between takes.

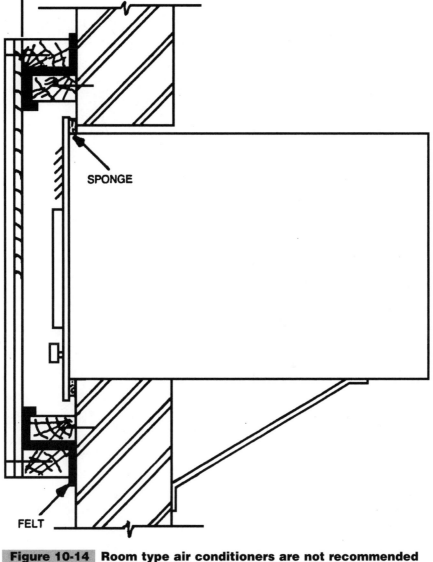

SPONGE

FELT

Figure 10-14 Room type air conditioners are not recommended for sound sensitive areas, but if they must be used, the above plan provides reasonable protection against intrusion of outside noise. The A/C unit is turned off and the door closed during actual recording.

A VIDEO MINI-STUDIO

Feature: Adjustable acoustics, saw-slot low peak resonators.

Recording studios of various types have been covered in previous chapters and it is time to consider one for video and television production. The client was not interested in commercial television production but instead wanted to train television technicians to operate the camera, lights, and control console as well as writers, producers and directors. This calls for a rather generous control room to accommodate the many observers in addition to those actually engaged in the task at hand. As there are no special requirements for the acoustics of the control room, this chapter deals only with the studio. Not only was television training stipulated, the studio also had to be suitable for recording musical groups and for speech in the form of dramatics, single narrator, and interviews.

The acoustical requirements for TV, music, and speech recording in the studio of Fig. 11-1 are divergent. The optimum reverberation time for speech for a studio of this size (7,418 cubic feet) is about 0.5 second; for music about 0.7 second. The requirements for television are less well defined as both speech and music are involved.

Considerable noise is produced in a TV studio as cameras are rolled around, cables dragged, and people move behind the camera. It is customary to use a boom-mounted microphone which can be brought only so close to those talking or it will dip into the picture. Even though the microphone is highly directional, the greater average source to microphone distance means that studio noise becomes a significant problem.

All of these things taken together have led the television production people to demand very absorbent walls and ceiling, the floor surface remaining hard to make rolling camera dollies and pedestals that much easier.

The requirements are much like the motion picture soundstage—make it as dead as possible. For both motion picture and television production, however, there is a saving grace in that the local acoustics are influenced strongly by reflections of sound from the walls of the setting. For example, if in the TV picture we see people in a library setting, reflections from the *flats* making up the visual bounds of the picture make the sound more or less what one would expect in a library in spite of one or two *open walls*. Movable absorptive flats are used to correct acoustical flaws which occur in such a fluid situation. A law of TV and motion picture production seems to be, "The next shot is entirely different!" Our goal, then, for television training work is to provide as dead a studio as is feasible.

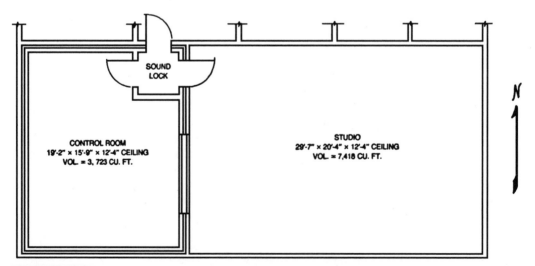

Figure 11-1 **Floor plan of studio suitable for television instruction and recording speech, drama, and music. The divergent requirements of these varied uses are met by adjustable acoustical elements in the studio.**

Louvered Absorbers

There are three acoustical conditions required: reverberation times of 0.5 and 0.7 second, and a third even more dead than the 0.5 second speech condition. To meet all of these requirements in a single room demands some method of adjusting room acoustics. There are numerous ways the acoustics of a room can be varied.[4]

In Chapter 5 the possibility of flipping panels having one reflective side and one absorptive side was explored. Hinged panels can expose deep absorptive layers when open, cover them when closed and so on. The application of adjustable louvered panels as described in Fig. 11-2 seems to fit the present case best of all. The hardware of the louvered windows commonly used in homes in the more temperate climates can easily be adapted by using $^{3}/_{16}$-inch tempered hardboard in place of the usual glass plates. Glass would be quite suitable acoustically, but far more expensive than hardboard. The simple movement of a lever opens and closes the panels of one segment, exposing the glass fiber behind when the panels are open (horizontal) and shielding it acoustically from the room when the panels are tightly closed. Adopting this means of adjustment of room absorption, the calculation of acoustical parameters of the room may proceed.

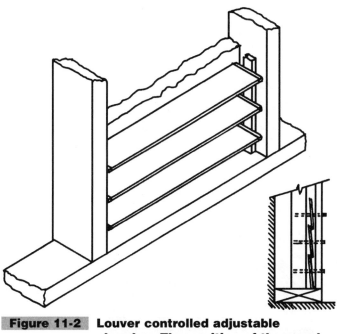

Figure 11-2   **Louver controlled adjustable absorber. The position of the panels (open or closed) controls the effectiveness of the 703 type of glass fiber behind. Partial closing of panels creates a Helmholtz resonator type of low peak absorber.**

Cyclorama Curtain

To provide the neutral background so often demanded in television work, a cyclorama curtain arranged along the east and south walls is used as shown in Fig. 11-3. This curtain, supported from a curved track, may be moved or retracted at will. Assuming that this curtain is made of 14-ounce cotton material, it is a significant absorber, especially at the higher audio frequencies. Besides being active as a visual background for TV, its deployment as shown in Fig. 11-3 also hides the low peak absorbers which are something less than beautiful. This does not cause an acoustical problem because the low frequencies upon which the low peak absorber acts are attenuated very little by the fabric.

Floor Covering

A carpet, 200 square feet in area, is placed at the west end of the studio next to the observation window. Speech and drama activities can take place at this end of the studio. If the

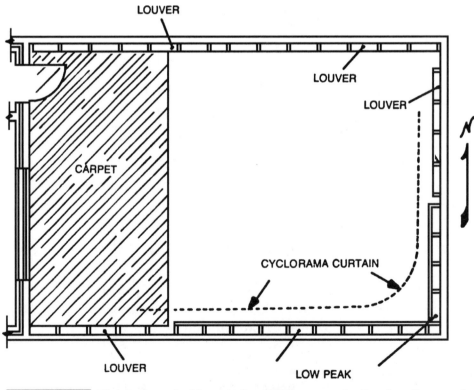

Figure 11-3 **Floor plan of video and multipurpose studio showing positions of carpet, cyclorama curtain, low peak and adjustable louver absorbers.**

adjacent louvers on the north and south walls are open, something approaching the dead end of a live-end-dead-end studio could be arranged.

The floor not covered with carpet is covered with vinyl tile which is excellent for rolling camera dollies. In fact, the entire floor should be covered with vinyl tile, even under the carpet. For unusual situations the carpet can be rolled up or its location shifted at will.

Ceiling Treatment

To keep the cost of acoustical treatment as low as possible a rather inelegant ceiling treatment is specified. A pattern of 31 panels of Owens-Corning Type 703 Fiberglas, 2 feet × 4 feet × 2 inches is cemented to the ceiling.

A suggested arrangement is given in the projected ceiling plan of Fig. 11-4. These semi-rigid boards (3 pounds per cubic foot density) should be relatively free from the sloughing off of troublesome glass fibers. If this is a problem, each panel could be wrapped in light-weight cloth before cementing in place.

Louver Absorbers

The basic section of louver absorber is shown in Fig. 11-5. The frame is of 2 × 6 lumber. The width of each section is 24 inches inside. The height of the two lower segments of each

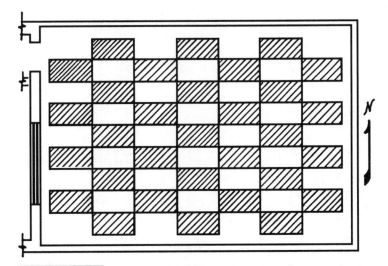

Figure 11-4 Projected ceiling plan showing suggested pattern of 24-inch × 48-inch × 2-inch panels of glass fiber of 3 pounds per cubic foot density cemented to ceiling. These may be covered with thin cloth before mounting if desired.

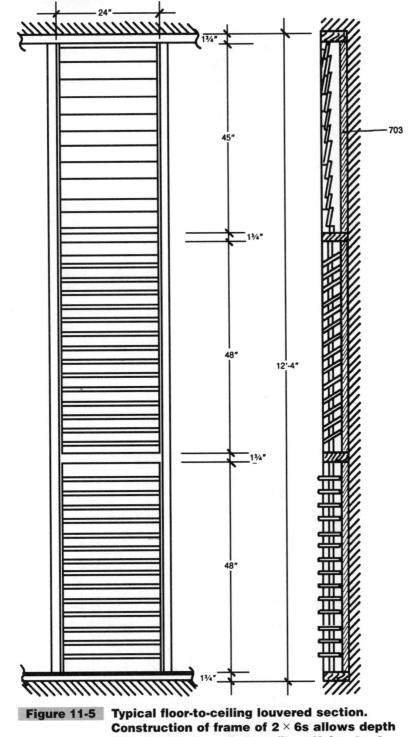

Figure 11-5 Typical floor-to-ceiling louvered section. Construction of frame of 2 × 6s allows depth for only 2 inches of glass fiber. Using 2 × 8s would provide space for 4 inches of glass fiber.

section is 48 inches inside. This allows mounting the 2 inch 703 glass fiber with no cutting in the lower two segments and only a single cut for the top segment. The louvers can be adapted to the width of section selected (24 inches) by cutting the louver panels of $\frac{3}{16}$-inch tempered hardboard the right length.

The louver sections are located in the studio as shown in Fig. 11-3. Thirteen louver sections are fixed to the north wall. By splitting the 13 into six and seven section groups a space is left near the center of the north wall for power and microphone outlets. Another group of four sections is similarly mounted on the east wall so that some wall surface is left exposed for electrical services in the NE corner of the studio. A fourth group of four louver sections is affixed to the west end of the south wall. A total of 21 floor-to-ceiling louver sections are thus distributed across three of the walls of the studio for a total of about 489 square feet.

LOW PEAK ABSORBERS

In certain foreign countries the specification of slat absorbers causes no problem because beautiful hardwoods are cheap and plentiful. Lumber costs in the United States have gone out of sight and buying straight-grained, good quality lumber is often ruled out by its cost.

To avoid such expense a low peak absorber using sheets of $\frac{3}{4}$-inch plywood with saw slots has been designed. This form of the familar Helmholtz low peak absorber is shown in Fig. 11-6. The framework is made of 2 × 10 lumber spaced 24 inches center to center to accommodate the 4-foot × 8-foot sheet of plywood without waste. The cross-bracing serves the double purpose of strengthening the structure and breaking the air space behind a single sheet of plywood into four cavities somewhat less than 2 feet × 4 feet each. This is important to discourage resonances in the cavity parallel to the plywood face. The slots are 4 inches center to center, arranged with respect to the 2 × 10 frame as shown in Fig. 11-6.

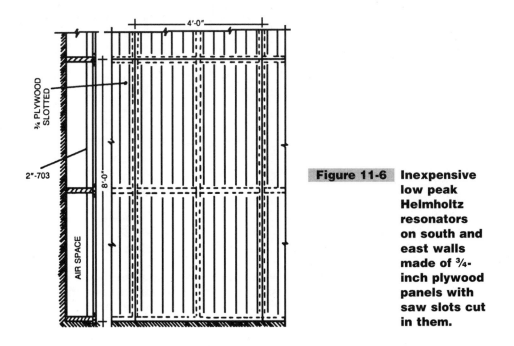

Figure 11-6 Inexpensive low peak Helmholtz resonators on south and east walls made of ¾-inch plywood panels with saw slots cut in them.

An uncertain factor in the design of this low peak absorber is the width of the saw slot, which affects tuning. A width of ⅛ inch is assumed in the present case, although increasing this to ³⁄₁₆ inch increases the frequency of resonance about 20 percent (from 144 Hz to 176 Hz for 4-inch spacing of slots). This is not too serious as the top of the resonance peak (the absorbence peak) is fairly broad if the 2 inches of 703 glass fiber is placed against the back of the slotted panel where the air particle velocity is great.

Keeping the saw slot width between ⅛ inch and ³⁄₁₆ inch should keep this shift of resonance peak within usable limits. The saw slots stop at the cross-bracing to keep the 4-foot × 8-foot panel strong.

The glass fiber may be cemented to the back of the panel, spots of cement being placed between, but not close to, the saw slots. The saw slots should be cleaned out as much as possible, removing slivers and rough protrusions in the slots. A stain and varnish finish for the slotted panels is recommended, taking care not to clog the saw slots in the process.

Figure 11-7 summarized the placement of acoustical materials and devices on the four studio walls. Because the cyclorama curtain hides most of the slotted panel absorbers, the visual impression on entering the room is that louvers cover most of the walls. This will, at least, give an impression of functional novelty, if not a thing of beauty and a joy forever.

LOUVER COMPLICATIONS

The practical aspects of louver-controlled wideband absorbers have been discussed, but there is more to this type of structure than meets the casual eye. The simplistic view is that the louver is only a cover for the glass fiberboards which may be closed or opened. But

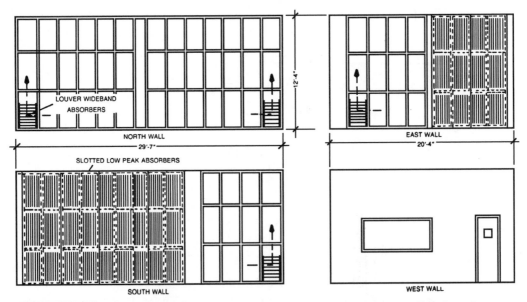

Figure 11-7 **Wall elevations of video and general-purpose studio showing placement of adjustable louvered absorbers and slotted low peak absorbers. The slotted absorbers are largely hidden by the cyclorama curtain.**

what happens when the louver slats are not tightly closed, when a narrow slit remains? The slit makes that segment into a low peak absorber of the Helmholtz type. This is a very indeterminate condition. How wide is the slit?

A slit formed by two louver slats approaching each other is a far more complex problem, mathematically, than having a slit of fixed width in a cover of fixed thickness. Therefore, the frequency at which the absorption peak occurs is uncertain in using the louver elements as low peak absorbers, but the possibility is intriguing.

Computations

Table 11-1 gives the step-by-step calculations to the following three conditions:

- TV condition (all louvers open)
- voice condition ($^1/_2$ of louvers open)
- music condition (all louvers closed)

The resulting reverberation times for the three conditions of Table 11-1 are shown graphically in Fig. 11-8. The unbalanced absorption provided by the carpet, cyclorama curtain, and the low frequency deficiency of the 2-inch glass fiber ceiling boards is compensated by the saw slot low peak absorbers in varying degrees for the above three conditions. Consequently, the bass rise of the reverberation time of Fig. 11-8 is least for the TV condition and greatest when all louvers are closed for the music condition.

Now a confession: the construction of the louver sections detailed in Figs. 11-2 and 11-5 for 2 × 6 lumber allows only room enough for 2 inches of 703. The computations of Table 11-1 and the heavy line graphs of Fig. 11-8 are for 4 inches of 703 in all louver segments and sections. If calculations are carried through for 2 inches of 703 in all louver sections, the bass rise of reverberation time is much greater, as shown by the broken lines of Fig. 11-8. Of course, changing the thickness of 703 has no effect on the music condition because all louvers are closed.

What is needed at this point is a series of detailed reverberation measurements for the specific studio in question. These should include measurements of reverberation time with different percentages of louvers in a narrow slit (low peak) and open condition.

Why go to the expense of 4 inches of 703 if 2 inches of 703 will work just as well using normally closed louvers as low peak slit resonators to compensate for the bass rise in reverberation time? This approach, obviously, requires some careful measuring and planning. It also requires some method, such as the use of a shim to set the slit width accurately in each segment. And it may be more straightforward to pay for 4 inches of 703 for the louver sections and live with the heavy graphs of Fig. 11-8. Of course, this would require a 2 × 8 framework instead of the 2 × 6 frame of Fig. 11-5.

However, the whole approach of using the louver sections as slit resonators could be reduced to simple operational procedures. If not followed, or carelessly followed, such operational procedures could result in some weird unbalanced acoustical conditions. To calibrate judgment, a rough calculation has been carried through to help evaluate the use of louvers as slit resonators for supplying low frequency absorption.

TABLE 11-1 TELEVISION STUDIO CALCULATIONS

SIZE   20'7" × 20'4", CEILING 12'4"
FLOOR   VINYL TILE, 200 SQ. FT. CARPET, 3/16" PILE, FOAM UNDERLAY
CEILING   31 PIECES 703 FIBERGLAS 2' X 4' X 2" CEMENTED TO CEILING
WALLS   LOW PEAK ABSORBERS, 278 SQ. FT.
     LOUVER ADJUSTABLE WIDEBAND, 489 SQ. FT. 4" 703
     CYCLORAMA CURTAIN 35 LIN. FT. 10' HIGH, 14 OZ MTL.
VOLUME   7,418 CU. FT.

MATERIAL	S AREA SQ. FT.	125 HZ A	SA	250 HZ A	SA	500 HZ A	SA	1 KHZ A	SA	2 KHZ A	SA	4 KHZ A	SA
Ceiling	200	0.05	10.0	0.10	20.0	0.10	20.0	0.30	60.0	0.40	80.0	0.50	100.0
Floor: vinyl tile	401	0.02	8.0	0.03	12.0	0.03	12.0	0.03	12.0	0.03	12.0	0.02	8.0
Ceiling: 31/703	248	0.18	44.6	0.76	188.5	0.99	245.5	0.99	245.5	0.99	245.5	0.99	245.5
Cyclorama	350	0.03	10.5	0.12	42.0	0.15	52.5	0.27	94.5	0.37	129.5	0.42	147.0
La peak absorb	278	0.90	250.2	0.84	233.5	0.64	177.9	0.36	100.1	0.17	47.3	0.06	16.7
Sa Subtotal			323.3		496.0		507.9		512.1		514.3		517.2
TV CONDITION													
All louvers open 4" 703	489	0.99	484.1	0.99	484.1	0.99	484.1	0.99	484.1	0.99	484.1	0.99	484.1
Total sabins, Sa			807.4		980.1		992.0		995.2		996.4		1001.3
Reverb. Time, Sec.			0.45		0.37		0.37		0.36		0.36		0.36
VOICE CONDITION													
33.6" louvers open	164	0.99	162.0	0.99	162.0	0.99	162.0	0.99	162.0	0.99	162.0	0.99	162.0
Total sabins, Sa			485.3		656.0		669.9		674.1		676.3		679.2
Reverb. Time, sec.			0.75		0.55		0.54		0.54		0.54		0.54
MUSIC CONDITION													
All louvers closed													
Total sabins, Sa			323.3		496.0		507.9		512.1		514.3		517.2
Reverb. Time, sec.			1.1		0.73		0.72		0.71		0.71		0.70

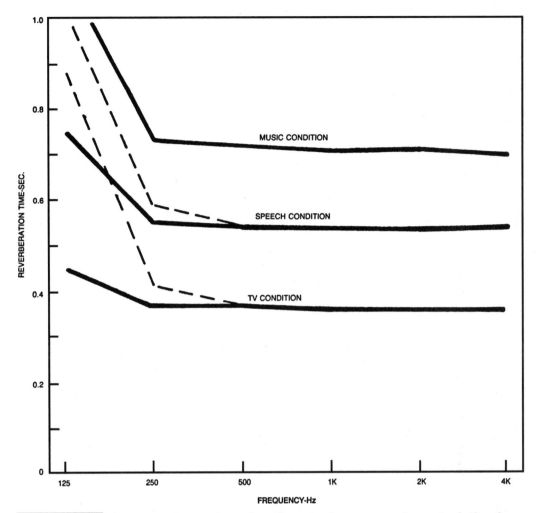

Figure 11-8   Calculated reverberation time vs. frequency characteristic of studio for three conditions of adjustable louvers: video condition (all louvers open), voice condition (⅓ of louvers open) and music condition (all louvers closed). The heavy lines apply to louvers with 4 inches of glass fiber, the broken lines apply to louvers with 2 inches of glass fiber.

In Fig. 11-9 the heavy graph A represents the reverberation time of the studio for voice condition with 2 inches of 703 in all louver sections. In this voice condition ⅓ of the louvers are open and ⅔ closed. Assuming that the slit width and depth (the overlap of the louvers), the effective cavity depth and all other parameters result in a low peak resonator comparable to the saw slot low peak absorbers, the absorption coefficients of the saw slot units can be used in this rough calculation (see Table 11-1). This is not a very secure assumption; hence the results may be off a considerable amount, but still be of help in a qualitative way.

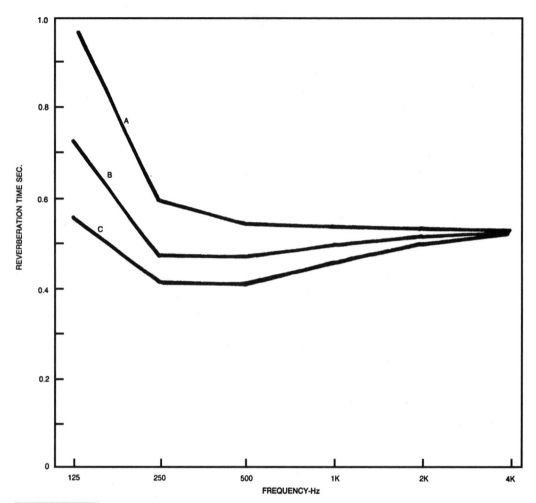

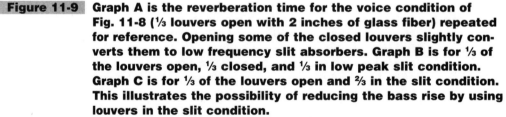

Figure 11-9 **Graph A is the reverberation time for the voice condition of Fig. 11-8 (⅓ louvers open with 2 inches of glass fiber) repeated for reference. Opening some of the closed louvers slightly converts them to low frequency slit absorbers. Graph B is for ⅓ of the louvers open, ⅓ closed, and ⅓ in low peak slit condition. Graph C is for ⅓ of the louvers open and ⅔ in the slit condition. This illustrates the possibility of reducing the bass rise by using louvers in the slit condition.**

With this assumption graphs B and C have been computed. Graph B is for ⅓ of the louvers open, ⅓ closed, and ⅓ in low peak slit condition. Graph C is for ⅓ of the louvers open and ⅔ in slit condition. Graph C flattens out fairly well, but falls to far below the 0.46 second goal. Having ⅓ of the louvers open is common to all three graphs. Other combinations of open and slit louvers would undoubtedly lift and straighten graph C but, it is emphasized, this should be done on the basis of measurements, not calculations with uncertain coefficients.

Television Facilities

The cyclorama certain has been discussed, perhaps a bit prematurely, in connection with its acoustical absorption effect. The relationship of this curtain to an overhead pipe grid for supporting lamps is shown in Figs. 11-10 and 11-11A.

The advantage of supporting lamps from above is that the floor is kept free of most lamp stands and cables. For full lighting it is sometimes necessary to provide a suitable *scoop* or

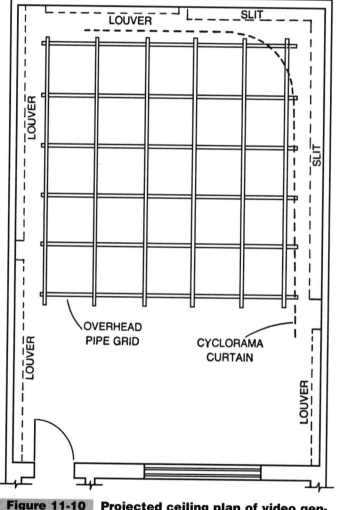

Figure 11-10 **Projected ceiling plan of video general-purpose studio showing relationship of overhead pipe grid for supporting lamps and the cyclorama curtain.**

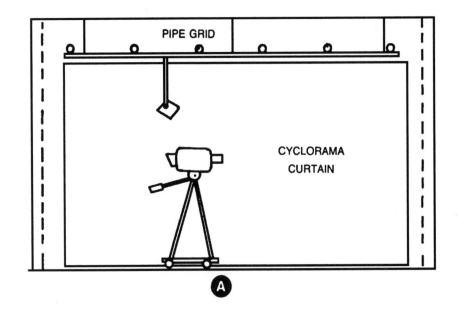

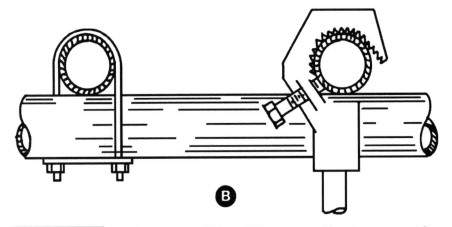

Figure 11-11 (A) Overhead pipe grid for supporting lamps and its relationship to cyclorama curtain. (B) U-bolt clamp method of securing pipe grid and conventional hook for lamp hangers.

broad on a stand, but a grid can care for most of the lighting equipment required for a small stage such as this.

The pipes, often 1-½ inches inside diameter, are secured at intersections with Ubolts as shown in Fig. 11-11B. The individual lamp units are suspended below the grid at an adjustable height on a vertical pipe secured to the grid by the toothed clamp also depicted in Fig. 11-11B. A safety chain from the pipe grid to the lamp unit is necessary as the adjustable devices sometimes slip.

Lamps are mounted, positioned, adjusted, and pointed from a ladder. In larger studios this is done from catwalks above. With one or more cameras, conventional acoustical treatment in the control room, videotape recorder, monitor bank, video and sound consoles, and the required supporting gear, the TV mini-studio described should be quite adequate for a meaningful training program. An illustrated trade reference catalog is helpful in setting up such a television training facility.[35]

A VIDEO AND
MULTITRACK STUDIO

Features: Adjustable acoustics, using a small studio as an isolation booth, service boxes for microphones and television equipment.

CONTENTS AT A GLANCE

Studio Plans

Conduits

Power Facilities

Studio Treatment
SWINGING PANELS

PLYWOOD WALL DIFFUSERS
WIDEBAND WALL ABSORBERS
CEILING TREATMENT
REVERBERATION TIME

Control Room Treatment
CEILING TREATMENT
FLOOR TREATMENT

What do video studios and multitrack recording studios have in common? Offhand, TV and multitrack seem like quite divergent activities. They both need ample space. In multitrack recording the acoustical separation between sources obtainable by screens and other devices is limited and it is soon discovered that a certain minimum of physical separation is necessary, which adds up to requiring a certain amount of floor space for each performer. Television's need for elbow room is more obvious, what with several cameras rolling around, overhead space for lights and multiple sets.

The studio selected for this chapter is, according to the general plan of this book, a budget project. This means compromises in many areas; the challenge is in keeping the magnitude of these compromises small. For example, the walls of this studio are 8-inch concrete block rather than double concrete with an air space. By plastering both faces and filling the block voids with concrete, a respectable STC 56 can be obtained which is within range of, say, STC 62. This illustrates a physical principle not too generally appreciated-that it is the last ounce that boosts the cost to excessive levels.

Put another way, the curve flattens out so that much more effort and money must be put in to get that last ounce than the earlier ounces. The message is simply that work of quality can be done with facilities somewhat less than the world's best. The best studio in the world comes far from guaranteeing the best product; skill and resourcefulness are still the indispensable ingredients.

Studio Plans

The layout of the studio to be studied is shown in Fig. 12-1. The plan includes two studio-control room suites. The large Studio A and its control room are for video and multitrack recording. The smaller Studio B and its control room are for general speech work.

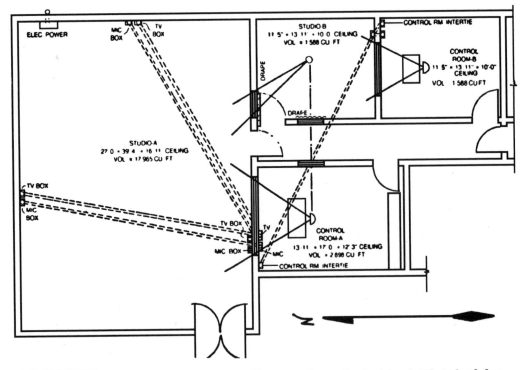

Figure 12-1 **Floor plan of larger studio complex adapted for both television and multitrack work. The use of Studio B for drum or vocal isolation booth is aided by small windows for visual contact and intertie lines between Control Rooms A and B.**

Studio B serves also as an isolation booth for vocals or drums when complete separation from what is happening in Studio A is required. A window in Control Room A lines up with a window in Studio B for coordination of activities via eye contact. Another window in Studio B looking into Studio A does the same between those two rooms. These incidental windows for eye contact are small but are built to the same double-glass standards as the larger observation windows.

Conduits

A suggested plan for conduit runs is indicated in Fig. 12-1. These conduits are laid before the floor slab is poured as shown in Fig. 12-2A. This gives the most direct path which pays off later in ease of snaking in the cables. Conduit terminal boxes are on the north and east walls. The boxes on the south wall under the control room window are fed directly through the wall from the control room as indicated in Fig. 12-2A. To avoid acoustical leaks these short conduits through the observation window wall must be very carefully caulked and after the cables are pulled through, glass fiber should be packed in the conduit from both ends.

One conduit serves video equipment and another holds the audio lines for each of the three terminal positions in Studio A. These conduits are fitted with boxes of the type shown

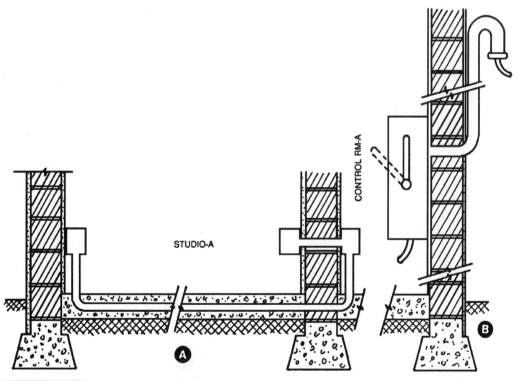

Figure 12-2 **(A) Duct runs laid before the concrete floor is poured provide separate ducts for television and audio use. (B) Electrical switch box arrangement for energizing portable cable runs for all set lighting. House lights are handled in the conventional way.**

in Fig. 12-3. Each audio pair terminates on a barrier strip from which flexible leads run to the professional type microphone connectors mounted on the lid. The video box must be adapted to the type of camera equipment to be used, but this is straightforward if the conduit and boxes are provided.

In the control room boxes larger than those in Studio A are required because they must terminate three conduits and the one going through the wall coming into the back of the box, if desired. Figure 12-2A and Fig. 12-4 illustrate one satisfactory method of arranging both video and microphone boxes and conduits in Control Room A.

A conduit should always tie control rooms together in a studio complex. By running six audio pairs in this conduit and terminating at both ends on a jackstrip, great flexibility results. Equipment in Control Room B can be used for a big job in Control Room A, or vice versa, without moving the equipment physically. Using the lines for microphone, cue foldback or talkback is also made easy with these intertie lines. The intertie box is shown in Fig. 12-4.

Power Facilities

Figure 12-2B gives details of the heavy duty electrical power switch and breaker box on the east wall of Studio A. This is to provide power only for set illumination. The general house lights are handled independently in the conventional manner. From this box portable cables run to spider distribution boxes and then to the individual lamp circuits on the pipe grid and on the floor. When this switch is open, all portable circuits are dead. For safety, great care must be exercised around these heavy duty lamp circuits and only experienced personnel should be allowed to work with them.

Studio Treatment

It has been pointed out that both television and multitrack recording require relatively *dead acoustics*. The client also expressed a desire to have Studio A acoustics suitable for recording musical groups in the more conventional manner.

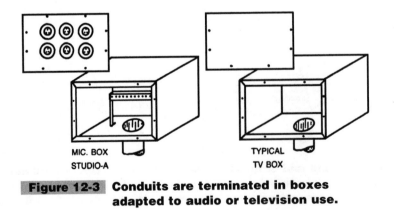

MIC. BOX
STUDIO-A

TYPICAL
TV BOX

Figure 12-3 Conduits are terminated in boxes adapted to audio or television use.

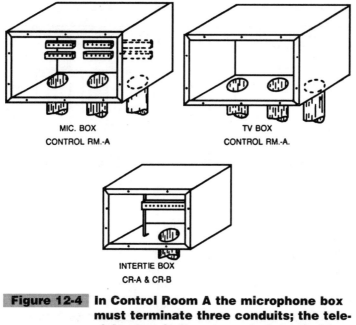

MIC. BOX
CONTROL RM.-A

TV BOX
CONTROL RM.-A.

INTERTIE BOX
CR-A & CR-B

Figure 12-4 **In Control Room A the microphone box must terminate three conduits; the television box is the same. An intertie box terminates the conduit connecting Control Room A and Control Room B.**

In fact, this use was in immediate demand while both video and multitrack were activities they hoped to enter soon. To minimize the degree of compromise between the three types of work a certain amount of adjustability has been built into Studio A acoustical treatment. This is accomplished by seven swinging panels on the north wall and five (rather, four full panels plus a half panel) on the east wall as shown in Fig. 12-5. By closing all panels approximately 334 square feet of 2-inch 703 are, in effect, removed from the room and replaced by about half that area of plywood. The closed panels also are physical protrusions which assist in diffusing sound.

SWINGING PANELS

The swinging panel construction is detailed in Fig. 12-6A and 12-6B. It is a simple frame of 1 × 4 lumber, with a plywood back (½-inch or ⅝-inch) for stiffening, which holds and protects 2-inch Type 703 Fiberglas absorbent. For efficient plywood cutting the frame is made of a size to accept 2-foot × 8-foot pieces of plywood. A cross-member at the midpoint adds strength, breaks the 8-foot length visually, and supplies a logical position for a third hinge.

Some acoustically transparent protective covera are desirable for the 703. This could be perforated metal sheets having at least 25 percent of the area in holes. It could be expanded metal, such as metal lath, or wire screen. Perhaps the simplest and most attractive, if not the most resistant to mechanical damage, is colorful cloth such as burlap or other loosely-woven fabric.

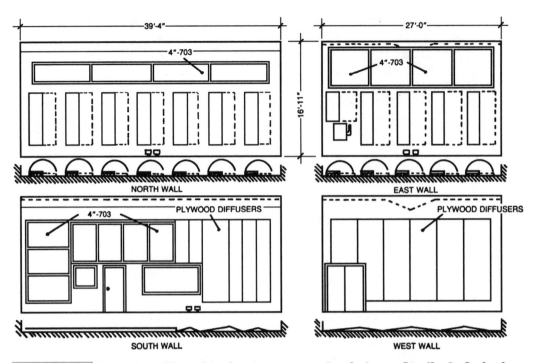

Figure 12-5 Wall elevations showing treatment plan in large Studio A. Swinging panels on the north and east walls allow considerable range in adjustment of acoustics, supplementing plywood diffusers/absorbers and wideband elements.

PLYWOOD WALL DIFFUSERS

Diffusers of ¾-inch plywood of triangular construction are used on the west wall and a portion of the south wall as shown in Fig. 12-5. On this figure the cross-sectional shapes are associated with each elevation. To utilize the plywood without waste each face is either 2 feet or 4 feet in width.

The 12-foot length uses plywood of that length, if available, or 8-foot plywood sheets may be extended because of the placement of the section dividers. A 2 × 4 ridge and frame provide the basic stiffening and nailing facilities required.

Figure 12-7 gives basic details of construction which apply to diffusers having either the 2-foot or 4-foot face width. These plywood diffusers, which are also fairly low frequency absorbers, can be painted without affecting their acoustical properties significantly.

WIDEBAND WALL ABSORBERS

The east end of the south wall (Fig. 12-5) is largely covered with wideband modules containing 4-inch thicknesses of Type 703 Fiberglas. Modules of similar construction but of different shape are also mounted above the plywood diffusers on the north and east walls. The exact dimensions of the individual modules are not significant. The total effective area is significant.

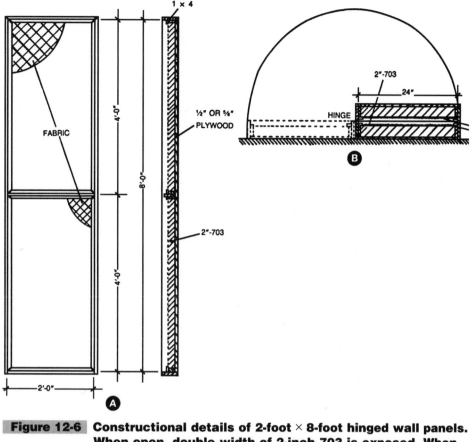

Figure 12-6 Constructional details of 2-foot × 8-foot hinged wall panels. When open, double width of 2-inch 703 is exposed. When closed, plywood protuberances contribute to diffusion and low frequency absorption.

As similar units have been described in earlier chapters, no space is given to their construction other than to mention the need for a frame of 1-inch lumber (plywood backs are optional) and a lightweight, loosely woven fabric for appearance and control of glass fiber particles. A perforated metal sheet (25 percent perforation minimum), screen, or expanded metal cover may be added for protection against physical abuse if desired.

CEILING TREATMENT

The Studio A projected ceiling plan of Fig. 12-8 reveals a triangular plywood diffuser approximately 8-feet wide running the length of the room down the center. The remainder of the ceiling surface is covered with a wideband absorber composed of the usual 4 inches of 703 Fiberglas.

A 2 × 4 gridwork holds the glass fiber semirigid boards which may be covered as discussed previously. In this case, expanded metal lath without the usual fabric is sufficient because of the height of the ceiling and the resulting distance from critical eyes and probing fingers.

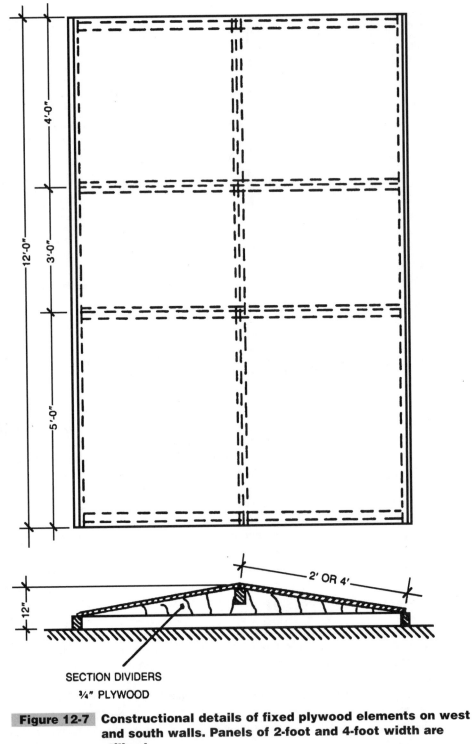

SECTION DIVIDERS
¾" PLYWOOD

Figure 12-7 Constructional details of fixed plywood elements on west and south walls. Panels of 2-foot and 4-foot width are utilized.

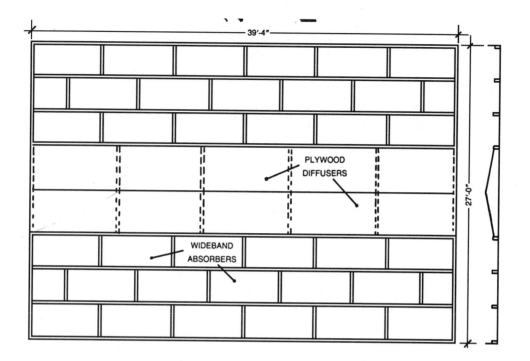

Figure 12-8 **Projected ceiling plan showing ceiling treatment. A triangular cross-section plywood element about 8-feet wide runs the full length of the room. The remainder of the ceiling area is covered with wideband sections of 4-inch 703 glass fiber.**

The triangular plywood ridge on the ceiling is constructed as shown in Fig. 12-9. The entire structure is built down from the 2 × 4 gridwork fastened securely to the roof slab. There is a significant amount of open space between the plywood skin and the concrete roof slab. This space can be put to good use for conduit runs or, possibly even as a duct for air circulation.

REVERBERATION TIME

Table 12-1 provides a summary of pertinent acoustical data and computations of reverberation time. This table has been simplified and abbreviated by neglecting the sound absorbed by the vinyl tile-covered concrete floor, the plastered concrete block wall areas not covered by plywood or wideband units, windows, doors, and air contained in the room. Admittedly, the absorbence of each of these is minor, yet overall accuracy would be improved by their inclusion. But the price paid in confusion and complication is too high. Measurements should be relied upon for final evaluation in any event.

Table 12-1 includes reverberation time calculations for two conditions:

■ All swinging panels open exposing the 2-inch 703.
■ All swinging panels closed, covering the 703 and exposing the plywood box of half width and double depth.

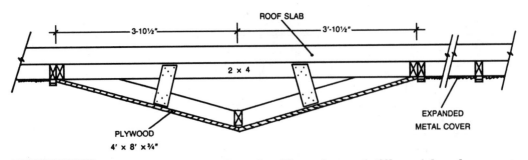

Figure 12-9 Details of construction of ceiling plywood diffuser/absorber.

The values of reverberation time from the table are plotted in Fig. 12-10 to give a graphic picture. It is interesting that at 125 Hz the reverberation time is essentially the same whether the panels are open or closed.

The 2 inches of 703 has an absorption coefficient of only 0.18 at 125 Hz, effective when the panels are open. When closed, the plywood shows a coefficient of 0.38, but this applies only to about half the area of the 703.

For frequencies of 250 Hz and above at which the 703 becomes essentially a perfect absorber there is a significant spread between the *all panels closed* and the *all panels open* graphs. When the panels are open a reverberation time of 0.58 second is obtained over most of the band. When closed, this increases to about 0.74 second.

Traditional music recording techniques normally require a reverberation time of about 0.9 second for a studio of this size (almost 18,000 cubic feet). However, 0.74 second constitutes a reasonable and usable compromise. For television and multitrack work the 0.58 second should be very favorable. Placing some of the musicians near the highly absorbent opened swinging panels, with adequate screens between, should result in excellent separation for multitrack work. Further, the thousand square feet of floor space should accommodate something like 20 musicians of either the traditional or multitrack types.

For practically infinite separation, sometimes required for soloist or drums, the use of Studio B as an isolation booth is possible. The arrangement of Fig. 12-1 makes this both possible and convenient, sending the microphone outputs from Studio B to Control Room A by way of the intertie lines between Control Rooms A and B. Studio B, in the present case, has a heavy work load in conventional recording of radio programs with a single narrator.

Control Room Treatment

The wall treatment of Control Room A to achieve a reverberation time of about 0.3 second is described in Fig. 12-11. By plugging 0.3 second reverberation time into Sabine's equation we find that about 473 sabins (absorption units) are required. How shall this be distributed between the three pairs of room surfaces? This is an excellent opportunity for illustrating the good acoustical practice of distributing each type of absorber between the N-S, E-W, and vertical pairs of surfaces in proportion to the areas of these pairs. In this control room with no carpet, we have only one kind of absorber-areas of 4-inch 703. In the case of Control Room A this distribution is shown in Table 12-2.

TABLE 12-1   STUDIO A ACOUSTICAL DATA

SIZE: 20'0" × 39'4", CEILING 16'11"
FLOOR: VINYL TILE ON CONCRETE
CEILING: 3/4" PLYWOOD DIFFUSER 304 SQ. FT.
WALLS: WIDEBAND 4' 703-722 SQ. FT.
SWINGING PANELS 2" 703-334 SQ. FT.
WIDEBAND 4' 703-449 SQ. FT.
3/4" PLYWOOD DIFFUSER 297 SQ. FT.
VOLUME: 17,965 CU. FT.

MATERIAL	S AREA SQ. FT.	125 HZ		250 HZ		500 HZ		1 KHZ		2 KHZ		4 KHZ	
		A	SA	A	SA	A	SA	A	SA	A	SA	A	SA
Plywood: wall-ceiling	601	0.38	228.4	0.19	114.2	0.06	36.1	0.05	30.1	0.04	24.0	0.04	24.0
Wideband 4' 703 walls-ceiling	1171	0.99	1159.3	0.99	1159.3	0.99	1159.3	0.99	1159.3	0.99	1159.3	0.99	1159.3
Swing panels OPEN 2 703	334	0.18	60.1	0.76	253.8	0.99	330.7	0.99	330.7	0.99	330.7	0.99	330.7
Total sabins, Sa			1447.8		1527.3		1526.1		1520.1		1514.0		1514.0
Reverb. Time, Sec.			0.61		0.58		0.58		0.58		0.58		0.58
Swinging Panels CLOSED Wideband, 4 703			1159.3		1159.3		1159.3		1159.3		1159.3		1159.3
Plywood: walls-ceiling +panels	768	0.38	291.8	0.19	145.9	0.06	46.1	0.05	38.4	0.04	30.7	0.04	30.7
Total sabins, Sa			1451.1		1305.2		1205.4		1197.7		1190.0		1190.0
Reverb. Time, Sec.			0.60		0.67		0.73		0.73		0.74		0.74

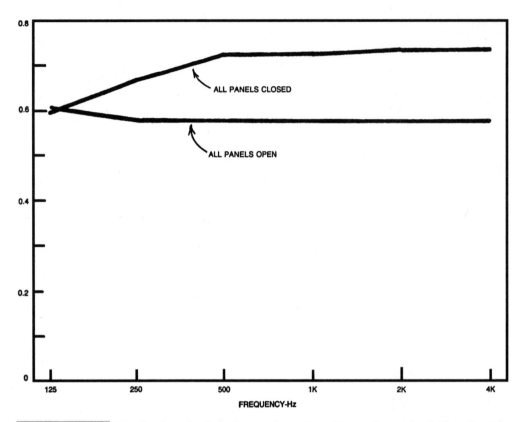

Figure 12-10 Studio A calculated reverberation time characteristics for all panels closed (conventional music condition) and all panels open (multitrack and television condition). Swinging the panels has practically no effect at 125 Hz because the lower absorption of the 2-inch thickness of 703 is offset by the plywood absorption of the closed boxes.

For the frequency range 125 Hz-4 kHz we can take the absorption coefficient of 4 inches of 703 as essentially unity. This means that one sabin of absorption is given by each square foot of 703. For Control Room A we then need 473 square feet of 703. This should be distributed approximately as shown in column (a) of Table 12-2. The distribution of the 703 shown in Fig. 12-11 accommodates only 390 square feet which yields a reverberation time of 0.36 second, which is acceptably close to 0.3 second. About 160 square feet of 703 should then be applied to north and/or south walls, but the north wall is almost filled with the observation window. The south wall has the work table and built-in drawers and cabinets. So, we do what we can and mount a frame of 20 sections, each 24 inches × 24 inches inside, on the south wall, totaling only 80 square feet. The other 80 square feet must be placed elsewhere. This is about all that can be done for the N-S mode.

How about the east and west pair of walls? On the west wall (Fig. 12-11) there are 20 sections, each 24 inches × 24 inches inside, yielding 80 square feet, and seven sections

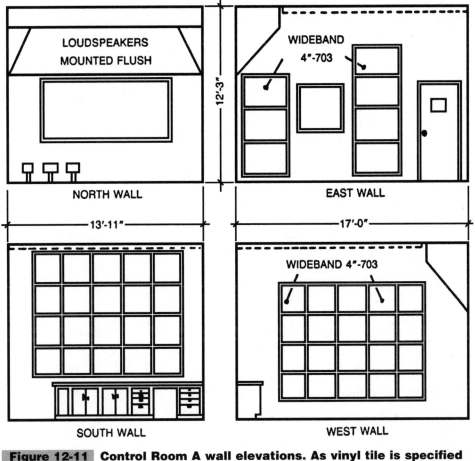

Figure 12-11 Control Room A wall elevations. As vinyl tile is specified for floor covering, the only material required for acoustical treatment is 4-inch 703 glass fiber.

24 inches × 36 inches inside on the east wall. This gives an effective area of 42 square feet for a total of 122 square feet. The E-W walls plus the N-S walls then offer a total or 80 plus 122 or 202 square feet of 4-inch 703.

CEILING TREATMENT

A total of 473 square feet of 703 is required and 202 square feet are applied to the walls; therefore, 271 square feet should be applied to the floor-ceiling pair of surfaces. Because 703 is not a very satisfactory floor covering, 188 square feet are placed on the ceiling. With a width of 13 feet 11 inches (13.9 feet), this means that a length of about 13.5 feet of 703 area yields an area of 188 square feet.

The frame takes up a respectable portion of this area; hence a length of 14 feet 6 inches is actually required to give the 188 square foot net.

TABLE 12-2 CONTROL ROOM A DISTRIBUTION OF ABSORPTION				
AXIAL MODES	AREA SQ. FT.	% TOTAL AREA	(A) EXACT DISTRIBUTION OF 473 SABINS	(B) PRACTICAL DISTRIBUTION OF 390 SABINS
N-S walls	416.5	33.8	160	80
E-W walls	341.0	27.7	131	122
Floor-ceiling	473.2	38.5	182	188
Totals	1230.7	100.0%	473	390

Figure 12-12 shows a grid of sections mounted on the ceiling, leaving some bare ceiling for the loudspeakers on the north end. The 24-inch inside dimension of each section is for the purpose of efficient cutting of the 24-inch × 48-inch sheets of 703. Of course, 24-inch × 48-inch sections require cutting only in the odd sections. It is the 703 area that counts. There are many ways to handle the mechanical mounting.

The rationale of Table 12-2 is based on distributing the 4-inch 703 glass fiber material in proportion to the areas of the surfaces associated with the three axial modes of Control Room A. Column (a) of this table gives the exact number of sabins for each pair of surfaces based on this premise. Column (b) lists the practical distribution of 703 areas of Figs. 12-11 and 12-12 yielding a reverberation time of 0.36 second. There are some unavoidable slips betwixt theory and practice. Observation windows, doors, hard-surfaced floor and work table have brought compromise. The principle is still a good initial guide, even if certain departures from it are necessary.

FLOOR TREATMENT

As indicated in the floor plan of Fig. 12-12, the floor covering specified is vinyl tile. Linoleum, wood parquet, or other hard surface is acceptable. By avoiding carpet with its unbalanced absorption, only 4-inch 703 is required in the acoustical treatment.

In the context of this chapter no detailed discussion of the treatment of Studio B and Control Room B is required because of their similarity to suites covered in previous chapters. For normal use of Studio B a reverberation time of the order of 0.3 second is required. The suitability of 0.3 second when used as an isolation booth depends on the type of sound source placed in it. As a vocal booth 0.3 second may be a bit low, but reverberation can always be added electronically. As a drum booth it should be satisfactory as it is, although idiosyncrasies of individual drummers often require temporary adjustments. For normal use of Studio B, short drapes can be drawn over the two unused windows to minimize reflection defects.

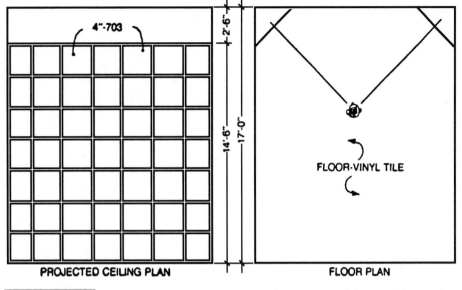

Figure 12-12   Projected ceiling plan and floor plan of Control Room B.

A SCREENING FACILITY
FOR FILM AND VIDEO

Features: Listening room characteristics, projection facilities, stepped seating area.

CONTENTS AT A GLANCE

Favorable conditions for viewing motion pictures must include both visual and aural factors. Visual quality depends on screen diffusion, screen brightness, and the steadiness of the projected image. Aural quality in regard to room acoustics is much the same problem as covered in previous discussions of studio and monitoring room quality with some relaxation of tolerances. From the viewpoint of sound reproduction, the motion picture projector sound head quality is vitally important, along with amplifiers and loudspeakers. Both the visual and the acoustical factors are treated in this chapter, as they are inseparably bound together in projecting motion pictures and in viewing video presentations effectively.

Floor Plan

For the best impression of a projected motion picture, the projection machinery must be in a separate room. This not only reduces the noise, it eliminates one more potential diversion of attention of those who should be paying full attention to the screen.

In Fig. 13-1 a very practical, low cost, high quality screening facility is described. It is small but effective accommodating up to 20 people very comfortably, a few more in a pinch. The individual seats may be upholstered, swivel-type, or simply canvas director's chairs. Arranged on carpeted wood risers of 4-foot width, a feeling of comfort and luxury can be imparted with modest outlay.

Door Arrangement

The background noise level standard adopted for this small theater (the NC 25 contour is commonly used, Fig. 2-5), and the environmental noise level outside the theater determine the type of walls required. This has been covered in Chapter 2.

Doors A and B of Fig. 13-1 provide entrance and egress for the screening room. If these are single doors, even though they have a solid core and are well weatherstripped, their insulation against noise from the outside is quite limited. If the external noise is low level, single doors may be adequate. Adding door C makes a sound lock between doors C and B. This would make little sense unless something similar is done to door A. In addition to the noise, another potential problem associated with door A is that, if opened while the theater is darkened, it may be very disruptive as light from the outside falls on the projected image, disturbing the viewers.

Screen

This theater should be equally valuable for projection of 35 mm or 16 mm film prints, or 70 mm or super 8 mm prints for that matter. The same screen can serve them all equally well. Aspect ratio is defined as the ratio of width to height of the film image and the screen proportions are determined by the proportions of the image on the film. Historically, an aspect ratio of four units wide by three units high (1.33:1) has been used.

When sound was introduced to 35 mm film, less room was available for the image of each frame, but a smaller 4×3 proportion persisted. Starting with the early introduction of *CinemaScope* and progressing on through many stages, the aspect ratio varied from 2.66:1 on down.

It is now general practice to provide for projection at proportions between 1.33:1 (old 35 mm film and normal 16 mm films) to 1.65 and 1.95:1 for flat, widescreen films and 2.35:1 for anamorphic projection. This means that ample screen area must be provided and that black masks are to be used to shape the screen to the format to be used, cropping the edges neatly. In Fig. 13-2 a screen 6-feet high and 14-feet wide meets the maximum proportions required, 2.35:1.

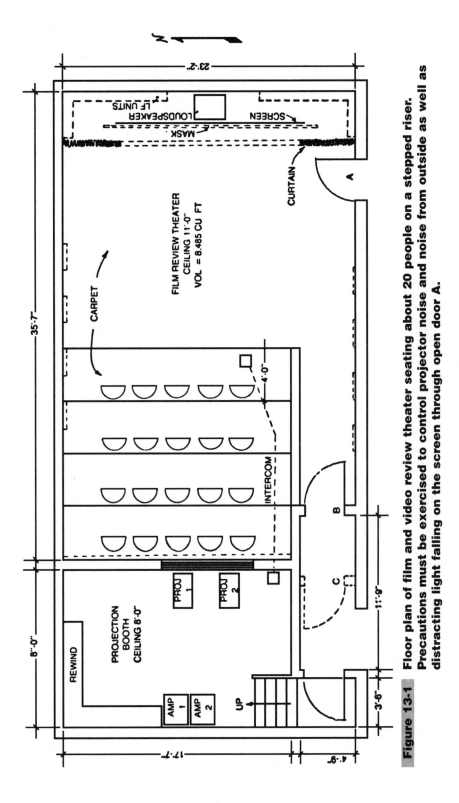

Figure 13-1 Floor plan of film and video review theater seating about 20 people on a stepped riser. Precautions must be exercised to control projector noise and noise from outside as well as distracting light falling on the screen through open door A.

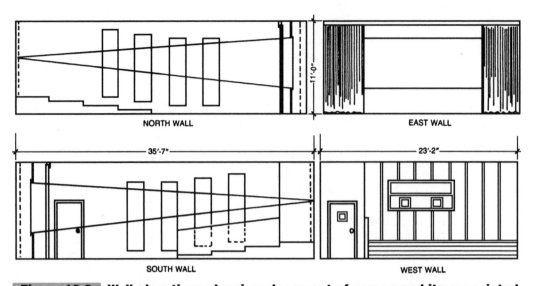

NORTH WALL   EAST WALL

SOUTH WALL   WEST WALL

Figure 13-2   **Wall elevations showing placement of screen and its associated elements as well as acoustical treatment. The rear wall and side wall panels are basically 4-inch 703 glass fiber covered with a perforated vinyl fabric.**

The masks are simply black cloth-covered frames of light wood or metal. These are arranged as a top and bottom horizontal pair, and a left and right pair for cropping the vertical edges. A fancy (and very convenient) installation would have these pairs adjustable by motor drive and remotely controlled from the projection booth so that one button would be pushed for 1.33:1, another for 1.65:1, etc. Other controls for highly professional projection would include motor-driven curtains and light dimmers, preferably synchronized.

The screen must be of the perforated type if the loudspeaker is positioned behind it as indicated in Fig. 13-1. The small perforations make the screen essentially transparent to sound in the audible band, yet are not visible at normal viewing distances.

Three types of screens are in common use, distinguished from each other by their surface.

- matte
- beaded
- metallized

A picture projected on a screen with a matte surface appears to have much the same brightness when viewed off to one side as when viewed from directly in front. The general screen brightness level of the matte screen is, however, quite low. Beaded and metallized screens have a more pronounced directional characteristic, throwing most of the light directly back toward the projector and giving a much dimmer picture off to the side. For quality projection giving a picture of equal brightness over the seating area of the small screening room, a perforated screen with a matte surface is probably the more suitable. It should be mounted in a frame with elastic cord so that its surface is very flat.

Projection Booth

The floor of the projection booth is at least 3 feet above the main floor of the screening room so that the projector beam clears the heads of those seated on the top riser. Actually, this beam should not be interrupted by persons walking anywhere in the theater, but with the limited 11-foot ceiling height this ideal cannot be attained with the riser plan shown.

The projection room is reached by five steps up from the alcove shared with door B. The projectionist's observation window is 7-feet long and about 18-inches wide and is of customary double-glass construction.

Glass in the two projector ports is quite a different problem because of possible color tint and refraction affecting the projected image. From the standpoint of theater noise, however, it is imperative that these small ports be fitted with at least one good thickness of glass. A work table is suggested for film rewinding and similar tasks.

The projection booth should be acoustically treated to reduce the effect of projector noise both in the booth and in the screening room. The surface area of the projection booth is about 691 square feet with an 8-foot ceiling. If the floor is vinyl tile and the walls and ceiling are bare gypsum board the average absorption coefficient might be about 0.05, giving a total absorption of $691 \times 0.05 = 34.55$ sabins at midband.

If the absorption coefficient was increased to 0.30 by the acoustical treatment of room surfaces, the total absorption in the room would be increased to $691 \times 0.30 = 207.3$ sabins. This would result in a decrease in projector noise level of $10 \log 34.55/207.3 = 7.78$ dB. This means that anywhere in the room, except in the immediate vicinity of the projector, the noise level is thus reduced almost 8 dB by the introduction of the absorbing material. This would make it much more comfortable for the projectionist and would reduce projector noise in the screening room as well.

Fire regulations pertaining to projection rooms must be determined before actual construction and acoustical treatment are begun, but acoustically the only requirement is to make the room as highly absorbent as feasible over the audible frequency range with no worry about uniformity.

Another detail which can contribute to smooth operation of the projection booth is some form of intercommunication with a control seat in the audience area. This link provides a professional touch to a projection event in such things as when to roll it, the adjustment of sound level, etc.

Theater Treatment

A screening theater of this type is basically a listening room and should be acoustically treated as such. There is the question of whether speech or music should be favored, but the understanding of narration and dialog is taken as the most fundamental requirement. For a room with an 8,485 cubic foot volume, the optimum reverberation time for speech is close to 0.5 second. The goal is 0.5 second, uniform 125 Hz-4 kHz, recognizing in advance that tolerances are not as tight for film or video viewing as they are for recording studios or critical monitoring rooms.

The entire floor area of the film review theater, including the riser seating area, is carpeted with a heavy carpet and pad as indicated in Table 13-1. This is in deference to the comfort and enjoyment of guests and in spite of the acoustical compensation required.

Another component of acoustical absorption having characteristics similar to carpet is the curtain. For this the area entered in the computation is that which the curtains offer when retracted to reveal the 14-foot width of screen (the 2.35:1 aspect ratio). The question arises, "should both sides of the partially retracted curtain be considered active absorbers?" This depends on how the curtain was placed when the coefficients were measured. The book (Appendix A) says, "Medium velour, 14 ounces per square yard draped to half area."

Although the inference is that the drape is not far from a wall (these are about 3 feet from a wall), the area of only one side of the curtain has been entered in Table 13-1 and the possible error of a few dozen sabins registered in the inner consciousness.

Wideband absorption, again with 4 inches of Owens-Corning 703 Fiberglas as the basic dissipative element, is applied to the north and south walls in the form of 2 foot × 8 foot wall panels and to the entire rear (west) wall. In both cases the 703 is covered with a decorative vinyl fabric perforated for a reasonable degree of sound transparency. The vinyl fabric supplied by L.E. Carpenter and Company (Appendix D), called *Vicrtex*, comes in many attractive patterns and colors, but is perforated at the factory only on order. The perforation percentage of the Vicrtex is estimated to be between 12 percent and 15 percent. This vinyl covering can be a main contributor to the decor of the room if carefully chosen.

The wideband wall panels are located as shown in Fig. 13-2, four to each side wall and positioned so that a panel on one wall faces bare wall between panels on the opposite wall. This should provide reasonable control of flutter echoes in the north-south mode, even as the carpet does for the vertical mode.

Figure 13-3 gives necessary details of construction of the wall panels. The perforated vinyl fabric comes in a 54-inch width which will provide two 24-inch panels. If desired, 1 × 6 lumber can be used for the external frame which results in an air space between the 703 and the back board. This increases low frequency absorption.

Figure 13-4 shows how the rear (west) wall is framed with 2 × 4s making spaces 24 inches wide to accommodate both the 24-inch width of the 703 and the 54-inch width of Vicrtex cover. The 2 × 4s are only about 3¾ inches and by shimming them out from the wall about ¼ inch the 703 does not tend to bulge the Vicrtex. A finish strip is nailed to each 2 × 4, covering the Vicrtex edges and overlap.

Low frequency absorption is required to compensate for low frequency deficiencies of carpet and curtains. The risers, constructed of ¾-inch plywood on a wood frame, contribute a significant amount of absorption, peaking about 125 Hz. In addition, the gypsum plaster board on walls and ceiling, even though covered by other elements, also absorbs well at low frequencies.

In Table 13-1 gypsum board of ½-inch thickness has been assumed, although relatively small changes would be expected if it were ⅝-inch or even double thickness.

As the riser and wall/ceiling surfaces do not give quite enough low frequency compensation, some perforated panel Helmholtz resonators are introduced to the room. These are not the most beautiful things in the world; hence they are hidden behind curtains and screen. By placing them in the corners, they take advantage of the fact that all modes terminate in corners.

TABLE 13-1 STUDIO A ACOUSTICAL DATA

SIZE 23'2" × 35'7", CEILING 11'0"

FLOOR & RISERS HEAVY CARPET AND PAD

WALLS SIDE WALLS 8 WIDEBAND PANELS 2'—8'
4" 703 COVERED WITH PERFORATED VINYL FABRIC (FIG. 13-3)
REAR 4" 703, PERFORATED VINYL COVER (FIG. 13-4)
BEHIND SCREEN LOW PEAK ABSORBERS IN CORNERS.
0.31" PERF, 8" DEEP 4" 703 (FIG. 13-5, 6)

VOLUME 8,485 CU. FT.

MATERIAL	S AREA SQ. FT.	125 HZ		250 HZ		500 HZ		1 KHZ		2 KHZ		4 KHZ	
		A	SA	A	SA	A	SA	A	SA	A	SA	A	SA
Carpet	841	0.05	42.1	0.15	126.2	0.30	252.3	0.40	336.4	0.50	420.0	0.60	504.6
Curtain, open	100	0.07	7.0	0.31	31.0	0.49	49.0	0.75	75.0	0.70	70.0	0.60	60.0
Wideband panels 8 at 14.8 sq. ft	119	0.99	118.0	0.99	118.0	0.99	118.0	0.99	118.0	0.99	118.0	0.99	118.00
Wideband east wall	120	0.99	119.0	0.99	119.0	0.99	119.0	0.99	119.0	0.99	119.0	0.99	119.0
Riser, 3/4" ply	324	0.35	123.1	0.19	61.6	0.06	19.4	0.05	16.2	0.04	13.0	0.04	13.0
Gypsum bd. 1/2"	1660	0.10	166.0	0.05	83.0	0.04	66.4	0.03	49.8	0.03	49.8	0.03	49.8
Low Peak	190	1.0	190.0	0.83	157.7	0.44	83.6	0.29	55.1	0.24	45.6	0.20	38.0
Total sabins, Sa			765.2		696.5		707.7		769.5		835.9		902.4
Reverb. Time, Sec.			0.54		0.60		0.59		0.54		0.50		0.46

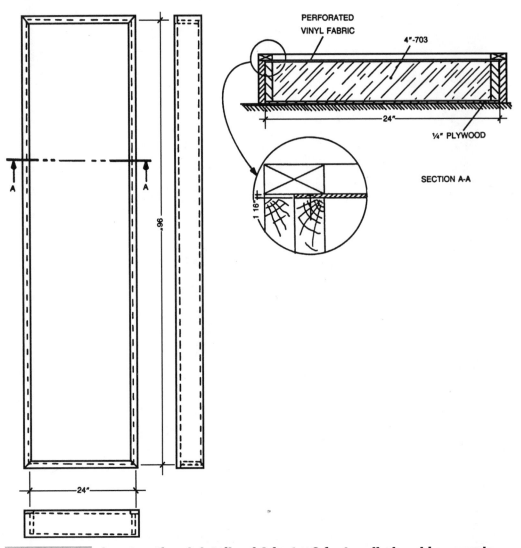

PERFORATED
VINYL FABRIC

4"-703

24"

¼" PLYWOOD

SECTION A-A

1 16"

96"

24"

Figure 13-3 Constructional details of 2-foot × 8-foot wall absorbing panels covered with vinyl fabric. The frame and spacer width must be adjusted to accommodate the 4-inch semirigid glass board without bulging of vinyl cover.

The positions of these Helmholtz resonators are shown in Fig. 13-5 and their construction is detailed in Fig. 13-6. The ³⁄₁₆-inch diameter holes on a square pattern 3 inches on centers turns out to be a perforation percentage of about 0.31 percent.

Other configurations of hole diameter and spacing yielding perforation percentages of about 0.31 percent ±10 percent are acceptable. Care should be exercised to assure that supporting 2 × 8s and dividing 1 × 8s fall between rows of holes. The ³⁄₁₆-inch Masonite sheets can be stacked for drilling to facilitate this chore.

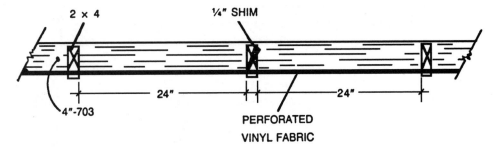

Figure 13-4   Detail of rear (west) wall treatment. The 2 × 4 frame is shimmed out about ¼ inch to avoid bulging of perforated vinyl fabric face by the glass fiberboard.

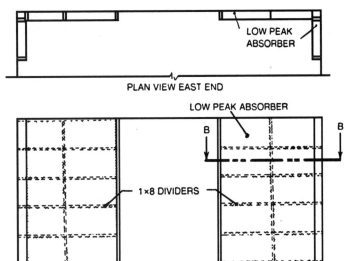

Figure 13-5   Behind the screen on the east wall is located a low peak absorber in each corner, floor to ceiling. Dividers of 1 × 8 lumber break up the air space to discourage modes of vibration parallel to the face of the absorber.

Reverberation Time

Following through on the calculations of Table 13-1 the reverberation time at each of the six frequencies is determined. The values from the table are plotted in Fig. 13-7. This graph varies from 0.46 to 0.60 second. However, as previously stated, the listening conditions should be quite acceptable.

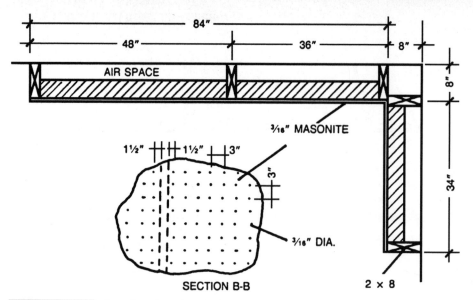

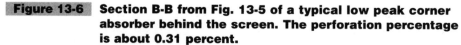

Figure 13-6 Section B-B from Fig. 13-5 of a typical low peak corner absorber behind the screen. The perforation percentage is about 0.31 percent.

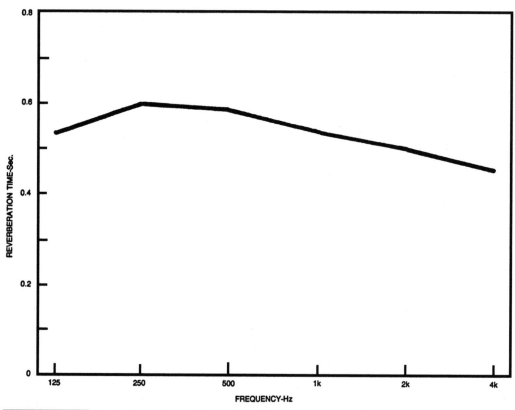

Figure 13-7 Calculated reverberation time of film review theater. This characteristic is satisfactory for this room which is considered primarily a listening room for speech.

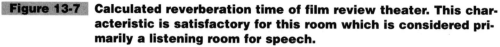

MULTIPLE STUDIOS

Features: Splaying of walls, space saving in studio suites, proprietary wall panels, sand for noise insulation, floating floors.

CONTENTS AT A GLANCE

All the good things should be maximized, all the bad things minimized. This message is repeated frequently enough to encourage maximum recall with minimum effort. A new building was to be built, but many activities were in competition for the maximum space a minimum budget would allow.

As for recording studios, talk at first swung toward a large number of very small studios. This was parried by solid information on the adverse effect of small studio spaces on sound quality and it was agreed to reduce the number of studios so that at least the 1500 cubic foot minimum volume could be realized. Most of the work is the recording of voice programs, but numerous languages are involved, necessitating a multiplicity of studio-control room suites.

It is desirable to make all these suites very similar so that language teams can go from one to another with no delay or inconvenience caused by lack of familiarity. Essentially identical acoustics of all speech studios would also allow complete freedom in intercutting.

Music recording of fairly large vocal and instrumental groups requires one music recording studio, but this larger studio must also be capable of being pressed into service for voice recording at times. This is a very challenging and interesting set of requirements, the solution of which poses some technical problems of general interest.

Heretofore in this book splayed walls have been conspicuous by their scarcity. One reason for this is that many of the studios studied here were located in existing buildings. In such cases splaying of walls required either reconstruction of great sections of the building or losing precious studio volume, or both. This is costly financially and acoustically. Splayed walls do reduce the chance of flutter echoes being produced, even as well-distributed absorbing materials do. In a new building, however, splaying represents little or no additional cost and no loss of room volume. Under such conditions it is most logical to include it.

Typical Recording Suite

Figure 14-1 represents a typical speech studio-control room suite having:

- Rooms of minimum volume (about 1500 cubic feet)
- Two splayed walls in each room
- Sound lock space shared by two or more suites
- An equipment storage space for each two suites

This is certainly maximizing function in limited space as two speech studios, their associated control rooms, an adequate sound lock and a shared storeroom are obtained in a rectangular area about 17 feet × 24 feet with a ceiling height of 10 feet. The plan of the single recording suite of Fig. 14-1 thus becomes an elemental building block of the larger studio complex. This sets the pattern for all control rooms and all speech studios; in fact, everything but the music studio.

Splaying Plan

Normally walls are splayed 1:10 or 1:5. Ratios in this form are readily understood by construction workers but they can also be expressed in degrees (5.7 degrees or 11.3 degrees)

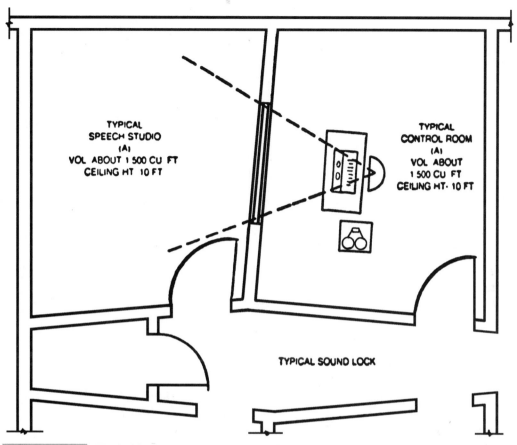

Figure 14-1 **Typical speech recording suite featuring rooms of about 1,500 cubic feet volume and splayed walls coordinated with a sound lock corridor. This is the basic unit of which the larger studio complex is composed.**

by plugging 0.1 or 0.2 into the trusty calculator and punching *arc-tan*. Due to a trigonometric inclination and a love of round figures an even angle of 5 degrees was adopted for the splay of these studio walls which was later realized to be a very odd ratio for the construction people, 2⅛:24.

The scheme of splaying all speech studios and control rooms is illustrated graphically in Fig. 14-2. The two walls to be splayed are simply rotated 5 degrees about their midpoint. As two walls are involved, and each wall can be rotated clockwise or counterclockwise about its midpoint, there are four possible combinations of room shapes with the 90-degree corner held in the same position. This 90-degree corner can be placed in other positions by rotating or flopping the sketches. Three of these four possibilities account for all the room shapes to be included.

The larger music studio splaying plan is shown in Fig. 14-3. Actually, it is the plan of Fig. 14-2A adapted to the different proportions and dimensions of the music studio, Studio C. It should be emphasized at this point that the four splaying plans of Fig. 14-2 have

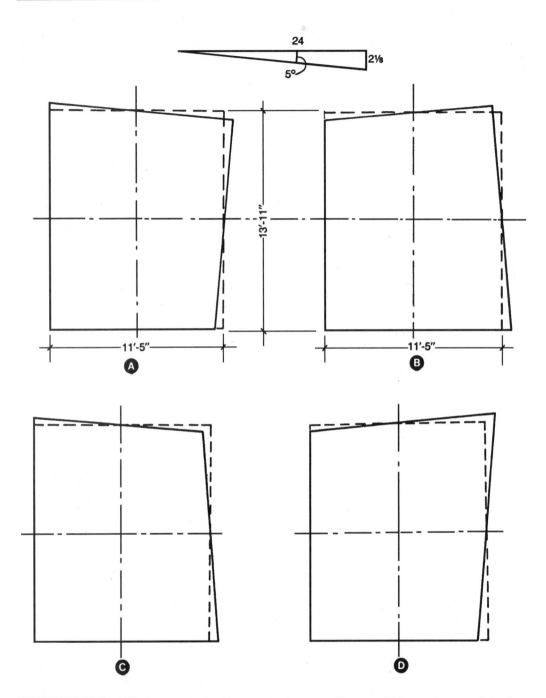

Figure 14-2 Splaying plan for the speech recording studios and associated control rooms. The two walls to be splayed in each room are rotated 5 degrees about their midpoints.

somewhat different areas, even though based upon identical rectangles. It should also be remembered that only the N-S and E-W modes are touched with the above wall splays. The vertical flutter echo must be cared for in some other way.

Room Proportions

To a first approximation the dimensions of the basic rectangle from which the splay pattern is derived can be used to establish proportions for the best distribution of axial modes. The actual modal frequencies, of course, differ slightly from these. One way of looking at it (Fig. 14-3) is that the sound energy reflected to a splayed wall from an opposing, but unsplayed, wall does not return to the same spot on the originating wall, a_1, but returns to a_2, a_3, etc. It "walks the slope" and tends toward becoming a tangential mode.

Another approach is to consider dimension d_1 to give one axial mode frequency and dimension d_2 another one slightly lower. Both outlooks are based on geometrical acoustics which fail miserably in the low frequency region giving us the most trouble. It is really a

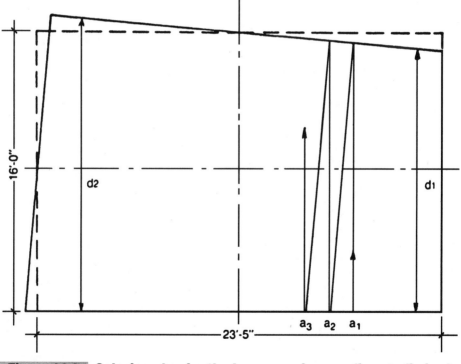

Figure 14-3 **Splaying plan for the larger music recording studio is also based on 5-degree rotation about the midpoint of each of the two walls.**

very complex problem and the mathematical tool of wave acoustics is a more powerful approach. Although only an approximation, establishing favorable room proportions according to the basic presplaying rectangle is a logical and practical approach and the only simple alternative.

Floor Plan

Using the typical speech studio-control room layout of Fig. 14-1 and the splaying plans of Figs. 14-2 and 14-3, the floor plan of Fig. 14-4 has been derived. It includes three speech studio suites, A, B, and D, and one music studio, Studio C, with its control room. Control Room C, serving the music studio, is comparable to the other smaller rooms. One sound lock corridor serves three studios and three control rooms. Were it not for the stairwell, all eight rooms might well have been served by a single sound lock. These studios are located in one corner of the top floor of the two-story building.

Traffic Noise

As the building housing the studios is on a well-traveled boulevard in a major city, traffic noise must be considered. Traffic noise varies greatly with the time of day and the only way to evaluate it properly is to run at least a 24-hour noise survey on the proposed site. Obviously, there is less traffic at night and noise conditions are at their lowest point in the early morning hours, but the 24-hour survey makes possible such statements as "The noise

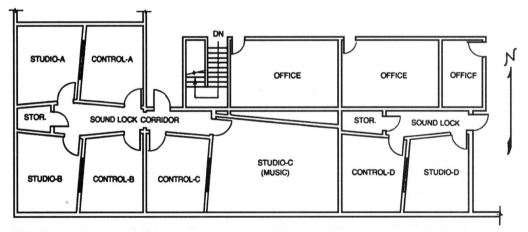

Figure 14-4 Floor plan incorporating three speech recording suites such as shown in Fig. 14-1 and a larger music recording studio. The control room for the music studio (Control Room C) is comparable to the speech studios and their control rooms.

level exceeds 75 dB(A) only 4 percent of the time." This is the sort of data required to support various types of decisions.

This ideal approach was not possible in this instance, therefore an octave analysis of peak boulevard traffic noise was made at the curb closest to the building site with the results shown in Fig. 14-5. This was done before the exact location of the building was known. Once the building location was set, the measured values at the curb were extrapolated to the nearest face of the building by assuming spherical divergence of sound with its resulting 6 dB reduction with each doubling of the distance from the line of traffic. This procedure yielded the broken line spectrum of Fig. 14-5, the estimated noise spectrum just outside the building.

The NC-15 contour is our goal for noise level within studios and control rooms which, of course, could very well have noise contributions from other activity within the building, air-conditioning equipment, etc., in addition to traffic noise. At the moment, however, only traffic noise is under consideration.

Special urging resulted in the placement of the studio corner of the building on the side away from the boulevard. This offers some protection from traffic noise. The building code requires ventilation of the attic space above the studio area which means that traffic noise of considerable magnitude pervades the attic space immediately above the studio ceilings.

With the wall construction to be described later, the greatest prospect for a traffic noise problem in the studios turns out to be via this ceiling path. If sound level measurements within the studios reveal traffic noise levels appreciably above the NC-15 contour, a layer of sand will be added between the ceiling joists above the double drywall ceiling. A 1-inch layer of sand "beefs up" the ceiling, acoustically speaking, 36 dB at 500 Hz on a mass basis and weighs only about 8 pounds per square foot. Each doubling of the sand thickness adds 3 dB transmission loss, but doubles the weight.

It is advisable to stop short of a thickness at which sand would break through the ceiling and pour down on unsuspecting personnel below. A modest amount of sand could add very substantially to the insulation strength of the ceiling against external noise.

External Walls

There is information in Fig. 14-5 which helps in deciding how heavy to make the external walls of the studio. The distance between the NC-15 contour and the broken line graph represents the minimum transmission loss the walls must provide. This loss requirement varies with frequency. It is maximum at about 1 kHz, decreasing for both lower and higher frequencies.

This is unfortunate in one sense because the sensitivity of human ears is greatest in this general frequency region.

It is fortunate in another sense because walls of common materials and normal construction can offer quite good transmission loss at 1 kHz, much greater than at much lower frequencies. The octave noise level of 64 dB at the face of the building is 47 dB higher than the NC-15 contour at 1 kHz which we are striving for within the studio. The external wall construction of Fig. 14-6 was considered to be adequate to provide this much transmission loss, especially knowing that the 64 dB applies to the front face of

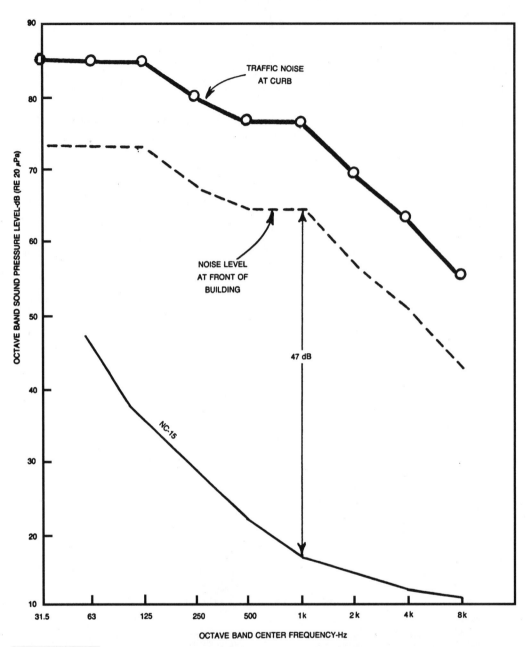

 Figure 14-5 Traffic noise peaks were measured at the boulevard curb and later (when the building position was established) extrapolated to the face of the building. The NC-15 contour is the goal for noise within the studio. The difference between these two graphs is the transmission loss the studio external walls must provide.

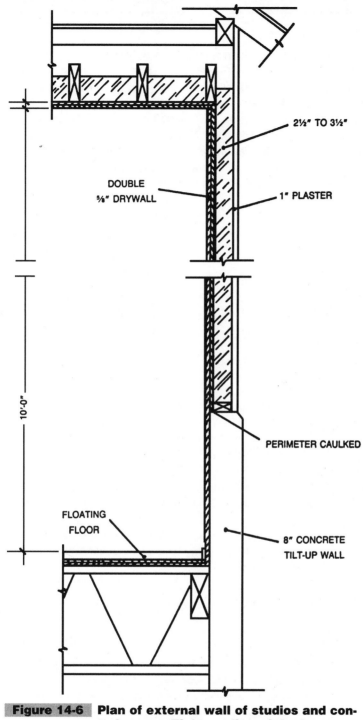

2½" TO 3½"

DOUBLE
⅝" DRYWALL

1" PLASTER

PERIMETER CAULKED

FLOATING
FLOOR

8" CONCRETE
TILT-UP WALL

10'-0"

Figure 14-6 Plan of external wall of studios and control rooms. Tight sealing of the base layer of gypsum board around its periphery contributes significantly to wall performance in protection of studios against external noise.

the building and the external walls of the studios are in the rear. The lower part of the wall of Fig. 14-6 is the top of the tilt-up panels. The upper part of frame construction is plastered outside and covered with double $^5/_8$-inch gypsum drywall panels inside. The insulation between wall studs serves the double purpose of thermal insulation and discouraging acoustical resonances in the cavity which degrade the effectiveness of the wall in attenuating external noise.

Ceiling construction is similar, except for the plaster. Should sand be added to the upper surface of the ceiling drywall at a later time, the insulation would first be removed, the sand applied and then the insulation would be replaced on top of the sand.

Internal Walls

The construction of typical internal walls is specified in Fig. 14-7. Double layers of $^5/_8$-inch gypsum board are standard on every wall or ceiling separating sound sensitive areas from outside noise. The gypsum panels of all studio walls (except external) are not nailed to the studs but are supported resiliently. The base layer board is secured to the resilient channels with screws. These resilient channels, U.S. Gypsum Rc-1 or equal, are first nailed horizontally to the studs, spaced 24 inches.

The vertical base layer board is then secured to the resilient channels with special screws of proper length so that the flexible action of the channel is not destroyed by screws hitting studs. The face layer of gypsum board is then applied horizontally with adhesive. All joints are then finished in the normal way.

Figure 14-8 illustrates the preferred method of staggering layers of gypsum board at corner intersections. The entire periphery of the base layer should be carefully caulked with nonhardening acoustical sealant. Such efforts toward hermetically sealing each room pays great dividends in reducing sound leaks and assuring maximum transmission loss of the wall. The resilient studio face of a wall resonates at a different frequency than the opposite nailed face.

Such resonance effects tend to reduce transmission loss at the resonance frequency and different handling of the two wall faces displaces one resonance point from the other, improving overall wall performance.

Floating Floor

To obtain sufficient protection against noise of other activities in ground floor rooms below the studio area, a floating concrete floor was required in studios and control rooms. There are numerous fancy mechanical ways of supporting floating floors on springs and rubber devices as well as proprietary impregnated glass fiberboards and strips which are excellent in floating floors, but expensive. An inexpensive method used in Germany a quarter of a century ago was pressed into service.[32]

A soft fiberboard is laid on the structural floor as a support for the concrete. The stiffness of the fiberboard is reduced by coating the underside of it with cork granules before

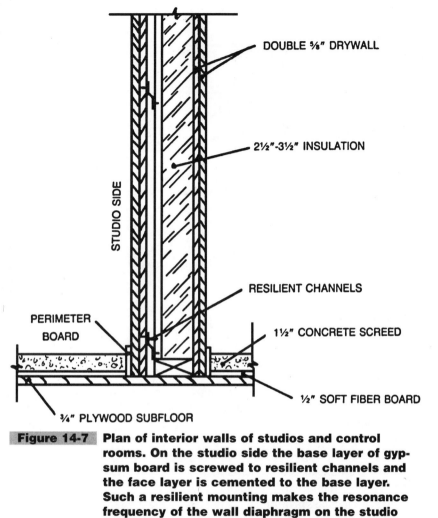

Figure 14-7 **Plan of interior walls of studios and control rooms. On the studio side the base layer of gypsum board is screwed to resilient channels and the face layer is cemented to the base layer. Such a resilient mounting makes the resonance frequency of the wall diaphragm on the studio side different from that of nailed panels on the other side, preventing coincidence and thus improving transmission loss of the wall.**

laying it on the structural floor. The fiberboard is then covered with plastic sheets and overlapped at least 3 inches. It also runs up over the perimeter board indicated in Fig. 14-7. The concrete screed is thus prevented from running into cracks between fiber boards which would form solid bridges between the structural floor and the floating floor, destroying the floating characteristics. The cork granules improve the impact sound insulation about 16 dB over the fiberboard alone. The 1½-inch concrete thickness is certainly minimum.

To reinforce such a floor (wise precaution) the thickness of the concrete must be 3 inches to 4 inches. In thinner layers it is almost impossible to keep the reinforcing screen in the center of the concrete layer during the pouring. If the reinforcing wires are on the bottom

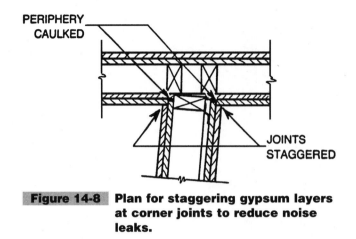

PERIPHERY CAULKED

JOINTS STAGGERED

Figure 14-8 **Plan for staggering gypsum layers at corner joints to reduce noise leaks.**

of the layer, little reinforcing results. The danger is that concentrated loads, such as the legs of a grand piano, may crack the concrete, reducing its sound insulating value.

Treatment of Studio A

The specification of carpet for all studios dominates their acoustical treatment. The drywall wall and ceiling surfaces provide a modest, but insufficient, amount of low frequency compensation for the carpet. Helmholtz resonators can easily supply the remainder of compensation required, but the problem is, where to put them? Thick boxes on the walls and ceilings are not an esthetic delight. It was decided to use a frame suspended from the ceiling to hold the low frequency boxes and illumination fixtures and to shield both from view with a lower frame face of plastic louver panels of either egg crate or honeycomb type openings.

Should the usual panels of 4 inches of 703 Fiberglas adorn the walls? In the interests of appearance it was decided to employ proprietary panels of glass fiber covered with decorative, perforated vinyl wallcovering manufactured by L.E. Carpenter and Company (Appendix D). *Vicracoustic panels* 2 feet × 8 feet × 2 inches were selected. The core of semirigid glass fiber, which does the absorbing, can be covered on one or both sides with ⅛-inch high density glass fiber substrate if required for protection against impact.

In this studio application the less expensive Type 80 panel, which consists only of the absorbing core wrapped on face and edges with perforated vinyl, was considered adequate. Panels (Fig. 14-9) are mounted on the walls by the use of Z-clips, one part of which is cemented to the backs of the Vicracoustic panel, the other screwed to the wall.

The low frequency absorption of these panels is increased from 0.47 to 0.57 at 125 Hz if the Z-clips are mounted on 1 × 3 strips, but the advantage of doing so at other frequencies is almost nil. The treatment of Studios A, B, C, and D is accomplished with carpet, low frequency Helmholtz resonators, and Vicracoustic panels added to the built-in absorption of the drywall surfaces.

The placement of the Vicracoustic wall panels in Studio-A is shown in Fig. 14-10. Even though the east and south walls are splayed, an attempt is made to place panels on one wall

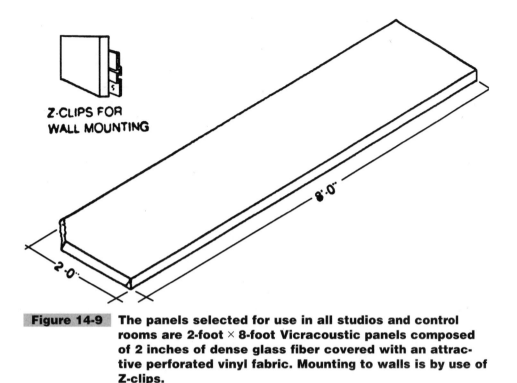

Figure 14-9 The panels selected for use in all studios and control rooms are 2-foot × 8-foot Vicracoustic panels composed of 2 inches of dense glass fiber covered with an attractive perforated vinyl fabric. Mounting to walls is by use of Z-clips.

to oppose bare wall (or window, or door) on the opposite wall. This can best be judged in the projected ceiling plan of Suite A in Fig. 14-11.

The constructional details of the low frequency absorbing boxes are given in Fig. 14-12. Similar to others considered in other chapters, the frame is of 1 × 8 lumber with a back of ½-inch plywood or particleboard. The face of ³⁄₁₆-inch masonite is filled with ³⁄₁₆-inch holes drilled on 3-inch centers. This gives a perforation percentage of about 0.3 percent and a resonance peak in the vicinity of 100 Hz.

The 4 inches of 703 glass fiber broadens this peak. The boxes should be spray painted with flat black paint to reduce their visibility.

The 7 foot × 10 foot suspended frame is placed in Studio A approximately as indicated in the projected ceiling plan of Fig. 14-11. Figure 14-13 shows the relationship of the 1 × 6 frame and the plastic louver layer. The three fluorescent fixtures assure that the plastic louver plane is the dominant visual feature of the room.

The 13 black low frequency boxes the frame contains will scarcely be visible. The placement of the low frequency boxes in the frame is important. Boxes 1, 2, 3, and 4 have their faces downward, resting on the open cells of the plastic louvers. Boxes 5 and 6 rest on their long edges and point north; 7 and 8 points south; 9 faces west; and 10 faces east. The three low frequency boxes 11, 12, and 13 resting on the fluorescent fixtures must, of course, be directed upward.

It is well that these last three have some soft material between the boxes and the metal reflectors to avoid sympathetic rattles when the room is filled with sound. In fact, an awareness of the possibility of rattles in the entire assembly is advised.

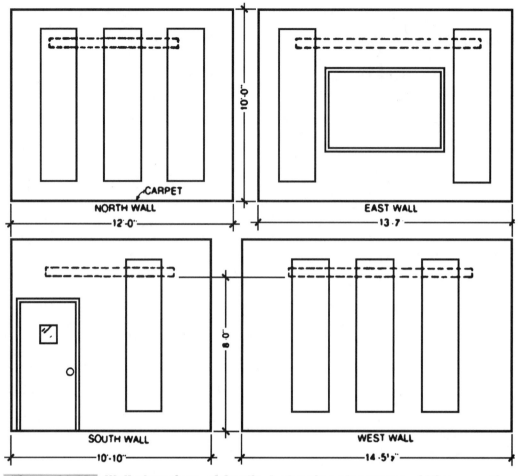

Figure 14-10 Wall elevations of Studio A showing placement of Vicracoustic panels. The broken lines indicate relative position of the suspended frame holding low peak absorbing boxes.

Reverberation Time of Studio A

Table 14-1 lays out the details of calculating (estimating) the reverberation time of Studio A. A reverberation time of 0.35 second was the goal and the plotted graph for Studio A in Fig. 14-14 shows that this goal has been approximated. The exceptionally high absorption of the Vicracoustic panels at 250 Hz and 500 Hz generates the characteristic dip noticed in the control rooms and Studio C as well. An irregularity of this magnitude in the calculation stage has little significance unless confirmed by subsequent measurements. At that time, and not before, some trimming might be necessary and justified. The beauty of the modular acoustical treatment is that such trimming, if required, can be easily carried out.

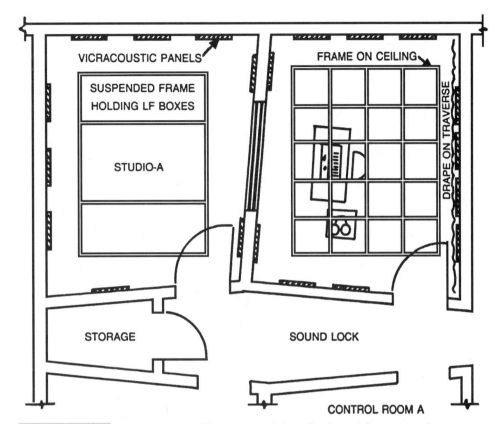

Figure 14-11 Projected ceiling plan of Studio A and its control room showing position of the following acoustical elements: Vicracoustic wall panels, frame suspended from ceiling in Studio A, frame fastened to ceiling in Control Room A and drapery on traverse in Control Room A.

Acoustical Treatment of Control Room A

The speech studios and control rooms are very similar in size and differ chiefly in the somewhat lower reverberation time goal of about 0.3 second for control rooms. Acoustically, a very major difference between the two types of rooms is that the floors of the control rooms are covered with vinyl tile instead of carpet. This factor means that the basic treatment will be accomplished primarily with wideband, or quasi-wideband, materials.

The grid of 1×6 lumber attached to control room ceilings makes 20 squares having inside dimensions of 24 inches $\times$ 24 inches which hold pads of 4 inches of 703 giving a total area of 80 square feet. The actual configuration is not too important, the effective area is.

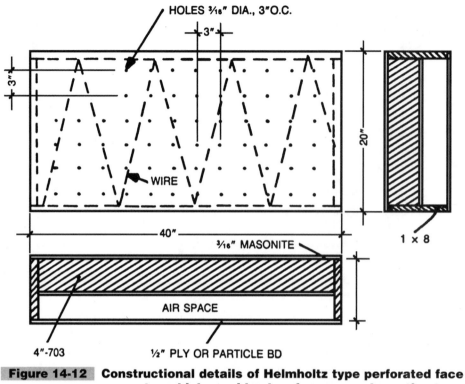

HOLES ³⁄₁₆" DIA., 3"O.C.

3"

3"

20"

40"

WIRE

³⁄₁₆" MASONITE

1 × 8

4"-703

AIR SPACE

½" PLY OR PARTICLE BD

Figure 14-12 **Constructional details of Helmholtz type perforated face resonator which provides low frequency absorption to compensate for carpet deficiency in the studios.**

Construction can follow similar frames described in earlier chapters with a fabric or screen facing. The air space between the 703 and the ceiling aids low frequency absorption of the material. The positioning of the ceiling frame is not critical, the position of Fig. 14-11 is suggested.

The placement of the Vicracoustic panels in typical Controls Room A is shown in Fig. 14-15. The four panels on the east wall are normally hidden behind a drapery which is retractable. This drape is included to flatten the reverberation time at 1 kHz and above is very close to 0.3 second as shown in Table 14-2 and Fig. 14-14. If the drape is retracted the reverberation time in the same high frequency region is close to 0.35 second, making the 250 Hz dip stand out a bit more. This drape may be considered an approved variable acoustical element, if desired.

With the drapes extended over the east wall, conditions are proper for listening to sounds from the speech studio with its reverberation time of 0.35 second. With the drapes retracted, the control room becomes more adaptable acoustically for recording an interview, for instance. The drapery material should not be too heavy (10 ounces per square yard material was used in the calculations).

Ordinary monk's cloth, running about 8 ounces per square yard, is acceptable. It should be hung as close to the wall as possible and still clear the panels. Only enough material should be hung to result in the fold almost disappearing when the drapes are extended. Having deeply folded drapes would introduce too much absorption.

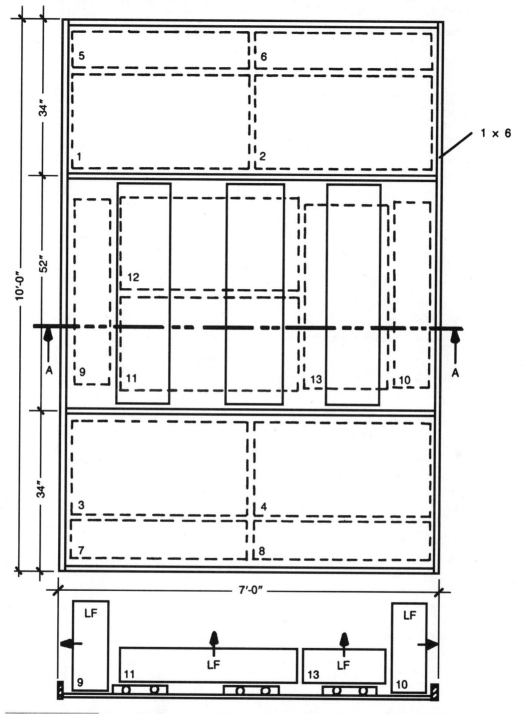

Figure 14-13 Frame of 1 × 6 lumber suspended from the ceiling of Studio A which holds illumination fixtures and the 13 low frequency boxes required in the room. Some boxes face downward, some point upward, and others point in the four horizontal directions.

TABLE 14-1   Speech Studio Calculations for Typical Studio A

SIZE   TWO WALLS SPLAYED FROM BASIC RECTANGLE
            11'5" × 13'11" CEILING 10'

FLOOR   CARPET, HEAVY WITH PAD, ON FLOATED FLOOR

CEILING   SUSPENDED FRAME HOLDING 13 LOW FREQUENCY
            HELMHOLTZ ABSORBERS 4.75 SQ FT EACH

WALLS   VICRACOUSTIC TYPE 80 PANELS (9) 2' × 8' × 2".
            PERFORATED VINYL COVERING 2" GLASS FIBER. FURRED OUT 1"

VOLUME   1,598 CUBIC FEET

MATERIAL	S AREA SQ. FT.	125 HZ A	SA	250 HZ A	SA	500 HZ A	SA	1 KHZ A	SA	2 KHZ A	SA	4 KHZ A	SA
Carpet	160	0.05	8.0	0.15	24.0	0.30	48.0	0.40	64.0	0.50	80.0	0.60	96.0
Drywall	668	0.08	53.4	0.05	33.4	0.03	20.0	0.03	20.0	0.03	20.0	0.03	20.0
Low Frequency	62	1.0	62.0	0.68	42.2	0.39	24.2	0.17	10.5	0.13	8.1	0.10	6.2
Vicracoustic panels	144	0.57	82.1	0.98	141.1	0.92	132.5	0.76	109.4	0.71	102.2	0.78	112.3
Total Sabins, Sa			205.5		240.7		224.7		203.9		210.3		234.5
Reverberation time, seconds			0.38		0.33		0.35		0.38		0.37		0.33

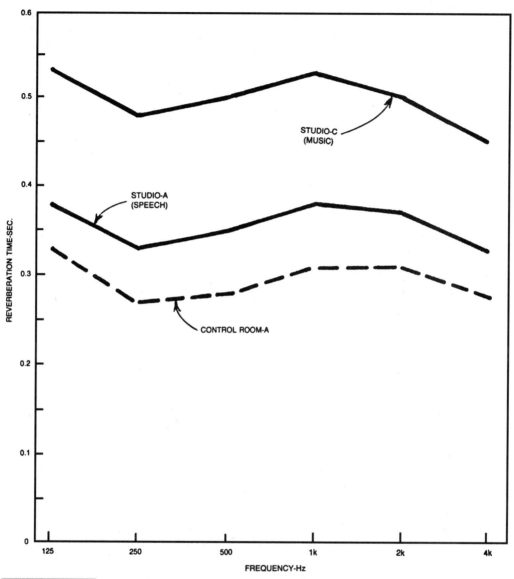

Figure 14-14 Calculated reverberation time for Studio A as shown which compares to the goal of 0.35 second. The goal for Control Room A is 0.3 second and for Studio C, 0.5 second. Measurements must verify such calculated estimates. The modular treatment plan allows trimming if required.

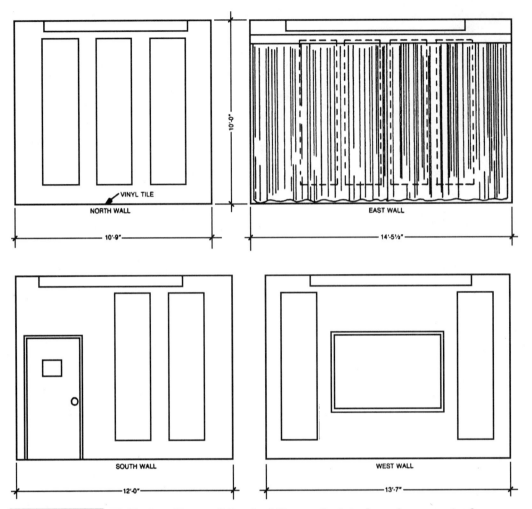

Figure 14-15 Wall elevations of Control Room A showing placement of Vicracoustic wall panels. The drape, normally covering the panels, is included to flatten the reverberation characteristic slightly. When retracted, the reverberation time in Control Room (A) is suitable for speech recording.

Music Studio Treatment

Music Studio C should have a longer reverberation time than the speech studios for two reasons, its greater volume and the fact that the music is better served by a longer reverberation time than used for speech. A goal of about 0.5 second was selected, compromising somewhat toward a speech requirement because the studio will be used for speech recording at times. In spite of its greater size, the music studio uses the same acoustical elements as the speech studios: carpet, Vicracoustic panels, and peaked low frequency absorbers.

TABLE 14-2 Typical Control Room Calculations

SIZE TWO WALLS SPLAYED FROM BASIC RECTANGLE
11'5" × 13'11" CEILING 10'

FLOOR VINYL TILE OVER FLOATED CONCRETE
CEILING WOOD FRAME HOLDING 80 SQ FT OF 4"-703
WALLS VICRACOUSTIC TYPE 80 PANELS (11).
2' × 8 × 2", PERFORATED VINYL COVERING
2" GLASS FIBER. FURRED OUT 1"
RETRACTABLE DRAPERY 9 × 14 10 OZ SQ. YD. EAST WALL

VOLUME 1,598 CUBIC FEET

MATERIAL	S AREA SQ. FT.	125 HZ		250 HZ		500 HZ		1 KHZ		2 KHZ		4 KHZ	
		A	SA	A	SA	A	SA	A	SA	A	SA	A	SA
Drywall	668	0.08	53.4	0.05	33.4	0.03	20.0	0.03	20.0	0.03	20.0	0.03	20.0
Wideband Ceiling	80	0.99	79.2	0.99	79.2	0.99	79.2	0.99	79.2	0.99	79.2	0.99	79.2
Vicracoustic panels	176	0.57	100.3	0.98	172.5	0.92	161.9	0.76	135.8	0.71	125.0	0.78	137.3
Drapery	126	0.03	3.8	0.04	5.0	0.11	13.9	0.17	21.4	0.24	30.2	0.35	44.1
Total Sabins, Sa			236.7		290.1		275.0		254.4		254.4		280.6
Reverberation Time, seconds			0.33		0.27		0.28		0.31		0.31		0.28

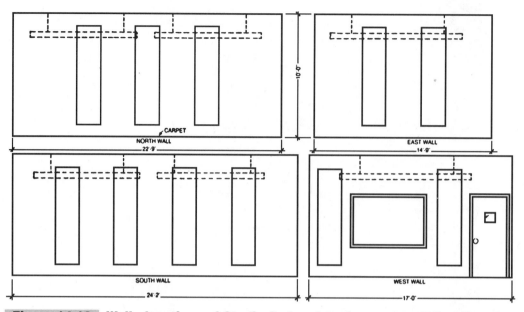

Figure 14-16   Wall elevations of Studio C showing placement of Vicracoustic panels in this music recording studio. Two suspended frames are required for illumination and to hold low frequency boxes.

Figure 14-16 shows the placement of the Vicracoustic panels on the walls of Music Studio C. The suspended ceiling frames are shown with broken lines, the lower edge being about 8 feet above the floor. These frames each hold 14 of the same low frequency resonator boxes used in the speech studios and described in Fig. 14-12.

Figure 14-17 shows the configuration of the 14 low frequency boxes in one of the ceiling frames. In general, the plan is the same as for the speech studio frame: the perforated side of those standing on edge facing outward (5, 6, 7, 8, 9,10): those resting on the fluorescent fixtures (11,12,13,14) facing upward; and the rest (1, 2, 3, 4) facing downward.

Table 14-3 reveals the details of the calculation of reverberation time for the music studio. The calculated values of reverberation time are plotted in Fig. 14-14. The deviations from the goal of 0.5 second are not significant, but measurements should verify the actual shape which will then be the basis of any trimming adjustments considered necessary.

A/C Duct Routing

In Chapter 2 the several basic principles concerning air conditioning ducts in studios were elucidated. Applying these principles to the present case of multiple studio suites, the plan of Fig. 14-18 resulted. Note that the maximum length of duct is placed between grilles of adjacent rooms or even rooms on the opposite side of the sound lock. Supply and return grilles in a given room should not be too close together to assure adequate circulation. In the Control Room C return duct, a U-shaped section was inserted to avoid a short path to the adjoining room.

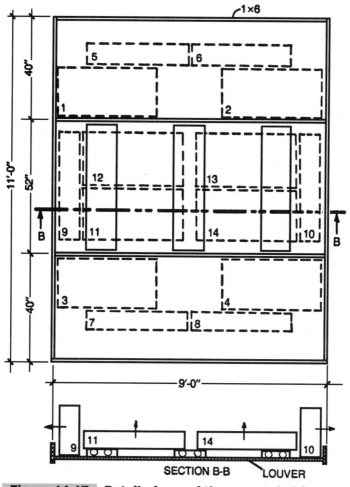

Figure 14-17   Detail of one of the suspended frames in the music studio. The positions of the 14 low frequency boxes are indicated.

TABLE 14-3 Calculations for the Music Studio C

SIZE: TWO WALLS SPLAYED 5" FROM BASIC RECTANGLE 16'-0" × 23-5" CEILING 10'

FLOOR: CARPET, HEAVY WITH PAD ON FLOATED CONCRETE.

CEILING: TWO SUSPENDED FRAMES 9' × 10' EACH HOLDING 14 LOW FREQUENCY HELMHOLTZ ABSORBERS, 4.75 SQ FT EACH

WALLS: VICRACOUSTIC TYPE 80 PANELS (11), 2' × 8' × 2" PERFORATED VINYL COVERING 2" GLASS FIBER. FURRED OUT 1"

VOLUME: 3,700 CUBIC FEET

MATERIAL	S AREA SQ. FT.	125 HZ		250 HZ		500 HZ		1 KHZ		2 KHZ		4 KHZ	
		A	SA	A	SA	A	SA	A	SA	A	SA	A	SA
Drywall	1,157	0.08	92.6	0.03	57.9	0.03	34.7	0.03	20.0	0.03	20.0	0.03	20.0
Carpet	370	0.05	18.5	0.15	55.5	0.30	148.0	0.50	79.2	0.99	79.2	0.99	79.2
Low Frequency absorbers	133	1.0	133.0	0.68	90.4	0.17	22.6	0.13	125.0	0.71	125.0	0.78	137.7
Vicracoustic panels	176	0.57	100.3 / 344.4	0.98	172.5 / 376.3	0.76	133.8	0.71	30.2	0.24	30.2	0.35	44.1
Total Sabins, Sa			376.3		359.5		339.1		254.4		254.4		280.6
Reverberation Time, seconds			0.53		0.48		0.50		0.53		0.50		0.45

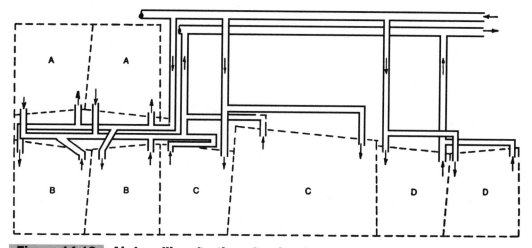

Figure 14-18 Air handling ducting plan for the studio complex. This plan places a maximum length of ducting between grilles of adjacent rooms to prevent crosstalk from room to room via the duct. Lined ducts attenuate sound in the ducts.

15

DIFFUSION CONFUSION

All that is required of the acoustical consultant is the design of studios and other rooms that engineers, musicians and the general public consider "good." This is a difficult and subjective evaluation and the job definitely does not fall into the neat categories of definite black-white, go-no-go things of this world. If a room is too reverberant or too dead it is judged "bad" and adjusting reverberation within relatively close limits is probably the greatest single factor in elevating a poor room to a good or at least a better condition. However, reverberation time is not the only factor involved. Another factor is the diffusion of sound in the room. Often two studios very similar in size with the same reverberation time have a very different sound. This difference can possibly be traced to diffusion of sound in the room.

The relationship of diffusion of sound in a studio to the general acoustical quality of that studio is something of a mystery that has baffled studio designers for the last half century. What is diffusion? The sound field in a studio is diffuse if at any given instant the intensity of sound is uniform everywhere in that room and at every point sound energy flows equally in all directions. It has to do with homogeneity of sound in a room. Such a diffuse condition is a basic assumption in the derivation of the reverberation time equations of Sabine and Eyring. It is apparent that a dominant standing wave condition or knowledge that sound conditions vary throughout a studio means that a diffuse condition does not exist.

Diffusion is not the problem in large rooms such as auditoriums as it is in small rooms such as recording studios and listening rooms. This is the result of the fact that the dimensions of the smaller rooms are comparable to the wavelength of sound to be recorded or reproduced in them.

In Chapter 1 the effect of room size was considered along with room proportions. It was noted that the more uniform the distribution of room resonance modes the better. This procedure contributes to the diffusion of sound in the room. Selecting a cubical space in which all axial modes pile up at certain frequencies with great empty spaces between these pile-up frequencies is a move away from reasonably diffuse conditions. It is impossible, by traditional methods at least, to attain truly diffuse sound conditions in a small space, but approaching it as closely as possible is a major goal in studio design.

Sound Decay Irregularity

In the measurement of reverberation time the modes of the room are excited, say, with high intensity random noise from a loudspeaker. When the loudspeaker sound is suddenly terminated, these room modal frequencies die away, each at its own frequency and own rate.

Figure 15-1 shows tracings from graphic level recorder records of five successive decays of an octave band of random noise centered at 125 Hz. The loudspeaker and microphone positions remained fixed. Figure 15-2 shows similar five successive decays under identical conditions except it is for an octave band of random noise centered on 4 kHz. The contrast in smoothness of decay is striking, yet these are typical of small studio decays at these frequencies. It is instructive to dig into this a bit further.

The 125 Hz and 4 kHz decays of Figs. 15-1 and 15-2 were made in a small multitrack studio 13 feet 5 inches × 18 feet 5 inches with a ceiling height of 7 feet 6 inches, volume 1,853 cubic feet. The object of the measurement was basically the determination of the reverberation time of the studio.

Establishing a best fit straight line average slope to the erratic decays at 125 Hz is far less precise than for the 4 kHz decays. To illustrate this, the *squint-eye slope* lines are included in Figs. 15-1 and 15-2 with the reverberation time (T_{60} seconds included at the top of each slope.

Of course, different observers establish slightly different slope fits, but the averaging of five slopes for each frequency and each microphone position gives a statistically significant mean value.

The mean value for the five 125 Hz decays for each of three mircophone positions in this studio is 0.291 second with a standard deviation of 0.025 second. The same for the 4

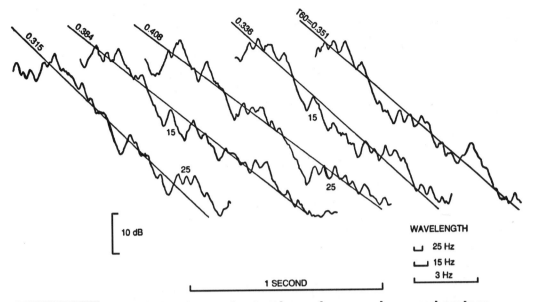

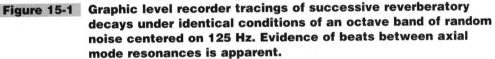

Figure 15-1 Graphic level recorder tracings of successive reverberatory decays under identical conditions of an octave band of random noise centered on 125 Hz. Evidence of beats between axial mode resonances is apparent.

kHz octave is 0.311 second with a standard deviation of 0.013 second. The standard deviation (the plus and minus deviations from the mean value which includes 67 percent of the measurements) for 125 Hz is twice that for 4 kHz which reflects the greater fluctuations in the 125 Hz measurements.

Of special interest in this chapter, the reverberation decays of Figs. 15-1 and 15-2 also reveal something of the sound diffusion conditions in this studio. Octave bands of random noise were used in both the 125 Hz and 4 kHz cases. An octave band centered on 125 Hz is considered to include energy from 88 to 177 Hz, the half-power (3 dB down) points. The 4 kHz octave covers 2,828 to 5,656 Hz, the one spanning 89 Hz, the other 2,828 Hz. The 125 Hz octave band includes relatively few modal frequencies of the room, the octave at 4 kHz many.

The axial mode frequencies for this small multitrack studio below 250 Hz are shown graphically in Fig. 15-3. Although there are no pile-ups several pairs are very close together and wide gaps (compared to the approximately 5 Hz bandwidth of each mode) occur.

In Fig. 15-3 the span of the 125 Hz octave includes six modal frequencies. Each of these modes has its own decay rate determined by the absorption material in the room involved in that particular mode. A single mode, if excited and allowed to decay without influence of any other mode, would decay exponentially which gives a nice straight line decay on a dB scale as shown in Fig. 15-4A. The octave containing the six axial modes of Fig. 15-3 might be considered a combination of B and C, as seen in Fig. 15-1.

Then how is the smoothness of the decay curves of the 4 kHz octave (Fig. 15-2) explained? The 125 Hz octave is only 89 Hz wide while the octave centered on 4 kHz is 2,828 Hz wide. The greater smoothness is explained by the greater width and thus the greater number of

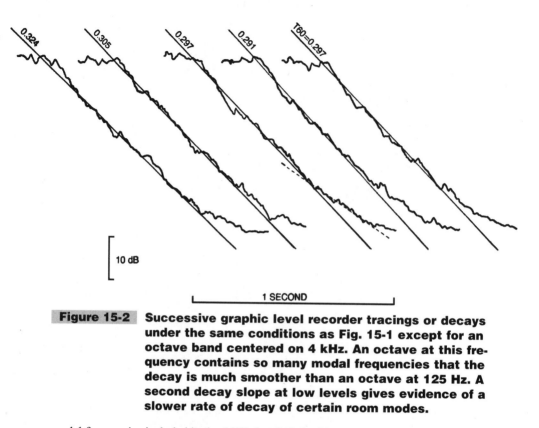

0.324   0.305   0.297   0.291   T60=0.297

10 dB

1 SECOND

Figure 15-2   **Successive graphic level recorder tracings or decays under the same conditions as Fig. 15-1 except for an octave band centered on 4 kHz. An octave at this frequency contains so many modal frequencies that the decay is much smoother than an octave at 125 Hz. A second decay slope at low levels gives evidence of a slower rate of decay of certain room modes.**

modal frequencies included in the 4 kHz band. Only the axial modes are plotted in Fig. 15-3. It would be impossible to show graphically even the numerous axial modes within the 4 kHz octave, let alone the tangential and oblique modes. In fact, considering all three types of natural frequencies of this multitrack studio, something of the order of 800,000 modal frequencies exist in the 4 kHz octave band while only 328 exist in the 125 Hz octave band.

Of course, as pointed out in Chapter 1, the tangential and oblique modes have less influence than the axial, but they do have some effect and this effect would be in the direction of smoother decays and better diffusion.

The effect of *difference beat frequencies* between axial modes of Fig. 15-3 can be detected in the decays of Fig. 15-1. The graphic level recorder paper speed for both Figs. 15-1 and 2 was 100 mm per second and a one second scale is indicated on both of these figures. If the 153.4 Hz and the 168.4 Hz modes beat together, a difference frequency of 15 Hz is produced.

One cycle of a 15 Hz signal is represented by the length of the line so indicated in the lower right-hand corner of Fig. 15-1. In the second and fourth 125 Hz decays there are fluctuations closely matching this frequency. The 126.3 Hz and 150.7 Hz modes beating together would yield a difference beat frequency of 24.4 Hz.

There are fluctuations in decay one and three which are close to 25 Hz. The closely spaced modes near 125 Hz and 150 Hz (Fig. 15-3) produce beats of 3.6 and 2.7 Hz. Variations corresponding to the more slowly varying beats near 3 Hz are more difficult to pinpoint, but there are even suggestions of these. In other words, the modal frequencies within the 125 Hz octave band account for the relatively great fluctuations of the 125 Hz decays.

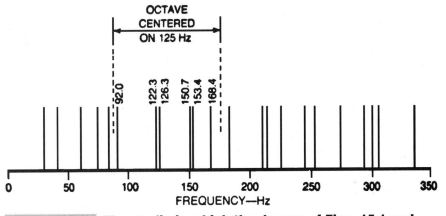

`Figure 15-3` **The studio in which the decays of Figs. 15-1 and 15-2 were taken has axial modal frequencies as shown. The octave centered on 125 Hz passes only the six indicated. The close pairs within this octave tend to beat with each other causing fluctuations in the decay trace at the difference frequency.**

The reason the five decays are not identical or similar can be explained by the fact that the different modal frequencies were not all excited to the same level. The random noise signal constantly changes in amplitude and frequency (within the octave limits). It is entirely fortuitous as to what instantaneous amplitude and frequency were at the time the sound was interrupted to begin the decay. A very smooth low frequency decay could result from a dominant single mode, although with octave bands this is unlikely.

An important indication of the diffusion of sound in this small multitrack studio is given in the low frequency reverberation decays as in Fig. 15-1. If the fluctuations are very great, the diffusion is poor. The smoother the decays, the better the diffusion.

Quantitative evaluation of diffusion conditions are not yet available from such decays, but good qualitative comparisons are not only possible, but part of the arsenal of informed workers in studios.

Diffusion information may also be gleaned from decay curves at higher frequencies. In Fig. 15-2 the broken line indicates a fairly definite suggestion of a second slope. This is probably the result of certain modal frequencies having less contact with the absorbing material in the room (i.e., modes that are less damped) or modal frequencies not fully excited as the decay begins. In this particular case, these modes do not affect things until the sound has decayed 30 dB, hence their effect would probably not be detectable in normal program material.

Variation of T_{60} with Position

Measuring reverberation time at different locations in a studio often reveals small but significant differences in reverberation time. These are usually averaged together for a better statistical description of conditions in the studio.

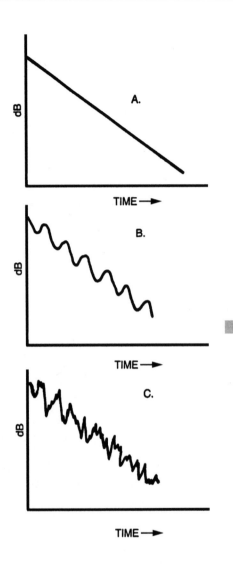

Figure 15-4 (A) A single mode decay exponentially, giving straight line decay on a logarithmic scale. (B) Two closely spaced modes, each having the same decay rate, beat with each other causing the decay to vary at the difference frequency. (C) Many closely spaced modes result in an erratic decay, the more modes the smoother the decay.

For example, Fig. 15-5 shows the reverberation time measured at three different microphone positions in the small multitrack studio mentioned in the previous section. It is noticed that the three graphs tend to draw together as frequency is increased. This suggests that such changes in reverberation time at different locations in the same studio are the result of a certain degree of nondiffuse conditions because we know that diffusion is better at high frequencies. Can this method then be used to evaluate the sound diffusion condition in a room?

The Engineering Research Department of the British Broadcasting Corporation asked the same question.[33] In their characteristically thorough way they measured reverberation time at 100 microphone positions in a 10-foot × 10-foot room. With no absorption material in the room (Fig. 15-6A), very diffuse conditions resulted in reverberation time long, but essentially constant throughout the room.

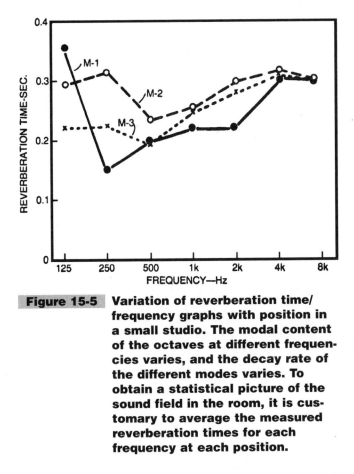

Figure 15-5 Variation of reverberation time/ frequency graphs with position in a small studio. The modal content of the octaves at different frequencies varies, and the decay rate of the different modes varies. To obtain a statistical picture of the sound field in the room, it is customary to average the measured reverberation times for each frequency at each position.

In Fig. 15-6B the reverberation time contours are shown when one wall was treated. The reverberation time is lower nearer the absorbing surface. They then demonstrated that geometrical diffusing elements on the untreated walls plus one absorbing wall resulted in quite complex contours. The laborious nature of this approach discourages further exploration of the method, although it shows some promise if special instrumentation were devised.

Directional Microphone Method

The signal output of a highly directional microphone in a perfectly diffuse room should be the same no matter where it is pointed, except when pointed at the source of sound. Ribbon microphones, with their figure-8 pattern, have been tried. Parabolic reflectors with a microphone at the focal point have been tried, as well as the line array type of directional microphone.

All of these methods have proved to be rather awkward to use and the results difficult to interpret. The greatest shortcoming of this method, from the point of view of small studios, is

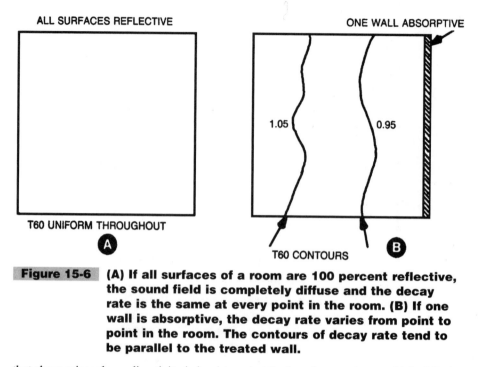

Figure 15-6 **(A) If all surfaces of a room are 100 percent reflective, the sound field is completely diffuse and the decay rate is the same at every point in the room. (B) If one wall is absorptive, the decay rate varies from point to point in the room. The contours of decay rate tend to be parallel to the treated wall.**

that sharp microphone directivity is hard to get at the low frequencies at which diffusion is the greatest problem. The prospect of this method of appraising diffusion in small studios is poor.

Frequency Irregularity

For the last half century there have been many serious efforts to evaluate diffusion in rooms by steady-state transmission measurements.[34,36] Microphone and loudspeaker positions remain fixed. The constant amplitude swept sine wave signal radiated from the loudspeaker and picked up by the microphone has the room effect impressed upon it. This room influence should reveal something about the room.

Figure 15-7 shows typical frequency response records taken in a music studio of 16,000 cubic foot volume having a reverberation time of about 0.6 second at the 100–300 Hz frequency region under investigation. Two things are very striking:

- the magnitude of the variations
- the differences from position to position in the room

The amazing thing is that fluctuations in point-to-point response of such magnitude occur in studios having the best acoustical treatment and those considered excellent in subjective evaluations. The loudspeaker response in the 100–300 Hz region is included in these recordings, but it remains constant through all the tests.

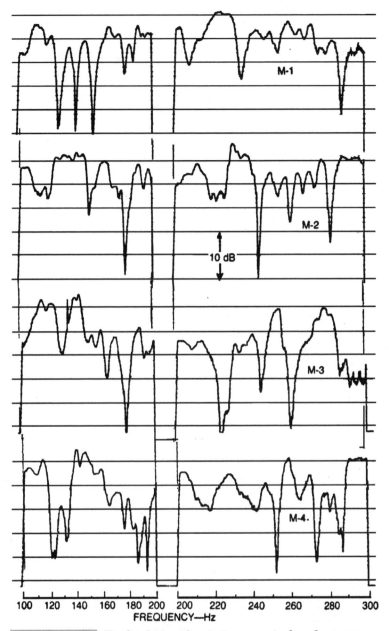

Figure 15-7 Typical steady state swept sine frequency response records taken at different positions in a music recording studio having a volume of 16,000 cubic feet and a reverberation time of 0.6 second. The striking variations are an indication that even a well-treated studio falls far short of truly diffuse conditions. At frequencies above 300 Hz, the curves becomes progressively smoother.

If such wild fluctuations are to be of any help in evaluating studios, it is necessary to find some method of reducing them to numbers. Bolt has suggested the term *frequency irregularity factor* obtained by adding all the peak levels, subtracting the sum of all the corresponding dip levels, and dividing the difference dB by the number of hertz swept. This frequency irregularity factor, or simply *FI factor*, is in dB/Hz.

Applying this procedure to Fig. 15-7 yields the FI factors tabulated in Table 15-1. Comparison of Fl factors for the different microphone positions would seem to tell us that conditions at M-2 are the best, considering the 100–300 Hz range, and that M-1 and M-2 are superior to M-3 and M-4.

Glancing back to Fig. 15-7 would seem to support this. What the FI factors of Table 15-1 tell us about sound diffusion in the room is not so clear. In measuring many studios it was noticed that larger FI factors were commonly associated with longer reverberation times. The 100–300 Hz FI factor for a dozen studios is plotted in Fig. 15-8 against their corresponding reverberation times. The broken line (not a least-squares fit) would seem to indicate a definite relationship.

In fact, theoretical studies and experimental results have shown that at high frequencies *frequency irregularity* is related only to reverberation time and that it gives no additional information on diffusion of sound in the room. Whether or not this is true in the 100–300 Hz region remains to be seen.

One thing seems to be clear, if at a certain microphone position the swept frequency response is within, say, ± 5 dB, that would be a good spot for a narrator to sit.

A corollary to that observation is that it is possible to compare microphone positions by a swept frequency signal test. For such a test a good place for the loudspeaker would be in a corner of the room. If standing waves such as indicated by the runs of Fig. 15-7 exist in what are called well-treated studios, conditions may be less bad in some spots than in others. This is of operational value only if recording in a studio can be done with a single microphone in a fixed position.

Size and Proportions of Room

A minimum studio or control room volume of 1,500 cubic feet has been urged. This is step one toward better diffusion. Rooms smaller than this are often plagued with coloration problems, impossible or impractical to correct. Of course, rooms having substantially greater volumes but still in the general small room category have plenty of diffusion problems also, but the chances of achieving satisfactory conditions by the application of the methods to be described are better.

By making a room large in terms of the wavelength of the sound to be recorded in it means that the modal frequencies will be closer together which means improved diffusion. Most of the methods of diffusing sound in a room to be considered later are most effective at the higher audio frequencies. Optimizing the proportions of the room is one of the most effective ways to improve diffusion at the low end of the audible band. There are a number of steps in the acoustical treatment of a room which tend toward better diffusion of sound in the room, once the major basic matter of room size and proportions are set (review Chapter 1 in this regard).

TABLE 15-1 FREQUENCY IRREGULARITY

MICROPHONE POSITION	FREQUENCY IRREGULARITY FACTOR, DB/HZ		
	100–200 HZ	200–300 HZ	100–300 HZ
M-1	0.777	0.556	0.667
M-2	0.497	0.775	0.626
M-3	0.663	0.852	0.758
M-4	0.764	0.771	0.768
Mean	0.675	0.739	0.705
Standard Deviation	0.129	0.127	0.069

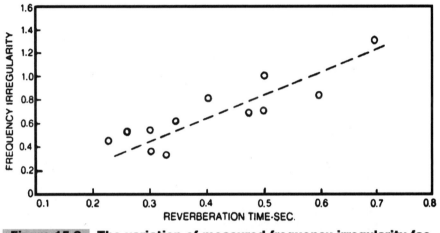

Figure 15-8 **The variation of measured frequency irregularity factor with studio reverberation time.**

Distribution of Absorbing Materials

Numerous controlled experiments and practical experience have demonstrated that concentrating the required absorbing material on one or two surfaces of a room is an acoustical abomination. Common sense emphasizes that this procedure often leaves some opposing or parallel walls untreated, producing some axial resonances.

The application of absorbing material in patches has been established as far superior to application in fewer large areas. This accounts for the proliferation of wall modules and sectionalized ceilings in studio designs in this book. The patches of absorbing material may be distributed by determining the areas of the N-S, E-W, and vertical pairs of surfaces and dividing the material between the three axial modes proportionally. At least, this is a respectable criterion to use as a rough guide, even though practical considerations like doors, observation windows, and floor coverings demand compromise.

Another important contribution of patches of absorbing material is that diffraction of sound, especially at the higher frequencies, takes place at the edges of each patch. Such diffraction contributes helpfully to diffusion. Placing absorbent where it acts on every axial, tangential and oblique mode is smart, remembering that all modes terminate in corners. Distributing absorbent in patches contributes to absorption efficiency as well as diffusion. Diffraction effects act as though the absorbent "sucks" sound energy from the surrounding reflective area which, in effect, increases its absorption coefficient.

Splayed Walls

The conventional wisdom among studio people has long held that splayed walls aid in diffusion of sound. Splayed walls do have the ability to help in the control of flutter echoes between opposite reflective surfaces, but do they really contribute significantly to diffusion? Model experiments have shown that the frequency irregularity factor is reduced with walls canted 5 percent but there is some question of this applying to practical studios with less smooth walls. The BBC made subjective tests in which experienced listeners listened with and without splayed walls with patches of absorbent on them and the results were inconclusive.

Rectangular and trapezoidal room shapes are illustrated in Fig. 15-9. The broken lines represent the sound pressure modal contours for a simple mode for both shapes. The small arrows represent the directions of particle motion in the two cases. The trapezoidal room shape most certainly has contributed something to diffusion, but the magnitude of the contribution is small for walls splayed the usual one part in 10.

It appears that justification of splayed walls must come from reduction of flutter echoes rather than improvement of diffusion. As there are other ways to prevent flutter echoes (such as patches of absorbent) it would seem that tearing an existing building apart to cant walls might be ill-advised. In new construction, however, inclining the walls might cost very little.

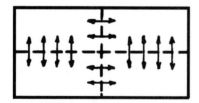

Figure 15-9 **The broken lines indicate the sound pressure modal contours for a simple mode in rectangular and trapezoidal rooms. The small arrows represent the direction of particle motion. It is obvious that the trapezoidal room shape contributes something to diffusion.**

Resonator Diffusion

In low frequency Helmholtz resonators, what happens to incident sound energy that is not absorbed by the system? It is scattered and scattering contributes to diffusion of sound energy in the room. This is not true of the porous type of absorbers in which energy not absorbed is reflected from the backing surface. Remember, however, that high frequency energy not affected by a perforated or slat low frequency absorber can be reflected and contribute to a flutter echo problem. This would suggest facing such units with a high frequency absorber or inclining its surface.

Geometrical Diffusers

There have been numerous, and presumably effective, goemetrical protuberances employed in studios to diffuse sound. Common among these are semicylindrical (poly) diffusers[37] and diffusers of rectangular and triangular cross section. The polycylindrical surface has been widely applied in studios, not only for its low frequency absorption, but for its ability to take sound arriving from a given direction and reradiate it through an angle of 100 degrees to 120 degrees. This contributes positively to diffusion in the room.

In spite of the salutory effect of polycylindrical elements, both controlled experiments[38] and theoretical studies[39] have demonstrated the marked superiority of rectangular protrusions over both cylindrical and triangular. The rectangular protrusions produce some effect when their depth is as shallow as one-seventh of the wavelength. Thus a rectangular element 6-inches deep has some effect down to approximately 325 Hz. This effect works to change the normal modes of a smooth-walled room. The cylindrical and triangular projections also do this, but to a lesser degree.

The acoustical distinguishing feature that sets the rectangular apart is the fact that it has finite portions perpendicular to the wall on which it is mounted. It has the ability of breaking up concentrations of modal frequencies better than cylindrical or triangular protrusions, of reducing the magnitude of dominant modes, and lowering the frequency irregularity in swept sine transmission. This provides some support for the proliferation of the not-too-beautiful wall modules in studio designs already considered.

Other diffusing elements found in every studio, control room, and listening room are people, tables, chairs, door and window frames, and equipment of every sort.

Diffraction Grating Diffusers

The optical diffraction grating can break down a beam of sunlight into all the colors of the rainbow. The type of grating used in optical studies has microscopically fine, parallel lines ruled on glass. Inexpensive plastic replicas are available as toys and the principle is applied in colorful advertising signs.

What has this to do with acoustics? Dr. Manfred Schroeder of the AT&T Bell Laboratories has made the connection in a very positive way.[44] Imagine a surface with a series of long, narrow, parallel wells, or grooves, on it. The wells are of either fixed or of varying depth, but of constant width. A sound ray impinging on this surface finds itself interacting with reflections from the bottoms of the wells. The phase (or time relationship) of the reflections from the wells varies with the depth of the wells; hence the arriving sound ray must interact with well reflections delayed varying amounts. Dr. Schroeder related the well depths to various mathematical sequences [45] which result in sound coming from a given direction to be scattered throughout 180 degrees. In this way mysterious mathematical names have come to identify different types of these diffusers: quadratic residue sequence, primitive root sequence, maximum length sequence, etc. They all scatter sound better than what was available in the past, but some are more efficient for specific tasks than others.

Figure 15-10 shows a specific type of diffraction grating sound diffuser, the kind Dr. Schroeder first tried. He needed different reflection coefficients of + and −1. A groove a quarter wavelength deep gives a reflection coefficient of −1 while no groove at all gives a reflection coefficient of +1. This emphasizes that a design frequency must first be selected. Let us take 1000 Hz as our design frequency. The wavelength of a 1000 Hz tone is 13.56 inches, a half wavelength is 6.68 inches, and a quarter wavelength is 3.39 inches. As shown in Fig. 15-10 the well depth must then be 3.39 inches and the unit well width 6.78 inches. It can be made of wood, metal, or plastic or any other substance which is a good reflector of sound. The question now is, "Will it diffuse only 1000 Hz sound?" This maximum length sequence diffuser will work well over about one octave, a half octave above and a half octave below 1000 Hz. The diffused energy is is confined to a hemidisk at right angles to the face of the diffuser.

The more complex grating diffusers perform much better. A quadratic residue diffuser is shown in Fig. 15-11. Here the sequence length is 17 and the relative well depths are as indicated and shown graphically. Two sequences are shown in the figure. Thin metal sep-

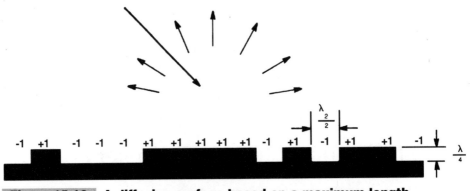

Figure 15-10   A diffusing surface based on a maximum length sequence. The grooves of quarter wavelength depth offer a reflection coefficient of −1, the high spots +1. Good diffusion a half octave above and below the design frequency is realized.

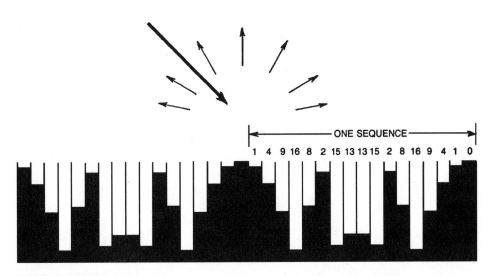

Figure 15-11 A diffusing surface based on a quadratic residue sequence. Good diffusion over most of the audible band can be achieved with this type of diffuser.

arators maintain the identity of each well. The width of the wells is about a half wavelength at the shortest wavelength to be scattered effectively. The maximum well depth is determined by the longest wavelength to be diffused. Although more difficult to construct, quadratic residue diffusers scatter sound effectively over most of the audible band.

Grating Diffuser Application

In the past the designer and builder of a studio, control room, or listening room had only reflection and absorption to work with; diffusion was largely out of reach. That has now changed. Standard diffusing units are now commercially available (RPG Diffusing Systems, Appendix D). Such units are being installed in recording studios and control rooms as well as concert halls, churches, and home listening rooms.

Diffraction grating sound diffusers are also being applied in small budget recording facilities. They are not a cure-all for the acoustical problems of small rooms but they do add a new tool to the designer's toolbox.

Are the principles elucidated in earlier chapters of this book outdated by the coming of effective diffusing elements? Not at all, but the prospects of better sound from small rooms are much improved. It will be a challenge to watch the growth of application of grating type of diffusers in the design of small recording and listening rooms.[46-53]

BITS AND PIECES
OF ACOUSTICAL LORE

Features: Biscuit tin modules, nursery tray modules, tuning resonators, LF compensation under the carpet, the cheapest absorber.

CONTENTS AT A GLANCE

Large absorbing modules of the wideband or low frequency peak type having appreciable air space behind a porous sheet are subject to a strange malady which degrades their absorption ability. There is the tendency for transverse modes of vibration being set up parallel to the face which decreases bass absorption. This can be prevented by breaking up the air space with dividers.

Partitioning Air Space

The modules described in earlier chapters include dividers so that the maximum transverse dimension is generally around 2 feet maximum. Partitioning an air space into an even greater number of small spaces can be quite expensive if lumber is used. Using lumber or metal sheets is not necessary because good performance can be obtained by breaking up the air space with corrugated cardboard dividers. Strips of corrugated board with width equal to the air space depth can be cut so that they fit together like egg crate dividers as shown in Fig. 16-1. Saw cuts with a buzz saw make quite serviceable slots.

Biscuit Tin Modules

Another way to partition the air space in absorbing modules is to make the modules the size of the partitioned space desired. Unit absorbers of modest size that can be readily arranged and attached to walls and ceilings also have many advantages in flexibility, sound diffusion, and general absorption properties.

However, the availability of such in finished, manufactured form is almost vanishingly small. The late Sandy Brown, British musician-turned-acoustician, reported a simple and cheap absorber built in a biscuit tin (cookie carton?).

Each tin box is about 8 inches × 8 inches × 5 inches deep with a lid having an opening almost as large as the box. A 1-inch thickness of rockwool with a wire screen cover is placed directly under the lid, leaving a 4-inch air space depth (Fig. 16-2).

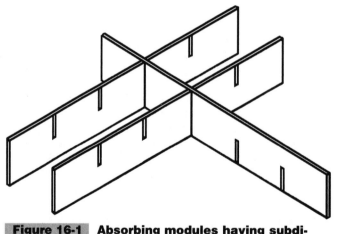

Figure 16-1 Absorbing modules having subdivided air space within are more efficient than those having undivided air space. Corrugated paper dividers are effective in discouraging modes of vibration in the air space parallel to the face of the module.

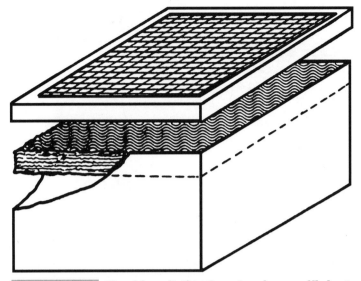

Figure 16-2   **The biscuit tin absorber is an efficient and inexpensive small module. Its use as a wideband absorber is limited because of its lack of low frequency absorption.**

This type of absorber was developed by the BBC and, apparently, was used for a time, but later abandoned because its low frequency absorption was too limited for the wideband absorber which it was supposed to be. This could be remedied only by thicker absorbent or deeper air space, both expensive, and the whole idea of the biscuit tins in the first place was performance at minimum cost. However, the idea might just be useful to someone who has a studio treatment problem and, simultanteously, falls heir to a few hundred containers like biscuit tins.

Nursery Tray Modules

After seeing and hearing of numerous horror stories concerning attempts to treat studios acoustically at minimum cost (egg cartons come to mind), the importance of truly budget-absorbing modules is emphasized. Although all cases treated in this book are of the budget type, there is a spread even within that category.

Constructing wooden frames to hold glass fiber to walls and ceilings is expensive and very inconvenient. It is expensive in both materials and time. Further, the temptation is to make the wood modules larger than desirable to reduce the number to be made and hung. A small module which could be arranged in groups forming the equivalent of larger panels, or in various patterns and designs, would take advantage of the smaller size and exploit it to advantage.

The molded plastic trays nurseries used for small bedding plants offer some promise. Figure 16-3 shows two types of trays in common use. The tray of Fig. 16-3 (A) happens to

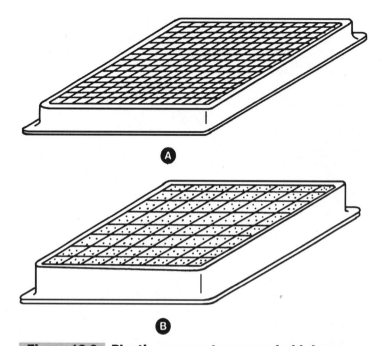

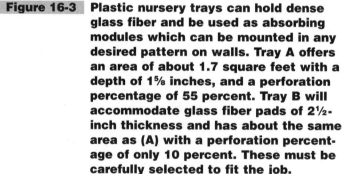

Figure 16-3   Plastic nursery trays can hold dense glass fiber and be used as absorbing modules which can be mounted in any desired pattern on walls. Tray A offers an area of about 1.7 square feet with a depth of 1⅝ inches, and a perforation percentage of 55 percent. Tray B will accommodate glass fiber pads of 2½-inch thickness and has about the same area as (A) with a perforation percentage of only 10 percent. These must be carefully selected to fit the job.

be 15½ inches square by 1⅛ inches deep. The large holes in the bottom give a perforation percentage of about 55 percent; that is, 55 percent of its bottom is in holes. These could be fitted with pads of glass fiber of 3 pounds per cubic foot density, 1½ inches thick and mounted to wall or ceiling surface with a few screws in the lip. The high perforation percentage means that this 1.7 square foot module would give the same absorption as 1½ inches of glass fiber without the plastic support. Any perforation percentage above about 15 percent or 20 percent would have essentially no acoustical effect. For much lower perforation percentages a Helmholtz resonator effect takes place. The 1½ inches of 3 pounds per cubic foot glass fiber would have very good absorption 1 kHz and above, and down to about 0.85 at 500 Hz, 0.4 at 250 Hz, and 0.1 at 125 Hz. Therefore, it is obvious that low frequency peak absorbers are required to provide adequate compensation if any appreciable area of such modules were used.

Figure 16-3B is a nursery tray of somewhat different dimensions. It is 15½ inches × 16½ inches × 2½ inches deep and has a perforation percentage of slightly less than 10 percent.

There are two advantages of this tray over the (A) tray if a wideband absorber is desired. It accommodates glass fiber of 2½ inch thickness, an inch thicker than (A), and the perforation percentage is down to where some Helmholtz effect takes place. Rough calculations indicate that it would resonate in the 350–400 Hz region which would tend to hold up absorption at the lower audio frequencies. Glass fiber of 3 pounds per cubic foot density of 2½-inch thickness would, without a cover or with a cover of high perforation percentage, be expected to yield excellent absorption down to 500 Hz, have an absorption coefficient of about 0.9 at 250 Hz, and 0.4 at 125 Hz. It is anyone's guess what the effect of the (B) cover would be, other than a general expectation of slower dropoff below 500 Hz.

The two trays considered require 3 pounds per cubic foot density glass fiber material of 1½ inch and 2½ inch thickness. The Owens-Corning Type 703 Fiberglas comes only in 1 inch and 2 inch thickness and splitting is not too convenient. The Johns-Manville Spinglass Series 1000 (3 pounds per cubic foot density) does come in 1 inch, 1½ inch, and 2½ inch thickness and would be equivalent in performance.

It is quite possible that nursery trays of many different sizes, depths, and perforation percentages are available, or even that plastic containers of quite different type could be adapted to this absorption module service. Their adaptability to a given acoustical treatment job would require a bit of planning and application of the principles given here and elsewhere. Their appearance, with an artistic eye and a spray gun (before the cores are inserted, of course), could be novel and interesting. Their performance could essentially be what is expected of the glass fiber of the same size and distribution without the tray covers.

There is one effect here, however, which tends to give small patches a higher absorbing efficiency than larger patches of the same area. When absorption coefficients are measured in reverberation rooms, coefficients greater than unity are encountered at times. These measurements are made by determining the reverberation time and from that the number of sabins absorption added to the room by the sample of material being tested. The absorption coefficients are determined by dividing the number of sabins at each frequency by the area of the sample. When coefficients greater than unity result (i.e., absorption greater than 100 percent) the advice is to consider it unity. The explanation usually given is that refraction effects at the edges of the sample under test make the sample appear larger, acoustically, than the physical size. By breaking the absorbent on a wall into small patches, refraction effects at the edges of all patches should give a greater absorption than calculated.

Tuning the Helmholtz Resonators

When a peak of sound absorption is needed at low audio frequencies to compensate for absorbers deficient at those frequencies (carpet, drapes, acoustical tile, etc.), or a sharp peak is needed to tame a troublesome modal frequency, attention naturally turns to Herr Doktor Professor Helmholtz. Panel type absorbers generally peak in the lows, but the absorption coefficient rarely exceeds 0.3 at the peak while Helmholtz resonators commonly come close to perfect absorption (a = 1.0). There is really no theoretical difference between the slit type and the perforated type; consider a long slit nothing more than a row of holes. In the acoustical design of a room it is very helpful to know where the peak of absorption occurs. One way to estimate this is to calculate the resonance frequency. Let us take the perforated resonator of Fig. 14-12 as an example:

Face panel .... ³⁄₁₆-inch masonite

Hole diameter .... ³⁄₁₆ inch

Hole spacing .... 3 inches on centers

Depth of airspace .... 7-⁵⁄₈ inches

The frequency of resonance of a perforated panel absorber backed by a subdivided air space is given approximately by[5]

$$f_0 = 200 \sqrt{\frac{p}{(d)\,(t)}}$$

where,

f_0 = frequency of resonance, Hz
p = perforation percentage
t = effective hole length, inches
= (panel thickness) + (0.8) (hole diameter)
d = air space depth, inches

Figuring perforation percentage is easily accomplished by reference to the sketch of Fig. 16-4 and is about 0.31 percent. The effective hole length is t = ³⁄₁₆ + (0.8) (³⁄₁₆) = 0.34 inches, approximately. With these numbers and the air space depth of d = 7.6 inches we can estimate the resonance frequency as follows:

$$f_0 = 200 \sqrt{\frac{0.31}{(7.6)\,(0.34)}} = 69 \text{ Hz}$$

The effective hole length is quite uncertain, being dependent upon the geometry, and the 0.8 (hole diameter) correction is only an approximation.

The indication of a frequency of resonance near 70 Hz is where the peak of absorption is expected to appear and the magnitude of that peak is close to a = 1.0.

Now, what is the shape of the absorption curve so that absorption calculations can be made at other frequencies? This information is best obtained from actual measurements on perforated absorbers, but few such measurements have been reported in this country. Some of the most complete measurements of this type have been reported by the Russian, V. S. Mankovsky.[40] From his excellent book are selected three Helmholtz type resonators with perforated faces. The physical dimensions (translated from metric to closest English equivalent) and other pertinent data are given in Table 16-1.

The measured absorption coefficients are plotted in Fig. 16-5 and it is tactily assumed that Russian measurement techniques are at least roughly equivalent to those in other parts of the world.

A comparison of graphs A and B dramatically show the effect of absorbent in the air cavity. The peak of graph B is so sharp it fell in the cracks between the 250 and 500 Hz measurement points but the broken line indicates a reasonable guess as to its shape. Without absorbent, this type of resonator has a tendency to "ring," i.e., to die away slowly when exciting sound ceases.

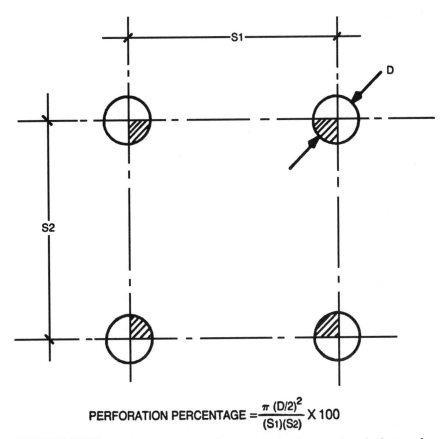

$$\text{PERFORATION PERCENTAGE} = \frac{\pi \, (D/2)^2}{(S1)(S2)} \times 100$$

Figure 16-4   **The perforation percentage of an extended area is easily calculated from its smallest repeat pattern.**

TABLE 16-1   DATA ON GRAPHS OF FIG. 16-5

GRAPH	DEPTH OF AIR SPACE	THICKNESS OF PERFORATED PLYWOOD	HOLE DIAMETER	HOLE SPACING	PERFORATION PERCENTAGE	FILLED WITH ROCKWOOL ?
A	2"	5/32"	5/32"	1-9/16"	0.785%	Yes
B	2"	5/32"	5/32"	1-9/16"	0.785%	No
C	2"	5/32"	25/32"	2-3/8"	8.5%	Yes

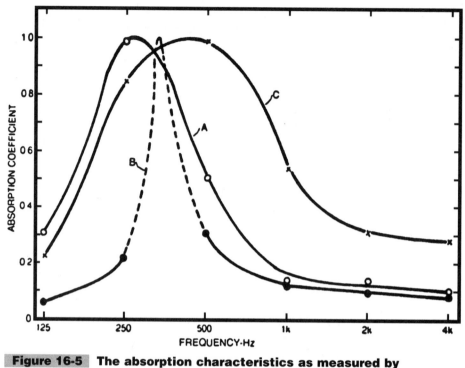

Figure 16-5 The absorption characteristics as measured by Mankovsky of three Helmholtz perforated face absorbers as specified in Table 16-1. Curve B is for a unit with no absorbent in the cavity. Curve A is for the identical unit with absorbent. Note the shift in resonance frequency and broadening of the peak resulting from introduction of the absorbent. Curve C is for a similar unit with much greater perforation percentage.

A great widening of the absorption peak and slight shift in resonance frequency results from the use of absorbent in the air space. Although the absorbent used (called PP-80) has not been identified by translator or editor, it is reasonably sure that it is common rockwool or glass fiber. The calculated resonance frequency applicable to the A and B graphs is about 467 Hz, which is not very close to the peaks determined experimentally.

Graph C is included to show the effect of a high perforation percentage, 8.5 percent. The depth is the same as A and B, panel thickness is the same, but hole diameter and spacing are different. The graph C in Fig. 16-5 is very broad. The calculated frequency of resonance in this case is 466 Hz which is in good agreement with the measured values of graph C. At such high perforation percentages the point is approached at which the cover has no Helmholtz effect. For example, 15 percent perforation percentage (or open space) is used for covers for wideband absorbers.

To apply data such as shown in Fig. 16-5 to any problem at hand is quite difficult in this form. Shapes vary as well as location of resonance peaks. These perforation percentages could well apply to absorbers tuned to other frequencies by using different air space depths. For example, the perforation percentage of 0.785 percent applied to a box made of 1 × 8s (7⅝-inch depth) would resonate in the vicinity of 120 Hz.

If this is where a peak is desired, how can graph A data be shifted down to this resonance frequency to give an estimate of coefficients to use based on the measured A values? In Fig. 16-6 the measured curves of Fig. 16-5 are brought together in what is called *normalized form*. These, in turn, can be applied to systems of other resonance frequencies. For example, graph A actually peaks at 270 Hz. At $2 f_0 = (2)(270) = 540$ Hz the absorption coefficient-efficient is read from graph A, Fig. 16-5, and found to be 0.42. This is plotted in Fig. 16-6 at $2 f_0$ as part of the normalized representation of graph A. The same is done to complete all three graphs in Fig. 16-6. An example of the use of the normalized graph A would be to build up an absorption graph using, for example, the $f_0 = 120$ Hz obtained for the boxes 8 inches deep and perforation percentage of 0.785 percent. At 120 Hz $a = 1.0$. At $2 f_0 = 240$ Hz $a = 0.42$, at $4 f_0 = 480$ Hz $a = 0.15$.

The same can be done for $1.5 f_0$ $2.8 f_0$ and other points in between, as well as for frequencies below the resonance peak. Another approach would be to trace graph A of Fig. 16-5 on a sheet of tracing paper and then slide the paper to the left until the peak coincides with 120 Hz. The absorption coefficients with the new tuning can then be read off on the assumption that the shape of the curve would not change for such a modest shift in frequency. These two approaches give, at least, something to use, even though the accuracy leaves something to be desired.

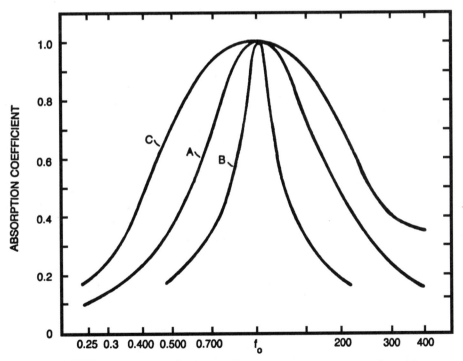

Figure 16-6 The three curves of Fig. 16-5 are repeated here in normalized form to assist in applying the curve shapes to other design tasks to estimate absorption coefficients to meet specific needs.

Some cases have been noted in which the calculated frequency of resonance does not agree very well with measured peaks of absorption. A relatively simple method of measuring the frequency of the absorption peak of a Helmholtz resonator is diagrammed in Fig. 16-7. A loudspeaker is driven by a sine wave oscillator. The resonator under test is placed 6 to 10 feet from the loudspeaker with the perforated face toward the loudspeaker. A microphone, placed inside the box, drives an indicator calibrated in decibels, preferably a sound level meter.

As the frequency of the oscillator is swept past the resonance point, a significant increase in reading will correspond to the great increase in sound pressure in the box at resonance.

If the oscillator output, the amplifier response, and the loudspeaker response are all constant with frequency, the box pressure indications are accurate. A second sound pressure measuring system with its microphone the same distance from the loudspeaker as the one inside the box could be used to measure the pressure outside the box and yield corrections for the other. However, if only the frequency of the peak is desired, just note the frequency at which the microphone in the box indicates the peak.

Such a test, made on the Helmholtz resonator of Fig. 14-12 with a graphic level recorder making a hard copy of the peak, is shown in Fig. 16-8. Surprisingly, a double peak was found. At first, transverse modes in the air space of the box were suspected, but this theory was rejected when the use of corrugated cardboard dividers in the air space had no effect on the double peaks. Panel resonance of the face and back of the box were then studied. The $^3/_{16}$-inch masonite cover resonates at a calculated frequency of about 69 Hz and the $^1/_2$-inch plywood back at about 51 Hz. The calculated resonance peak of the resonator

LF UNIT

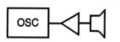
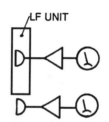

Figure 16-7 The resonance frequency of a Helmholtz type absorber may be obtained experimentally outdoors as shown. The rise of pressure inside the box is detected by a microphone placed in the box as the frequency of the sound from the loudspeaker is swept through the range of interest. A second system measuring the sound pressure outside the box is a possible refinement to separate out instrumental variations in sound pressure.

action is 69 Hz. Checking the accuracy of such calculated resonance points is what this section is all about, but pinpointing just what causes which peaks in Fig. 16-8 is somewhat doubtful, although the higher peak at 100 Hz is probably the Helmholtz action and the peak at about 63 Hz is due to panel action, possibly of both front and back acting together.

The sound pressure within the resonator box is increased about 22 dB at resonance. This corresponds to a 13-fold increase in pressure. The double peak characteristic may actually be an advantage in that high absorption prevails over close to an octave.

Low Frequency Compensation in the Floor

This is an idea whose time may not yet have come, at least for general application. However, it seems good enough to consider here as it stretches the imagination and points to a possible wider use in the future. There is no denying that it is nice to have carpets in studios, control rooms, and other listening rooms. Nor is there any way to deny their acoustical effect: high absorption in the highs, low absorption in the low frequencies.

If only a special carpet pad were available that would absorb well in the lows and very little in the highs. Just compensating for the carpet, the problem would be solved. The Japan Victor Company expended great effort in one of their small Tokyo studios to sweep the solution to the problem under the carpet, so to speak.[41] The novel way they did this was to lay the carpet on top of a Helmholtz resonator array. Under the carpet and pad of 1.6 inch thickness is a perforated board and cemented excelsior board of 3.5 inch thickness. Under that is 5.5 inches of air space. A neat solution, to say the least. This eliminated unsightly boxes, or other special ceiling or wall treatment to gain the required low frequency absorption.

The JVC solution is effective, although rather costly for general use. Is there a less expensive way of placing low frequency absorption under the carpet? Recent improvements in the Proudfoot Soundblox (The Proudfoot Company, Inc., P. O. Box 9, Greenwich,

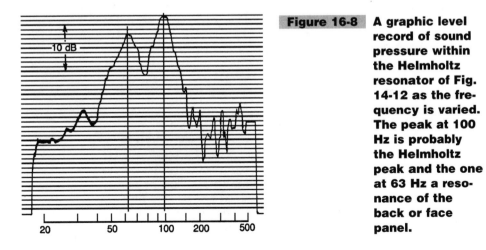

Figure 16-8 A graphic level record of sound pressure within the Helmholtz resonator of Fig. 14-12 as the frequency is varied. The peak at 100 Hz is probably the Helmholtz peak and the one at 63 Hz a resonance of the back or face panel.

Connecticut 06830) suggest a possible solution. Soundblox are proprietary concrete blocks utilizing the cavity inside plus carefully designed slots, metal septa, and fibrous filler as Helmholtz resonators. Figure 16-9 shows the appearance and cross section of one of the many types available, the 8-inch Type R block.

All the elements of a Helmholtz resonator are there along with the compartmentation effect of the metal septum and the fibrous filler to improve and widen the absorption effect. In Fig. 16-10 is a plot of the measured absorption coefficients of the 8-inch Type R Soundblox. The broken line graph shows the absorption coefficients of a typical heavy carpet to show that the carpet absorbs well where the Soundblox is deficient, and vice versa.

If 100 square feet of carpet and 100 square feet of Soundblox were placed in separate locations in a room, the coefficients would be proportional to the sabins absorption of each patch and the problem would be one of simple addition to get their combined effect. If carpet is used over the Soundblox only the sound not absorbed by the carpet reaches the Soundblox. The Soundblox can act only on the sound energy reaching it, which is less than that falling on the carpet by the amount the carpet absorbs.

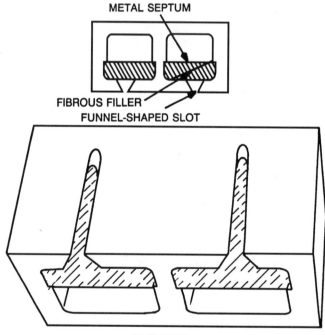

METAL SEPTUM

FIBROUS FILLER

FUNNEL-SHAPED SLOT

8″ TYPE R

Figure 16-9 The proprietary 8-inch Type R Soundblox concrete block which is a Helmholtz type sound absorber having the absorption characteristics shown in Fig. 16-10. The sound absorption is increased by dividing the cavity with a metal septum and the use of funnel-shaped slots.

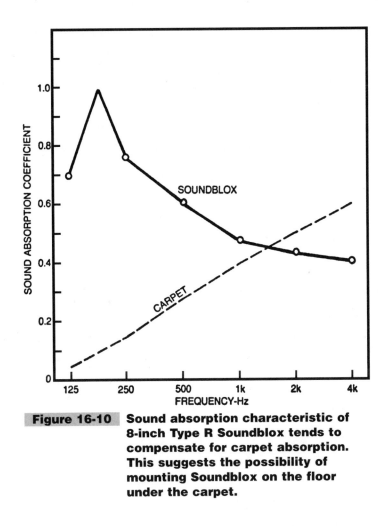

Figure 16-10 **Sound absorption characteristic of 8-inch Type R Soundblox tends to compensate for carpet absorption. This suggests the possibility of mounting Soundblox on the floor under the carpet.**

However, at the low frequencies carpet is almost transparent to sound where the Soundblox have their peak of absorption. In this case, the blocks absorb what the carpet does not. At 4 kHz the reverse tends to be true and in less definite form. The carpet allows 40 percent of the energy to pass through it and the Soundblox absorb only 40 percent of this. Thus the combined effect is a bit complicated: Certainly adding coefficients to obtain the combined effect is not legal.

Placing a layer of Soundblox, slots up, on the studio floor under a carpet is then a possibility to compensate for the acoustical effect of the carpet. There is the problem of mechanical strength of the Soundblox to withstand a concentrated load such as a grand piano leg. Certainly some load distributing layer is necessary between the carpet and the Soundblox. The JVC people solved this by using a perforated board and cemented excelsior board. The perforated board, of course, was part of their Helmholtz system. A perforated plywood layer having a high percentage of its area in holes might provide adequate load distribution.

Their cemented excelsior board apparently has its counterpart in this country in such products as National Gypsum's Tectum which might serve satisfactorily as a mechanical protection for the Soundblox. Here are the elements of a possible studio floor with acoustical advantages, but in strong need of further study and experimental verification.

The Cheapest Wideband Absorbers

For a truly budget studio there is need for an absorbing panel having respectable characteristics and of very low cost. Glass fiber panels 24 inches × 48 inches × 2 inches and 3 pounds per cubic foot density cost something like $3.00 each. The crudest and cheapest approach would be to slap a few gobs of acoustical tile cement on the back of the raw panel and stick it to the wall. By expending a modicum of imagination and effort, the panel can be covered with cloth of some sort which will not only control the irritating glass fibers but add a touch of class and color.

All budget approaches have their compromises. Stretching cloth over the panel tends to round off the corners and destroy the neat shape. The panels of glass fiber are fragile. If gentle handling can be guaranteed, no problem, but the first elbow blow will be memorialized permanently. All of these disadvantages can be at least minimized by building the panels along the line of that described in Fig. 16-11.

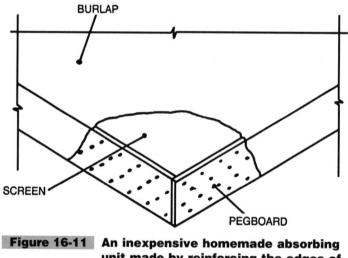

Figure 16-11 An inexpensive homemade absorbing unit made by reinforcing the edges of a panel of 3 pounds per cubic foot density glass fiber with pegboard and stretching burlap over it, then cementing the cloth on the back of the panel. A screen may be used if impact resistance is required.

Edge absorption can be largely retained by using edge boards of pegboards as thin as available. Impact insurance is in the form of a wire screen, perforated plastic sheet (20 percent perforation or greater), or plastic fly screen. The cloth cover should be of light weight and loosely woven. Burlap is a possibility and it is available in many colors or you can tie-dye your own! The cloth is wrapped tightly around the edge boards and cemented on the back with acoustical tile cement. Mounting the panel to the wall can be done in a rather unrelenting way with cement. The panel is so light, 4 pounds for the glass fiber plus edge boards and cloth, it could be supported easily by a couple of rings attached to the panel by a cemented cloth tape.

The 2-inch depth of the panel carries with it poorer absorption coefficients at 125 Hz (0.18) and 250 Hz (0.76). If panels are built of two 2-inch thicknesses, good absorption could be expected down to and including 125 Hz.

ACOUSTIC EQUATIONS

Feature: Understanding the room resonance problem.

While these equations relate to noise reduction, reverberation time, sound pressure level, sound power level, etc., they are not very complicated or for that matter difficult to use with a small pocket calculator. They are provided here for convenience, rather than as anything new or original. The main reason for using these equations is to get ballpark answers, not definitive answers. Take, for example, the first equation for the speed of sound in air. For every degree of variation in temperature, the change in speed is slight. At 100 Hz a variation in temperature of 10 degrees (65 degrees–75 degrees) results in only 11 feet per second difference (less than 1 percent). In terms of wavelength, the 10 degree temperature change results in a change of only 1¼ inch out of 11 foot wavelength at 100 Hz (again less than 1 percent deviation). On the other hand, manufacturers' optimistic absorption coefficient ratings or a variation between test lab and real-world mounting methods can cause huge discrepancies. As an example, one manufacturer has an absorption coefficient rating for their ceiling and wall panels of 0.40 and 0.24 at 125 Hz when mounted on 1½ inch and ¾ inch furring strips respectively. If the end user does not know or understand the effect of spacing from the wall, the predicted reverberation time would be very inaccurate. The difference in sabins absorption for a 200 square foot ceiling (80 and 48) would cause the reverb time (in a theoretical room of 2900 cu. ft. volume having no other sabins) to climb from the predicted 1.8 seconds to 3 seconds. A good rule of thumb is to get information on test methods and the mounting method utilized whenever possible. Again don't expect exact answers or predictions from any acoustical equation, as there are always variables that will throw the results off a little. Instead, use them as clue finders that not only give you an idea of what to look for and where to look but also what treatment is called for and where to position it for the best results.

Speed of Sound in Air

$$C = 49.03 \sqrt{460 + T} \text{ in feet per second}$$
$$C = 20.05 \sqrt{273 + Tc} \text{ in meters per second}$$

where:

C = the speed of sound in air in feet per sec or meters per second
T = temperature in degrees Fahrenheit
Tc = temperature in degrees Celsius   (1)

Example: $C = 49.03 \sqrt{460 + 70} \times 70 = 73.212$

Wavelength

Audible sound's wavelength at lower frequencies can be comparable to a small room's dimensions; for example, the wavelength at 20 Hz and 70 degrees is 56.4 feet. At low frequencies the sound build-up problems may be very difficult (read expensive) to correct. Therefore, the ability to determine wavelength and relate it to a room's dimensions is important for avoiding low frequency room resonance problems. In addition it is helpful in finding the particular dimension causing build-up at some specific frequency at which corrective measures should take place.

$$W = \frac{C}{F} \qquad (2)$$

where:

W = wavelength in feet
C = the speed of sound in air in feet per second (1130 ft/sec for normal, average conditions)
f = frequency of sound

Example: $\frac{1128.8}{125 \text{ Hz}} = 9$ feet

Absorption of Sound (Absorption Units)

It's easy to become confused by the absorption coefficients given by manufacturers of acoustical materials. The following two equations relate the absorption of sound in sabins to absorption coefficient and room size.

$$a = \frac{S}{S_A} \tag{3}$$

where:
 a = the absorption coefficient
 S = absorption in sabins
 S_A = the total surface area of the room in square feet

Example: a wall of 500 square feet had a rating of 14 sabins:

$$a = \frac{14}{500} = 0.028$$

also,

$$S = (S_A)\,(a)$$

Example:

$$(500 \text{ square ft.}) \, (0.028) = S = 14 \text{ sabins}$$

Another method of very roughly rating the sound absorbancy of materials in a single-number rating is the Noise Reduction Coefficient (NRC). This is the average, to the nearest 0.05, of the sound absorption coefficients in octave bands centered at 250, 500, 1000, and 2000 Hz. Unfortunately, there is no way to obtain information on the variation of absorption coefficient with frequency from the NRC. In this case the manufacturer should be consulted in order to get sound absorption coefficients at the standard frequencies.

Total Number of Sabins

To find the total absorption in sabins of a room, use the following equation:

$$(S)(a) = (S_1)(a_1) + (S_2)(a_2) + (S_3)(a_3) \ldots \tag{4}$$

where:
 a = average absorption coefficient
 S = total surface area of the room, sq. ft.
 S_a = the total absorption in sabins
 S_1 = the first surface area
 a_1 = the absorption coefficient of S_1
 S_2 = the second surface area
 a_2 = the absorption coefficient of S_2, etc . . .

Example:

$$\text{Room dimensions} = 20' \, L \times 15' \, W \times 10' \, H$$

ceiling = acoustic tile rated at 0.04 absorption coefficient
floor = carpeting rated at 0.05
two L walls = brick at 0.03
two W walls = concrete block rated at 0.1

$$
\begin{aligned}
S_1 &= (20)\,(15)\,(0.04) &&= 12 \\
S_2 &= (20)\,(15)\,(0.05) &&= 15 \\
S_3 &= (2)\,(20)\,(10)\,(0.03) &&= 12 \\
S_4 &= (2)\,(15)\,(10)\,(0.1) &&= 30 \\
\hline
\text{Total} && &= 69 \text{ sabins}
\end{aligned}
$$

Going back to equation (3) with a total area of 1300 sq. ft. we find that this room has an average absorption coefficient of 0.053.

Reverberation Time

After doing a few of these equations, especially when the average absorption coefficient is unknown or at least imprecise, the value of an electronic reverberation timer becomes obvious. The data given by their usage along with the following equations makes matters much simpler.

$$
a = \frac{0.049\ V}{(S)\,(T_{60})}
$$

or

$$
T_{60} \quad \frac{0.049\ V}{(S)(a)} \tag{5}
$$

where:

a = the average absorption coefficient
V = the volume of the room in cubic feet
T_{60} = the reverberation time, seconds
S = the total surface area of the room in square feet

Example:

$$
a = .075,\ V = 3000 \text{ cu. ft.},\ T_{60} = 1.5 \text{ sec.},\ S = 1300 \text{ sq. ft.}
$$

$$
a = \frac{(0.049)\,(3000)}{(1300)\,(1.5)} = \frac{147}{1950}
$$

$$
a = .075
$$

$$
T_{60} = \frac{(0.049)\,(3000)}{(1300)\,(.075)} = \frac{147}{97.5}
$$

$$
T_{60} = 1.5 \text{ sec.}
$$

To further illustrate the relationship of reverberation time and the absorptivity of surface materials we have:

$$T_{60} = 0.0498 \left(\frac{V}{A} \right)$$ (6)

where:
  T_{60} = the reverberation time, seconds
  V = the volume in cubic feet
  A = the total absorption in sabins

Example:

$$T_{60} = 0.049 \left(\frac{3000}{98} \right) = 1.5 \text{ seconds}$$

By increasing the absorptiveness of the room by adding absorption materials, you can reduce the room's reverberant or reflective energy by the following ratio (~ means "equivalent to"):

$$\frac{E_1}{E_2} \sim \frac{A_2}{A_1}$$ (7)

where:
  E_1 = the reflective sound energy before treatment
  E_2 = the reflective sound energy after treatment
  A_1 = the total amount of room absorption in sabins before treatment
  A_2 = the total amount of room absorption in sabins after treatment

Sound Pressure Level (Spl)

The reduction in sound pressure level (SPL) of a steady noise or other sound in a room resulting from the addition of absorbing material from A_1 to A_2 sabins may be found from:

$$\text{Reduction in SPL} = 10 \log \frac{E_1}{E_2}$$ (8)

$$= 10 \log \frac{A_2}{A_1}$$

The more reduction required in dB, the more sabins required until in order to achieve a 10 dB reduction 10 times the original sabins is needed. An overly optimistic if not impossible task.

If you are dealing with absorption coefficients, you can avoid converting to sabins by using the following equation;

$$D = 10 \log \frac{1}{1 - a}$$

where:
  D = the decrease in SPL in dB
  a = the average absorption coefficient

Example A:

$$D = 10 \log \frac{1}{1 - 0.5} = 3 \text{ dB}$$

Example B:

$$D = 10 \log \frac{1}{1 - 0.09} = 0.4 \text{ dB}$$

Sound Pressure Level and Sound Power Level

Sound pressure and power are rated in decibels and are therefore referenced to standard levels. They are most often determined by electronic measurement. The following equations should help to shed some light on their meaning.

$$Lp = 10 \log \left(\frac{P}{P_{\text{ref}}} \right)^2 \tag{10}$$

$$Lp = 20 \log \frac{P}{P_{\text{ref}}}$$

where:
    L_p = the sound pressure level, dB
    P = the root mean-square sound pressure as measured
  P_{ref} = the reference pressure
      = 2×10^{-5} (N/M^2)
      = 0.0002 dynes per centimeter squared
      = 0.002 microbars
      = 20 micropascals (RMS) μPa

$$Lw = 10 \log \left(\frac{W}{10^{-12}} \right) \text{dB} \tag{11}$$

$$Lw = 10 \log w + 120 \text{ dB}$$

where:
  Lw = the sound power level, dB
  W = the acoustic power in watts

Even with the most advanced calibrated electronic equipment automatically taking care of both measuring and referencing sound power and pressure levels it is still important to determine the difference between direct and reflected energy.

Direct Sound in dB

$Lp = Lw - 20 \log R + 2.3$ (for feet)   (12)
$Lp = Lw - 20 \log R - 8$ (for meters)

where:

R = the distance between the sound source and the receiver in feet or meters
L_p = the sound pressure level at the receiver in dB (re. 0.0002 μbars)
L_w = the sound source power level in dB (re. 10^{-12} watts)

Reverberant Sound In dB

$$LP = Lw - 10 \log A + 16.3 \text{ (for feet)} \tag{13}$$

$$LP = Lw - 10 \log A + 6.0 \text{ (for meters)}$$

where:

A = The total room absorption in sabins (square feet) or metric sabins (square meters)
L_P = The sound pressure level at the receiver in dB from reverberant sound only (re. 0.0002 μbars)
L_w = The sound power level at the source (re. 10^{-12} watts)

Room Cutoff Frequency

When dealing with room dimensions and troublesome modes it is important to know the room's cutoff frequency. Anything above that frequency is free of build-up problems and thus gives a smooth response. Below the cutoff frequency is where trouble may lurk in the form of bunched up modes and large gaps between modes giving an uneven response.

$$Cf = 20,000 \sqrt{\frac{T}{V}} \tag{14}$$

where:

Cf = The room's cutoff frequency, Hz
T_{60} = The reverberation time in seconds
V = The volume of the room in cubic feet

Examples:

$$Cf = 20,000 \sqrt{\frac{1.5}{3000}} = 447 \text{ Hz}$$

$$Cf = 20,000 \sqrt{\frac{0.5}{3000}} = 258 \text{ Hz}$$

$$Cf = 20,000 \sqrt{\frac{2.5}{3000}} = 557 \text{ Hz}$$

$$Cf = 20,000 \sqrt{\frac{1.5}{5000}} = 346 \text{ Hz}$$

These examples show that the longer the reverberation time the higher the room cutoff frequency while the larger the room, the lower the cutoff frequency.

Helmholtz Resonators

These equations will be helpful if you want to construct your own Helmholtz resonators or diaphragmatic absorbers to correct low frequency build-up problems. Unfortunately, while the three for Helmholtz give resonance frequencies, the amount of actual absorption at that frequency as well as its bandwidth are most easily determined experimentally.

Resonance Frequency of Perforated Helmholtz Resonator

$$f = \frac{C}{2\pi} \frac{P}{d(L + 1.7R)} \tag{15}$$

where:

f = the resonance frequency in Hz
c = the speed of sound in feet per second
P = the perforation ratio (hole area to panel area)
d = the distance of the panel from the wall in inches
L = the panel thickness in inches
R = the radius of the holes in inches

$$f = 200 \sqrt{\frac{P}{(D)(T)}}$$

where:

f = the resonance frequency in Hz
P = the perforation percentage (percentage of area in holes)
D = the air space depth in inches
T = the effective hole length in inches = the panel thickness + (0.8 × the hole diameter)

While these two equations yield similar results, they are not identical, due to the fact that actual measurements vary greatly. Therefore, both are included for comparison and possible averaging of results.

Resonance Frequency in Hz of Slat-Type Helmholtz Resonators

$$f = 2160 \sqrt{\frac{r}{(d)\,(D) + (W + r)}} \tag{17}$$

where:

f = the resonance frequency in Hz
r = the slot width in inches
w = the slat width in inches
D = the air space depth in inches
d = the effective depth of the slot in inches = 1.2 × the thickness of the slat

The Resonance Frequency of Diaphragmatic Absorbers

$$f = \frac{170}{\sqrt{(m)(d)}} \tag{18}$$

where:

f = the resonance frequency in Hz
m = the surface density of the panel in pounds per square foot
d = the depth of the air space in *inches*

$$f = \frac{49}{\sqrt{(m)(d)}} \tag{19}$$

where:

f = the resonance frequency in Hz
m = the surface density of the panel in pounds per square foot
d = the depth of the air space in *feet*

Noise Reduction

In terms of real-world studio, auditorium or listening room design, sound transmission either to or from the outside is of the utmost importance. Barriers, walls, enclosures, etc., are rated according to transmission loss which is related to the transmission coefficient by the following equation:

$$TL = 10 \log \frac{1}{T} \tag{20}$$

where:

TL = The transmission loss

T = The transmission coefficient, the ratio of transmitted to incident sound

To find the noise reduction in dB of a structure, use the following equation:

$$NR = TL + 10 \log \frac{(a_2)\,(A_2)}{S} \tag{21}$$

where:

NR = noise reduction, dB

TL = the transmission loss

a_2 = the average sound absorption coefficient in the receiving room

A_2 = the total area of the receiving room

S = the surface area of the partition separating the source and receiving room

The significance of this equation should be obvious. Here the effect of treating the receiving room is given importance. While other equations point only to the separating partition as a factor in noise reduction, this equation points out the importance of the absorptive qualities of the receiving room. Often it is much more cost-effective to treat a room's wall surfaces rather than rebuild one or more of the walls. The determining factor as to which of these two routes to take will be found by the following equations which examine the transmission loss-effectiveness of the partition.

Noise Reduction As It Relates to Transmission Loss

$$NR = L_1 - L_2 \tag{22}$$

$$TL = NR + 10 \log \frac{S}{A} \tag{23}$$

$$TL = NR + 10 \log S - 10 \log A \tag{24}$$

where:

NR = the noise reduction in decibels

TL = the transmission loss of the partition in dB

L_1 = the average sound pressure level in dB in the source room

L_2 = the average sound pressure level in dB in the receiving room

S = the area of sound transmitting surface of the partition or other test specimen in square feet

A = the total absorption in the receiving room in sabins

$$NR = TL - 10 \log \left(\frac{1}{4} + \frac{S}{R} \right)$$

(25)

where:

NR = the noise reduction in dB

TL = the transmission loss in dB

S = the surface area of one side of the partition in square feet

R = the room constant of the receiving room

Room Constant

The room constant is related to the reverberant nature of the room and can be found with the following equations:

$$R = \frac{S}{\left(\dfrac{TS}{0.049(v)} \right)^{-1}}$$

(26)

where:

R = the room constant

S = the total room surface area in square feet

V = the volume of the room in cubic feet

T = the room's reverberation time in seconds

$$R = \frac{Sa}{1 - a}$$

(27)

where:

S = the total surface area of the room in square feet

a = the average absorption coefficient of the room

Any opening in a partition, even if very slight, degrades its noise reduction capability. The following equation shows this:

$$TL = 10 \log \left(\frac{1}{OA} \right)$$

(28)

where:
    TL = the maximum transmission loss
    OA = the ratio of opening to the total surface area

Limp Mass Partitions

Diaphragmatic or limp mass partition transmission loss is dependent on frequency as can be seen from the following equations:

$$TL = 20 \log Wa + 20 \log f - 33.5$$

$$TL = 20 \log Wb + 20 \log f - 47.5 \tag{29}$$

where:
    TL = the transmission loss in decibels
    Wa = the surface mass of the partition in pounds per square foot
    Wb = the surface mass of the partition in pounds per square meter
    f = the frequency in Hz

ACOUSTIC MATERIALS

Acoustics is a combination of art, science, and applied experience. Sound coming into contact with an object is either reflected off, absorbed by, transmitted through the object, or some combination of all three. This is the whole ball of wax in terms of everyday acoustics. However, the acoustician's job involves not only the design stage, but also specifying, and detailing the acoustical materials to be used and, during the construction, taking care of specific details and making sure that certain specifications are met. A thorough knowledge of acoustical materials and their properties is required.

It is necessary to know about a material's properties so as to determine how vibration might be generated by them and how those vibrations or airborne sounds will be transmitted, dampened, and attenuated by them. A material's stiffness or "modulus of elasticity" and its resonant frequency are important to its ability to isolate vibration while its peak STL is important as it indicates its ability to stop airborne sound transmission.

In order to understand how acoustic materials function, you must first understand the physics of sound. Energy that causes a variation in pressure is a result of molecular movement. Sound energy starts at some source, travels via some path and reaches the receiver. The path must be an elastic medium. Molecular movement is constant for each medium. The magnitude of the movement depends on the pressure and the temperature of the medium. In any medium molecules are constantly in motion, smashing into each other and bouncing off. Sound energy is superimposed on this random motion. To see what effect sound has on molecular motion you must freeze or stop all that continuous motion for an instant. Figure 18-1 shows what happens to a molecule when force is exerted upon it by sound.

Molecule A is displaced to one side and strikes B setting it in motion. More importantly than the bouncing is the fact that A not only returns to its original position, but continues past it to a location opposite to its initial direction of movement. Then it returns to its original position. If this movement is repeated at regular intervals there is a vibration. The areas where A + B, B + C, C + D, and D + E are close together are called compressions. The pressure voids this compressed state causes are called rarefaction. The distance between two adjacent peaks of rarefaction or for that matter two adjacent peaks of compression is the wavelength of that frequency in that particular medium.

Sticking to the molecular example of Figure 18-1, the sequence of A's movement (from top to bottom) is one cycle, or a complete excursion of the molecule. Frequency is the number of these cycles that occurs during the period of one second. The preceding example holds true for invisable sound power in air and mechanical vibrations in solids. The maximum displacement of A is equal to the amplitude of motion. In audible sound the excursion or movement is a very small amount.

The important aspects of the path or the elastic medium the sound is passing through are elasticity and density. Greater mass causes greater inertia. Elasticity or stiffness is also a factor along with hysteresis or internal friction. Therefore, in any consistent medium the speed of the sound traveling through it is fixed. The velocity or speed of sound is proportional to elasticity and inversely proportional to the density of the material it is passing through. In other words the more a material weighs the slower sound passes through it while the stiffer it is the faster sound will pass through it. These two seemingly simple relationships can become complex. As examples, the speed of sound in lead very dense is approximately 4000 feet per second while in less dense wood it is 11,000 feet per second,

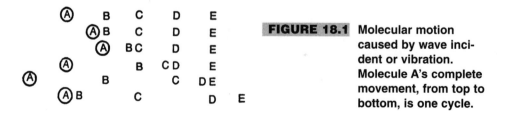

FIGURE 18.1 Molecular motion caused by wave incident or vibration. Molecule A's complete movement, from top to bottom, is one cycle.

and still less dense air, 1130 feet per second. These facts do not appear to add up until the hardness of the material is taken into concideration.

Force is equal to mass × acceleration. All materials have mass. In order for this mass to be moved it requires acceleration, especially when moved back and forth in a cyclical (vibrating or oscillating) motion with its changes in direction and velocity. Thus the process requires some amount of force and force acting through a distance is energy. The transfer or conversion of this energy into heat for example is one of the functional properties we seek in acoustical absorbent materials. Thus materials have to be considered on the molecular scale along with the larger scales of construction. Amplitude, wavelength, acoustic energy, sound pressure, and speed must all be considered as factors in choosing the proper material to solve a specific problem. By using various equations you can gain an insight into the workings of acoustical materials and therefore would never be gullible enough to consider a sand-filled paint as having significant acoustical absorption or believe that lining a room's walls with any one treatment would solve problems at all frequencies.

Acoustical properties are discussed in terms of ratios and coefficients. The two help in the comparison of properties. The range of normal hearing in humans covers a huge variation in pressure from 0 to 1,000,000. Instead of dealing with all these numbers, the decibel, a logarithm, of a power ratio, is utilized. The resulting range of 1 to 120B is a bit easier to deal with. However as a log of a ratio [10 log (value a/value b)] the decibel has to be referenced to some value. In sound pressure levels the dB is referenced to dynes per square centimeter, with sound power to watts. Thus single number comparisons between the two cannot be made. To further complicate matters since it is a log even the addition of decibels requires complicated equations.

The human ear's response to sound is not uniform with frequency. It is more sensitive to higher pitched sounds than to lower. The crossover point is around 1000 Hz; below this point the ear becomes increasingly less sensitive. This fact has caused the use of weighing scales when measuring sound pressure. In order to accurately compare test results it must therefore be known if, and what, scales were utilized during testing. As a matter of interest: The A scale heavily discriminates against low frequency sounds and closely aprroximates the characteristics of the human ear; the B scale moderately discriminates against sounds below 500 Hz; and the C scale is essentially flat. Sound pressure level comparison may be made as long as the same weighing scale was used for deriving both. To help avoid confusion, many values will be given followed by dB(A), dB(B), or dB(C). If only dB is given you must check out the test methods used with the manufacturer. We know that sound energy sets a sound wave in motion and that power is the amount of energy that produces a certain amount of sound pressure. Sound pressure falls off with distance. Sound pressure measured at the receiver's location is usually the parameter desired. Sound power cannot easily be measured directly. Sound pressure is what is normally measured.

A very important aspect of understanding the way in which sound absorbing material works is how pressure affects them. Take for instance a hollow rubber ball in a pressure chamber. If the pressure was increased beyond the internal pressure of the ball it would collapse. If the pressure in the chamber was decreased well below that of normal atmospheric pressure the ball would expand. Note here that the air which constitutes the pressure did not have to enter or escape from the ball to achieve either effect. Sound absorbent materials offer resistance to sound pressure variations but unlike the simple rubber ball this resistance varies with the material. If a material offers low resistance there is little or no attenuation of the sound and it will easily pass through the material unless it is very thick or dense. Sound barriers offer high resistance to sound pressure, enough so that most of the energy is reflected back and little or no sound escapes through to the other side. It is precisely for these two reasons that many acoustic materials utilize a combination of both high and low resistive materials to achieve both sound absorption and sound transmission losses.

Acoustic materials act as transducers, usually converting sound energy into thermal. Materials can be tested and the repeatability of these test results yields a high level of confidence in the laboratory data but equate to their characteristics in real life only to a certain degree. These test results are in the form of relative measures and/or coefficients. They are not absolute values and should not be considered as such. Their only usefulness is for a comparative look at various materials. As an example note the problems encountered in Chapter 5.

Does this mean that the labs or the manufacturers of acoustic materials lie? No, and even coefficients with values higher than theoretically possible can be obtained in honest lab tests.

Another form of confusion and inaccuracy results from the use of different bandwidths during testing. Measurements are not always made at individual pure tone frequencies even though one (center) frequency is given, but often with a grouping of frequencies such as a full octave or $\frac{1}{3}$ octave bands. The center frequency given along with the number of bands tested will point to which of these two was used and should be noted.

Octave band center frequencies are; 31.5, 63, 125, 250, 500, 1k, 2k, 4k, 8k, and 16k Hz. One-third octave band center frequencies are 25, 31, 40, 50, 63, 80, 100, 125, 160, 200, 250, 315, 400, 500, 630, 800, 1000, 1250, 1600, 2000, 3150, 4000, 5000, 6300, 8000, 10,000, 12,500, 16,000, and 20,000. A material's acoustic properties relates directly to its physical properties of stiffness and density. Yet this relationship is not linear. As an example fiberglass density if changed from 1.6 pounds per cubic foot to even up to 6 pounds per cubic foot would show little difference in absorption. All materials have some acoustical properties since they all absorb, reflect, or radiate or dampen sound to some degree. In the strictest sense, however, acoustic materials are those whose physical properties particularly suit them for some acoustical purpose.

Sound Absorptive Materials

Absorptive materials come in three basic types: porous, diaphragmatic, and resonant or reactive. There are literally hundreds of porous absorbant material types with the most common being mineral or glass wool, molded or felted tiles, perforated panels and boards, plasters, sprayed-on fibers and binders, foamed open-celled plastics, elastomers, glass or mineral fibers, and even normal room furniture such as cushions, upholstery, carpets, drapes fabrics, curtains, etc. All are fuzzy and/or soft and when sound pressure is exerted upon them they move. This movement causes friction which converts the pressure to heat and thus depletes the acoustic energy. In order to determine the amount of this depletion or the attenuation to be expected, one must delve a bit deeper into the inner workings of sound absorption.

Air inside a porous network of fibers, cells, granules, or particles is pumped or oscillated back and forth within the material when sound pressure is exerted on its facing. The network offers some resistance to this flow and this causes the network to vibrate which causes frictional losses. The frictional losses occur as heat and the acoustic energy within the material is reduced accordingly. The energy converted gives the absorption of the material and is expressed in coefficients that relate the converted energy to the original energy. If a material absorbs 100 percent of the sound such as an open window would, it would be 100 percent absorptive. The unit of one square foot of totally absorbent surface is equal to a sabin and all acoustically absorptive materials are compared to this. Thus if an acoustical material absorbs 45 percent of sound energy at a particular frequency it would have an absorption coefficient of 0.45. This rating is not perfect for real-world usage, but works well when comparing one material's absorptiveness to another's.

The internal structure of porous materials can be one or more of several basic types shown in Fig. 18-2. In actual materials the network is rarely this regular. What these networks do is introduce an amount of resistance, or more correctly impedance, to the flow of the air inside the material. If the resistance is too low, no frictional loss occurs. If too high, the air flows will be too restricted causing little fiber movement and therefore no frictional loss. Inbetween these two points lies a fairly broad area and an optimum flow resistance is not an exact value for most materials.

The internal structure of absorptive materials must have openings. As an example, while two foams may look the same, an open cell foam is sound absorptive while a closed cell foam is sound reflective. The fibers, particles, or cells should be random in orientation and

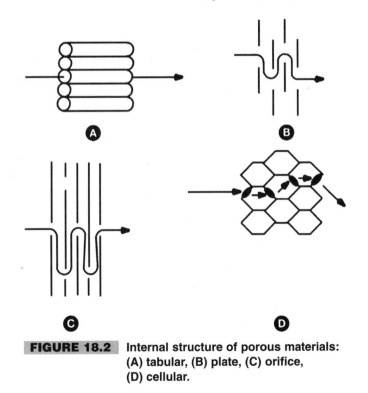

FIGURE 18.2 Internal structure of porous materials:
(A) tabular, (B) plate, (C) orifice,
(D) cellular.

size and shaped irregularly so that air travel takes as complicated a path as possible. Should the individual fibers or components be too large they will reflect and if too small they will be too dense and offer too high a resistance to air flow, producing little frictional loss. Remember that the movement or oscillation of molecules is very small and that it is not important that air enter or leave a material, only that it moves inside it. Air movement can be in any direction within the material. Porosity of the material is related to density and is unique to that material and on its own density is not an accurate indication of the absorptive ability of even a group of similar materials.

Fiber, pore, cell, or particle size is important as stated, but so is their orientation. To picture this, think of many layers of ordinary screening materials. This structure would be absorptive and its absorptivity would vary with the spacing of the strands, orientation of strands, spacing of between strands, and the spacing between screens. Absorption is also related to the thickness of the material. Absorption increases with thickness but not in a 1:1 ratio as at some point added thickness yields no more absorption.

Absorptive materials are used to lessen the amount of sound reflections in a room and thus lower reverberation time and reduce the sound pressure level within the room. Absorptive materials should be placed as close to the offending sound source as possible or directly on the surfaces that have unwanted reflections. Absorptives should be placed on ceilings, walls, floors, and on panels surrounding noise sources as well as within partition cavities and backed with a reflective material. The reduction of sound pressure or reverberation time due to absorbent materials added to a room depends on the size and geometry of the room, the absorption coefficient of the existing materials, the location of the noise source or sources, amount of absorptive material added to the area, and even the mounting method of the added absorptive material. It is impossible to predict the results to be expected from acoustical treatment without taking all of the above into account. To do so one must understand the difference between the two main absorption coefficients used.

Most acoustical tests of absorptive material are performed using either random incidence or normal incidence test. Incidence is the term used to describe the direction of a sound wave striking a surface. Normal is perpendicular and random is from all angles. Both test methods are strictly defined and adhered to. Random is governed by ASTM Standard C423 "Standard Method of Test for Sound Absorption of Acoustical Materials in Reverberation Rooms." Here the absorption coefficient of a material is determined by the change its introduction causes in the rate of decay of a sound in a reverberant room. First the total absorption of the room is measured without the test specimen. A sound source is turned on until its level reaches a steady state. The sound source is stopped and the rate of reverberation decay measured. The test specimen is brought into the room and the test repeated. The difference is noted and the total amount of absorption caused by the specimen is divided by the area of the specimen in square feet and the resulting value is the absorption coefficient. The laboratory room must not only be hard, reverberant, and diffused, but large enough so that this diffused area is not broken up by the specimen. Yet while the specimen also has to be small enough so as not to break up the diffusion of the reverberant field it must be large enough to give accurate results. Small specimens can also give higher absorption values than larger ones. Here diffraction can increase absorption. Since sound waves that strike the material are bent or diffracted at the edges they see an effective area that is greater than the area of the test specimen.

This can often yield coefficients higher than 1 even though this value cannot be exceeded in theory. This however poses no problem for comparing the absorption coefficients of two materials as the standard size of 72 square feet is always used. However, expecting these results in the field is not practical. Should a lab round off or adjust the test results so that the peak of 1 is not exceeded they must according to C423 specify exactly how this was done. Additionally due to the fact that an absorber's low frequency absorption can be increased due to air space behind the specimen, standard repeatable mountings are utilized and that mounting will be specified in the lab reports (see Fig. 2-12). As you can see, it is always wise to obtain a copy of the lab report on a given material from the manufacturer.

Normal incidence coefficients utilize an impedance tube as per ASTM Standard C384 "Test for Impedance and Absorption of Acoustical Materials by the Tube Method." Here a small sample of the material is placed at one end of a sealed tube and a tone is generated within. A small movable microphone measures the minimum and maximum sound pressure levels and from these the absorption coefficient is found. The impedance tube method uses a sound source radiating pure tones usually at octave intervals. The random method is determined by testing at six different $\frac{1}{3}$ octave bands—centered at 125, 250, 500, 1K, 2K, and 4 kHz. Comparing normal incidence test results to random incidence is not easy. One rule of thumb is that normal is $\frac{1}{2}$ random if random is a small value; as random gets large, normal increases to almost reach the same value. In general normal incidence results are always smaller than random results by $\frac{1}{4}$ to $\frac{1}{3}$.

Absorption is also related to frequency. The particle oscillation is related to frequency and that is why tests are performed at several ranges or frequencies. Furthermore, the material's dimensions as compared to the wavelength of sound falling upon it are also important. Theory states that maximum absorption occurs when material thickness is $\frac{1}{4}$ wavelength of the lowest frequency to be absorbed. Spending a little time on equation number 2 in Chapter 17 will graphically show how foolish it is to think that any paint with a bit of sand added to it can cause any appreciable absorption. Most foams used for absorption range from at least $\frac{1}{2}$ inch to 4 inches to 8 inches in thickness. "Acoustical Foams" that are only $\frac{1}{8}$ inch to $\frac{1}{4}$ inch thick are pretty much a joke, and a bad one at that.

Two more points should also be made. Real rooms are seldom perfectly diffused environments and often have standing wave problems between any two parallel surfaces. Absorptive material placed on other walls will not yield the desired amount of attenuation. Even when placed on the correct walls the effective amount of absorption is often lower than calculated. By distributing the absorption widely and randomly, a closer to calculated result can be reached. This is due to the edge effect or the diffraction of sound at the edges of the material as well as the face. Another way of rating the absorptive effect of materials is by their acoustic impedance. This is the ratio of the sound pressure to the particle velocity of the sound wave at the surface of the material. As can be expected this is a very complex quantity but it does allow for accurate reverb time predictions in rooms with uneven absorption.

Examples of Absorptive Materials

Foam rubber is made by whipping latex, adding a gelling agent, and pouring the mixture into molds. The result is an open-celled material of variable density. Plastic foams foamed by chemicals causing gas bubbles yield closed-celled structures before modification. Foams can transmit, absorb, or reflect sound depending on their permeability. Permeability is measured by the pressure drop of a gas passing through it. The pressure drops as pore size increases, but is also affected by the foam's surface roughness and the number of closed cells within. Plastic foams normally have a thin skin interconnecting many of the pores (nonreticulated) which can block the passage of sound and act as a barrier. For absorption more open cells are required, thus allowing for more air movement within the material. This is accomplished by reticulation. Thermal reticulation is performed by passing a flame through the foam and producing a fire polishing. In chemical reticulation the foam is immersed in a caustic chemical. Both methods remove some of the cell wall, leaving the strands between the walls intact. The air within the foam now moves more freely, and more vibration or friction occurs at the strands and is thus dissipated. The magnitude of the strand's oscillation has little effect on the absorption. Thus the strand's rigidity is not important. Should all the cell wall be removed the foam can become virtually transparent to sound. An example of this is the foam used on microphones or as speaker grilles. It is important for foam to be backed by a hard reflective surface. This not only aids in stopping sound transmission through it but the sound pressure increases as reflection takes place. The phase of the reflected energy is reversed from that of the energy. Varying the pore size allows for the tuning of absorption at a particular frequency. Therefore the efficiency of foam can be easily varied. As an example, one-inch width foam with 20 pores per linear inch would be 20 percent efficient in absorbing frequency X. With 60 pores per linear inch it would improve to 40 percent efficient and at 91 pores would be 80 percent. The permeability of foam and the resistance to air flow is measured in rayls. Units of flow resistivity show flow resistance per thickness. If flow resistance is high, permeability will be low and if flow resistance is low, permeability will be high. Even though foams may have the same thickness, density, number of cells, and cell size their absorption can be very different due to the amount of reticulation which affects permeability. The flow resistance is proportional to the number of membranes or skins reticulated. To a certain degree the following holds true: While less than 100 rayls means almost complete reticulation and more than 400 rayls yields high flow resistance, absorption in between these two points is not exactly predictable. However, lower rayls tend to have better high frequency absorption and higher rayls better low.

The load compression of foam denotes its ability to support itself as well as loads. In general, polyester-based foams are less flammable than polyether. Additives can give both more flame-extinguishing features but also tend to change flow resistance. Polyester-based additives are more durable in highly humid environments while polyether can survive very dry climates. Polyester is also not as affected by ultraviolet light. Polyether can be produced with greater than 90 percent open cells while polyester about 60 percent. Here, a higher percentage equals lower air flow resistance and thus makes for better absorptive wedges, while 60 percent makes better sheet foam absorpters. Polyether is also less susceptible to mechanical fatigue. Low cut-off frequency is a function of air flow resistance. Values for polyurethane-polyether should range from 40 to 50 percent and for polyurethane-polyester from 30 to 40 percent. As far as foam anechoic wedges, those can yield close to 100 percent absorbency to below 100 Hz when using wedges of 36-inch depth. Thus it can work as well as fiberglass, and actually end up doing a better job in the long run because this material's composition has less variation per unit and remains stable much longer.

The shock absorption and compression capabilities of foam are functions of its resiliency. The density of foam should range from 1 to 2.5 pounds per square foot as anything over that won't increase the foam's acoustical absorption. If correctly formulated, they have good sound absorption characteristics, are only fair at dampening, and yield some vibration isolation, but, unless very thick, do not make such great sound barriers. In addition to wedged and convoluted shaping, some manufacturers provide foams with fuzzylike surface coatings that tend to increase absorption at higher frequencies.

So much information exists on fiberglass and mineral absorbers that it is not necessary to rehash it again here. Stated as simply as possible, fiberous materials bonded together often by resins, handle high temperatures better than foams and are about equally absorptive but are irritating to both work and live with. Felts also have good absorptive characteristics. They can be woven, nonwoven, or knitted. Their fibers are often bonded via pressure, chemical action, and/or heat. They demonstrate better wear resistance than glass fibers and are denser than both foams or fiberglass.

Wood fibers can also be bonded together. They are much more rigid and are used for wall paneling and roofing. As roofing, wood fibers are bonded together with cement giving it a higher transmission loss especially with the common additional thermal layers of tar and stone pebbles that are often used. Spray-on absorptive materials can be very convenient to reach hard to get at areas or when covering irregular surfaces. These combine fibrous materials and bonding. If the bond is too tight they yield a high resistance. If bonded too loosely not only will the resistance be too low but the material could fall apart. When mixed properly these materials can stay up even at 2" thicknesses. Yet this will generally only attenuate the higher frequencies and is only a little better than putting carpeting on a ceiling. However, if applied to hung panels these materials can exhibit broader band absorption. Thus spray-ons are dependent not only on the thickness of the material applied but also on the base they are sprayed upon. It is essential to get manufacturer's specifications on application methods when using these materials.

Functional absorbers are panels, baffles, drums, cones, double cones, freestanding room dividers, blocks, etc., and all are easily installed. Panels, for instance, can be hung from the ceiling exposing all six sides to the sound. Since most are three dimensional, they give higher absorption than single surface absorbers. They are extremely convenient where existing surfaces are not suitable for wall panels or where ceilings are blocked by pipes, lights, ducts, etc. Because they are often three dimensional they are rated in sabins per unit as opposed to sabins per square foot. Hanging can be a little tricky as their absorptiveness is affected by both the number of them used and their spacing. As Equation 9 (Chapter 17) shows, at some point you'll get the maximum absorptivity no matter how many additional units are added. Space absorbers are pretty much the same as functional absorbers. They are either freestanding or hung where other absorptive materials are not practical to mount. These again can be thick panels of various materials but more often are of a more spectacular three-dimensional shape such as hollow triangular pyramids of fibrous material, wood fiber cones, or two cones joined at the base and thus enclosing a large amount of air. Others are hollow cylinders of glass fiber with perforated metal or screen facings. One manufacturer, Acoustic Science Corporation, offers absorptive cylinders that include an aimable limp mass reflector which effectively allows the end user to either tune in or turn off mid- and high-frequency absorption while still taking advantage of low- to low mid-frequency absorption. Additionally their units can be ordered with tuned resonator built into them giving high absorption at a specific frequency down to below 40 Hz. All functional and space absorbers offer very high amounts of absorptivity, but the "Tube Trap" from ASC offers several unique problem-solving ideas and extreme versatility all in a single package.

Most curtains, fabrics, and upholstered seat cushions make good absorbers. When building absorbers the resultant performance can be calculated from the known parameters of the materials utilized. However, full-scaled finished product tests are best as mounting methods, edge, coincidence, and random incidence effects have a great deal to do with performance. Some very good absorbers have graduated flow resistances. Here the flow resistance is minimal at the surface where the sound wave falls upon it and becomes gradually more resistive as the energy moves within until the maximum resistance occurs at the back allowing no energy to pass through the unit.

Finally absorptive systems are also offered. Roof deck systems are structural materials that have absorptive undersides. Whether perforated metal with fibrous material behind it or strictly a sound absorptive panel, roof decks not only reduce reverberation and level within, but aid in transmission loss. Aside from their use as roofs many can be used as combination ceiling and floor units. Wall treatment panel systems are generally decoratively covered fibrous panels that can be mounted to walls to increase a room's absorptiveness. Mountings vary greatly with any space behind the panels adding to the low frequency absorption almost exactly as if adding thickness. Thus these systems offer broad band absorption and can completely control a room's acoustics. Sound absorptive systems can be either freestanding or structurally mounted—freestanding such as hung units or screens with feet and structural as built-in such as ceiling, certain walls or roofs.

Often absorptive materials are chosen to meet environmental conditions. Temperature, moisture, dust, abrasion, high-pressure levels, and force cause some absorptive materials to become useless in a short period of time. There are however materials or combinations of materials that can be utilized even in the worst conditions. Metal wools for instance are unaffected by humidity or temperature, while foamed plastics or ceramics can at least handle moisture. High sound pressure can destroy most materials yet urethane foams work well as do porous masonary blocks, porous rough faced metals and dampened metal panels.

Modifications to absorptive materials are also helpful. Various facings can be used as long as their porosity allows the sound wave incident to cause air movement within the absorptive materials. Impact and abrasion problems can be eliminated by the use of facings. These protective coverings can also lend strength, hardness, ease of cleaning, weight, maintainability, moisture resistance, durability, and a more pleasing appearance to any absorptive material. One modification is to cover materials with a perforated material such as metal that is much more easily maintained. These facings affect the absorptivity of the material in some way usually by decreasing the high end absorption and increasing the low end absorption. When the reflective areas between the perforations are large, a great deal of high frequency content is reflected. On the other hand, low frequencies with their longer wavelengths diffract around the reflected areas and still cause pressure behind the surface thus activating the air inside the absorber. Perforated facings generally vary from roughly 10 to 50 percent open area depending on the thickness of the facing material, hole size, and spacings. More porous materials such as speaker grill cloth or screens can also be used.

Unperforated films can also be used so long as they are light and flexible enough to offer little resistance to the incident wave. Thin plastic bags, for instance, can completely surround a material. In the case of mineral and glass fiber, plastic bags protect against degradation by moisture, pollutants, and oils while giving them the ability to be cleaned. Additionally, they allow for them to be moved without detrimental effects to the materials or those moving them. These films or bags must not be attached firmly or consistantly as this could affect the unit's overall absorptive effect.

Painting absorbers such as acoustic tiles can be tricky and the manufacturer should always be consulted. Even if a manufacturer states that a material can be painted the results of this on absorption should be known beforehand. Paint is, in effect, a rigid nonlimp film and will cause many materials to reflect rather than absorb sound. Some tiles are purposely made with large openings to allow for painting. They often have 15 to 20% open area which prevents the bridging or filling in of the hole by normal painting.

Maintenance of absorptive materials is very important. Spray-ons can crumble and are not easy to clean. Thin perforations can clog from oil and other pollutants. Obviously anywhere that impact or abrasion is possible the absorptive material should be checked regularly. Mountings are very important. Manufacturer's ratings are calculated using very carefully followed mounting methods and thus they should be consulted in regard to the proper mounting for desired effect. Most materials may be attached directly to a hard impervious surface but are often suspended (hung ceiling) on runners or furring strips leaving an air space behind them. This air space has the effect of increasing low frequency absorption but can also decrease high frequency absorption with a crossover point at around 500 Hz. Depending on the materials used, air space can vary from 2 to 16 inches. After 16 inches there is little added effect. Due to this lab tests are most often performed with mounting method #7 which is a mechanical suspension system with 16 inches of air space behind it.

Diaphragmatic Absorbers

When a sound wave falls upon a panel, the panel vibrates at its natural period. Panels are moved more by low frequencies than high and in fact most high frequencies are reflected off the panel, thus the term bass trap. Because no panel is perfectly elastic some friction takes place. In addition, energy is also lost due to internal damping in the panel as well as damping due to the means of panel support. All panels have a resonance frequency and in a resonant condition the amplitude at which they vibrate is greatest. Depending on the panel's size and other charcteristics affecting its damping it will transmit into the trap most of the energy hitting it at its resonant frequency and higher order modes and are most effective between 50 and 500 Hz. Panel absorbers can include sheets of gypsum board, wood paneling, windows, suspended ceilings, ceiling reflectors, and even wood platforms (such as staging). Thin sheets of plywood, plastic, metal, or even paper can be used as diaphragmatic absorbers. There are vacuum-formed ceiling panels of vinyl, damped sheet metal, and plywood assemblies used today. One unit is of molded fiberglass no more than $\frac{1}{8}$-inch thick. It is shaped into pyramids, is approximately 2 feet by 2 feet, and used as a hung ceiling providing air space behind them. They have good broadband absorption with equal coefficients from 125 to 4000 Hz. Because diaphragmatic absorbers are more absorptive at lower frequencies they are used to supplement other materials or to affect specific low frequency problems. Since it is often too expensive or impractical to utilize very thick amounts of fibrous materials for low frequency absorption, these bass traps are highly cost-effective. Combinations of diaphragmatic panels and fibrous materials on their facings often give exceptional absorption. The distance a diaphragmatic panel is placed in front of a hard-walled surface is an important factor. Here the air space acts like a spring boosting both its absorption and its resonance frequency range can be extended via the insertion of porous absorptive material in the air space between the diaphragm and the hard surface. It is important to know the effects of diaphragmatic absorbers as a seemingly unabsorptive wall during the design stage can cause a great deal of low frequency loss to a studio or a control room. Along with air space the absorptivity of diaphragmatic panels varies with mass and elasticity of the panel and is hard to calculate. While resonance frequency can be predicted (see Equations 18 and 19, Chapter 17) most units are built, tested, and adjusted on an experimental basis for optimum absorption.

Reactive or Resonant Absorbers

Resonators like diaphragmatic panels are most effective at low frequencies. They are basically perforated panels with an enclosed air space behind them which may or may not include absorptive materials. In order to understand how reactive panels work, it's simpler to describe the function of a single opening (see Fig. 18-3). When a sound with a large wavelength hits the resonator, it causes the air in the neck of the enclosed cavity to vibrate back and forth. The air in the cavity acts like a spring. The cavity/channel combination tunes to a specific frequency much as a bottle sounds when air is blown across the top. The resonance effect has a very narrow bandwidth.

There are many ways to both vary the resonance frequency and broaden the bandwidth. The cavity size and hole diameter as well as length of the hole affects the resonance frequency as can be seen from Equations 15 and 16 (Chapter 17) for perforated resonators and Equation 17 for slotted resonators. In addition, by placing absorptive material inside the cavity, just behind the panel, the bandwidth is effectively broadened. Placing absorptive material inside the holes or channels will increase this broadening even further. Panels have been made with both varying thickness causing different channel lengths and tunings as well as varying hole sizes. Here the larger holes and thicker panel sections tune to lower frequencies while the opposite is true for the smaller holes and thinner panel sections. Even with these variations the frequency range of resonant absorbers will not exceed much over 400 Hz.

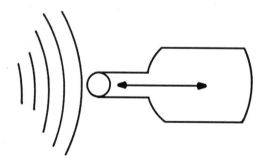

FIGURE 18.3 Resonant or reactive (Helmholtz) absorpter. When wave coinsidence occurs, air in the enclosed cavity vibrates back and forth moving in and out of the channel tuning to a specific frequency much like air blown across the top of a bottle.

In situations requiring a large amount of narrow-band, low frequency absorption or when utilized to supplement other absorptive materials, resonators are very useful. Resonators can be attached to ducts or other structures where some low frequency emmission needs attenuation. The resultant sound reduction can be augmented by the use of a layer of porous materials. The design of such resonators while aided by the previously given equations for resonance frequency is still basically experimental as the exact bandwidth and amount of attenuation are difficult to predict. Concrete blocks with slots cut into their facing form a channel that links the sound to the cavity, which in this case is the hollow section of the block. Most concrete blocks are already absorptive and this slot augments any peak in low frequency absorption. Additionally, if absorptive material is placed inside the cavity the absorption is broadened. These units generally work well from 100 to 300 Hz. However, many have fairly wideband absorption.

Soundblox from the Proudfoot Co. are manufactured near the job site by Proudfoot franchises thus eliminating expensive shipping. The molds are provided by Proudfoot and fit standard automatic block machines. Therefore, uniformity is insured and units are available across the country. Soundblox have a cavity/slot resonator construction and in some cases utilize metal septa or dividers to split the cavity and porous materials can also be added. Because the cavities are closed at the top they are effectively resonators. Some slots are funnel shaped for better acoustical coupling to the cavities. The metal septum divides the cavity into two, each an octave apart in tuning, broadening the absorber's bandwidth. Soundblox are load-bearing structural units and are installed as common cinderblocks but must be laid on a full bed of mortar for their rated transmission loss performance. These units come in an array of sizes, facings, slot, and cavity sizes. Perforated ceramic tiles which are often filled with porous absorptive materials also offer a broadened frequency range.

Noise Control

So far we've only discussed the adjustment or acoustical treatment of sound within a confined area. Dealing with unwanted sound entering or escaping from a room is considered noise control. Again we are dealing with the same source-path-receiver situation, but here, since the source is causing unwanted sound, first try to control it there, either by enclosing, decoupling or isolating it vibration-wise. Next, should this not prove effective enough, move on to the path, dealing with the direct sound via barriers and the reflected sound with absorption. Finally if all else fails the receiver can be modified (as in situations such as with factory workers) via earmuffs or actually enclosing the receiver. This process involves a strict step-by-step procedure. Octave band measurements of the area are made to determine the present amount of absorption vs. frequency. Next the source or sources of the offending sound must be located. It is then determined if the sound can be controlled at the source, along the path or at the receiver as well. Normally sound takes many paths and therefore there are equally as many solutions. Airborne sound can be direct or reflected and can pass through or around boundaries and objects fairly easily. Airborne sound can even sneak through pipes and air ducts. Vibratory sound can readily pass through floors and walls. Thus before one can begin building a studio the desired degree of quietness must be known as well as all factors about the location in terms of outside noise, the buildings structural limitations, all piping, duct, and electrical paths as well as any mechanical equipment nearby.

Remember the sound absorption ratio was determined by the ratio of incident minus reflected energy to the original energy. Now the portion of incident not reflected or absorbed but transmitted is important. This fraction of the incident energy is equal to the transmission coefficient and from this you can derive the amount of transmission loss to be expected (see Equations 20-25 in Chapter 17).

Sound barriers are acoustic materials that reflect, contain, or isolate. Their function is to stop or block acoustic energy and the transmission of airborne sound from one place to another. Barriers should have high density, be impervious, and limp. Common barrier materials include heavily loaded plastics, lead sheet, steel sheet, concrete, gypsum board, rubber synthetics, and sheet asphalt. Barriers have natural vibration modes or resonances which are determined by the structure's size, mass, elasticity, and means of fixation at the edges. While all resonant modes are important the fundamental is the point of least isolation. Below this the barrier is essentially a spring so that as frequency increases, isolation increases. Walls, ceilings, floors or partitions must provide enough isolation to achieve the desired level at the receiving area. Since a control room must have a background noise sound pressure level of only 20 to 30 dB and there are expected to be exterior noise levels of 110 dB, the amount of transmission loss provided by the structures inbetween the two must reach 70 to 80 dB! A good sound barrier must not only be heavy but also airtight. If only 10 percent of a wall is open it will make absolutely no difference what the rest of the wall is made out of. Thus sealing is most important in order for a barrier to reach its full capabilities. Noise will flow out of any opening and much noise can flow out of a very small opening (see Equation 28 in Chapter 17). Barriers should be overlapped at the seams and should be caulked and filled. Remember partitions are seldom just walls: Electrical outlets, television cables, telephone cables, and access doors can wreak havoc with a walls' airtightness. All caulking used should be nonstaining, nonshrinking, and have good bond strength and a long elastic life.

More force is needed to accelerate a wall at higher frequencies than at lower. The mass law predicts that transmission loss increases 6 dB per each doubling of mass or frequency (see Equation 29, Chapter 17). In real life this increase is closer to 4 dB and is primarily effective at high frequencies. Limpness or stiffness on the other hand is also important as the resonance of a material is determined by mass and limpness. Noise level entering a room decreases with mass and increases with stiffness of the walls. There are definite frequency limits to the applicability of the mass law to real-life materials. At low frequencies a wall cannot be considered on the basis of mass alone: Its rigidity also has to be taken into account. In addition, most walls are constrained at the edges with most movement occurring in the center. Not only do stiffness and resonant modes of vibration cause the wall to depart from the mass law but so does the coincidence effect.

A wall's movement, though slight, is very complicated. While in motion a shear wave occurs. When this shear wave is a component of the wavelength of the incident sound wave transmission loss decreases. If a wall is driven at its resonance it transmits sound, additionally, at every frequency above a certain critical frequency. There is a particular phase angle of incidence where even a very large panel vibrates as if it were in resonance. Thus walls have a coincidence drop in transmission loss at a critical frequency where the wavelength in air equals the bending wave of the structure set in motion. There is always some angle and frequency at which a sound wave can match a bending wave in a solid. Double walls avoid this coincidence effect and their transmission loss approaches the sum of both if well separated. These walls should not be exactly the same as then there can be resonance between them.

The Sound Transmission Class (STC) rating is usually reasonably accurate and is based upon careful measurements in controlled lab environments. Yet, like absorption coefficients, dependence on these values in the field can lead to problems as buildings are a collection of systems susceptible to the weakest link in the chain. For an STC of 40, not only do the walls have to meet this rating but also the ceiling, too, or sound will simply travel up and over the wall. Partitions should be carried from the structural floor all the way to the structural ceiling. Where this cannot be achieved the plenum space must have equal integrity. Most important in these ceiling passages is the framing around pipes, conduits, and ducts that pass through from one area to another.

Predicting transmission loss is very difficult: Actual partitions must be tested in reverberant rooms and this is even more straightforward than the tests for absorption coefficient. ASTM Standard E 90 "Standard Recommended Practice for Laboratory Measurement of Airborne Sound Transmission Loss of Building Partitions" dictates the test procedures. The specimen wall is mounted between two reverberant rooms with a great deal of precision so the only sound path between the two is through the specimen. One reverberant room is excited with sound and the sound pressure level in both rooms is measured at $16\frac{1}{3}$ octave bands from 125 to 4000 Hz. The transmission loss is then calculated (see Equations 21-24, Chapter 17). Many people prefer to compare a single number as opposed to a graph of 16. Here the STC comes into play. ASTM Standard 413 "Standard for Determining of Sound Transmission Class" rules. The resultant TL data is compared to a standard curve and the way in which the deviations from the standard are handled makes this single number rating a convenience. Yet it would not be used for choosing one barrier over another. Transmission loss results from ASTM E 90 are much more accurate as it calls for the specimen to be a minimum of 8 feet with expected doors, windows, etc., of normal size. In other words the specimen should be exactly the same as used in the field. While most manufacturers publish reliable data it still is difficult to achieve those ratings in the field. These ratings are an accurate comparison, however. Ratings may vary if different mountings are used. Even different metal studs can cause different ratings. For example, all gypsum wallboards may look alike but their performance does vary.

Barrier Materials

Lead is primarily a barrier. Lead sheet and composites are dense, limp, and impermeable giving them the ability to block sound. Sheet lead is maleable, readily bent, folded, and formed. It easily conforms to irregular surfaces, does not spring back once formed, and is quickly cut with no special tools. Joints can be folded by hand, taped, stapled, or crimped. Lead readily accepts adhesive cements, tapes, or pressure-sensitive tapes. In cramped plenum spaces it can be difficult to fit conventional barrier materials around pipes, ducts, conduits, etc. The ease of use in terms of cutting and forming sheet lead make it very simple to install. Thus a quality installation becomes less expensive labor-wise. Because all openings are detrimental, lead's formability greatly helps block even the smallest leaks around the smallest of wires which often penetrate through plenums. More rigid barrier materials are not as conducive to this work. Sheet lead is specified in pounds per square foot and ranges from ½ pound (1/128-inch thick) to 8 pounds (⅛-inch thick). Lead-loaded vinyl or lead powder-loaded vinyl can be coated with glass fiber, nylon, cotton, or other fabric backings and are again specified in pounds per square foot (½, ¾, 1, 1½, 3, or more) and are acoustically completely limp. Hung as curtains, folding doors, blanketing, or pipe and duct wrap, they are more expensive than plain sheet lead. Their advantage in terms of colors, textures, and decorative patterns with more durable finishes make them very helpful to designers. Lead-loaded plastic materials make good barriers and are easy to install since they are generally very flexible. They can be made into curtains by simply making a frame to hang them on. They can even be transparent. Lead is not expensive in sheet form and due to its limpness and natural damping qualities it is not affected by coincidence and follows the mass law relatively closely.

Sheet lead or loaded vinyls whether lead-loaded or loaded with other heavy materials, can also have foams added thereby adding sound absorption. As can be seen from equation #21, additional absorption material in the receiving area is very beneficial to transmission loss. As an example Carrol George, Inc. manufactures a loaded vinyl with foam adhered to both sides. While only 0.6 inches thick it yields impressive Tl ratings of,

14 dB at 100 Hz

14 dB at 125 Hz

14 dB at 160 Hz

13 dB at 200 Hz

18 dB at 250 Hz

17 dB at 315 Hz

21 dB at 400 Hz

20 dB at 500 Hz

22 dB at 630 Hz

26 dB at 800 Hz

27 dB at 1000 Hz

29 dB at 1250 Hz

32 dB at 1600 Hz

35 dB at 2000 Hz

38 dB at 2500 Hz

38 dB at 3150 Hz

41 dB at 4000 Hz

44 dB at 5000 Hz

The above ratings were measured by Riverbank Acoustical Laboratories and as an example of the depth of information provided by lab reports is shown by the following quote.

Description of the Specimen

"The test specimen, 48 inches (1.22 m) wide and 96 inches (2.44 m) high was mounted directly into the laboratory test opening, attached to a frame, and then sealed with a mastic over the entire perimeter to the test opening. The specimen was constructed as follows: a 1 pound per ft.^{2} (4.9 kg/m^2) filled vinyl noise barrier septum faced on both sides with a $\frac{1}{4}$-inch (6.4 mm) thick 2 pounds per ft.^{3} (32.0 kg/m^3) density urethane foam adhered to the septum. The specimen had an overall thickness of 0.6 inches (15.2 mm) and weighed an average of 1.09 pounds per ft.^{2} (5.32 kg/m^2). The transmission area, S, used in the computations was 32 ft.^{2} (3.0 m^2)."

When using lead or loaded composites one should avoid rigidly fastening to stiff surfaces. Viscoelastic adhesives or intermittent fastening should be used. It is more effective to hang lead sheet limply between two walls as opposed to rigidly fastening it to one.

Other barrier materials include brick, gypsum board, wood, concrete, steel, and glass. In all barriers, the more porous the material the less transmission loss yielded. Mastic cements are heavy, dense, flexible asphalted materials with elasticity ranging from semirigid to flexible. They are used for barriers and damping in hollow doors, appliances, walls, etc. Spray-on materials can be applied into a cavity, reducing noise by damping the structure and filling in any leaks. Adhesives can bond any material as long as the manufacturer's recommendations are met. Sealants like adhesives come in room temperature cure, hot melt elastomers, and tapes with preapplied pressure sensitive adhesives. Window glass also offers barrier properties that vary according to strength, thickness, number of panes, and construction methods. As an example, single-strength windows offer an STC of 19; double strength 21; $\frac{1}{4}$ inch, 26; $\frac{1}{2}$ inch, 31; and 1 inch, 34 to 37. Along with these properties one must also consider edge restraints. Resilient edge restraints can provide as much as a 6 dB improvement over stiff edge attachment.

As noted in the section covering observation windows (Chapter 1), the strips of rubber or neoprene underneath and bearing the weight of the glass should not compress more than 20 percent under the load, but the side and top resilient mounts can be more pliant. After witnessing many window-seating experiences that turned out badly, I can offer the following tips. Beveling the edges of the glass can save a lot of trouble. Since it takes several individuals to hoist the window into place (often this includes any musicians who are available at the time), beveling is a good idea as it helps avoid cut hands. Those same sharp edges can also slice and possibly defeat the advantages of using resilient isolation strips.

The best bet is to install edgework such as those used for decorative protective trims, sealing channels, weather stripping, or even the type of U-shaped edge capping normally found surrounding the edges of helmets on those glass panels from the moment you accept delivery. These are simply pushed on, right over the raw edges. They come in dual durimeter sponge or solid rubber and have flexible wire or steel-segmented cores which make them strong and resilient. Additionally they'll grip and hold to any contour without your needing to pull or stretch them around those dangerous edges. They render the added benefit, upon installation, of helping to avoid the "crowding" of the glass within the framework which can cause cracks and actual breakage. This usually occurs some time after the installation when the system has shifted or settled into place. More than once I've witnessed a studio owner in the midst of showing off a newly installed control room or isolation booth window only to sadly find the glass panel cracked. It's important to note that latex foam takes less of a compression set and for that reason it is more forgiving than sponge rubber seals, especially if your framework is a little distorted, that is to say, not perfectly plumb.

Obtaining these seals poses no challenge as they are readily available from most automotive parts suppliers, yet by getting in touch with manufacturers such as Cooper-Standard Automotive or Lauren Manufacturing Company (both are listed in the manufacturers of acoustic materials appendix in the back of this book) you will get some very educational material. The latter company also manufactures complete doors and windows, and both offer door seals, gaskets, and various elastomeric products such as strips and flexible cellular materials that can aid in solving vibration isolation problems.

Normal glass may not have the best transmission loss ratings, but when double or triple glazing is used the TL increases. Here an air space of 1 inch or more is used, but any more than 4 inches of air space provides negligible results. Laminated "acoustic glass" provides good isolation as does safety glass which has multiple layers of thin glass laminated with thick soft layers of poly vinyl-butyrate plastic.

For example, 1-inch glass with $\frac{1}{2}$-inch airspace provides an STC rating of 32, $\frac{1}{4}$-inch safety glass, 31; $\frac{1}{4}$-inch laminated (with plastic) 38 to 40. Two panels of different thicknesses can prevent resonant and coincidence effects. Sealing is also important. Spacings of greater than 1 inch or laminated glass can provide good TL ratings; however if the window is openable TL could decrease as much as 10 dB, even when closed. If windows must be openable, use dual cam or catch methods, apply two to three edge gaskets, and close them tightly with as much pressure as possible. In fact, if windows are openable only with the use of a special wrench or tool it's even better.

If a barrier or wall is made up of different sections with different TL ratings such as doors and windows it is important to have an idea of its average transmission loss. This averaging takes the transmission coefficient of each section multiplied by the area of that section and divides the sum by the total area. This method can lack some accuracy due to both real-life deviations from lab conditions and the effect of the very poor TL of the weakest link. Curtains which are often loaded plastics or vinyls offer good strength and barrier performance while remaining flexible. This flexibility aids sound TL by reducing coincidence and causing them to behave close to the limp mass law. However, since their tops are often open, noise reduction is not maximized. Furthermore, during lab testing their edges are sealed while in actual usage this may be impractical as access is often important. Therefore even with curtain edges overlapping TL ratings will be lower in the field.

Partition systems are used to divide a room temporarily or semipermanently. They are rigid and come in all types of materials, sizes and finishes. They are most often used in open plan offices. Operable rigid partitions ride on rollers in a track mounted on the ceiling. Thus they can be easily moved, folded against a wall, or extended out to divide a room. Semipermanent panels can be erected or disassembled on-site but generally require tools. They are made of all types of absorptive panel materials often with air space within the panels and interlocking edges to prevent sound leakage. They come in decorative finishes and allow for various changes in work space usage. Some sound panels for noise control have heavy barrier material in back of a porous material faced with perforated metal. Thus they fullfill two tasks by reducing reverberation on one side and TL to the other. The thicker they are the more absorptivity and TL they have. Often these panels are 2 to 8 inches in depth. To further increase TL some use heavily loaded vinyl or lead in between but spaced away from the absorptive and backing layers. To further aid in TL the rigid backing side can also have a damping layer added to it thereby reducing radiation from the back side and also help eliminate problems with coincidence.

Enclosures are used to completely or partially surround an offending noise source such as machinery. They are made up of interlocking panels. They provide protection from dirt, oil, and water by adding a thin facing or totally enclosing their porous material in a plastic bag. Enclosures have rigid backing and can include windows and access doors for machine operation and maintenance. They are usually rated in TL per machine as most are custom-made. Due to heat buildup (low sound transmission means low heat transmission) ventilation with attenuated ducting is also often provided.

Quiet rooms are the opposite of enclosures. Here an area is provided for operators, which effectively separates them from the noise of (for example) a factory floor. Thus they can also be used for audiometric (hearing) tests or as recording isolation booths and are similar to telephone booths. They are generally not very effective in shielding against low frequencies however.

Since sound can easily travel along a plenum above ceilings from one room to another, ceiling systems, while mostly absorptive, must also be used as barriers. Therefore many are rated for transmission loss. Acoustical ceilings may not be highly thought of, but offer a wide range of uses. They can be intended for anything from simple decorative to industrial or heavy duty usage. They vary in density and thickness. They can be made of anything from mineral or glass fiber to wood fiber. Their absorption coefficient ranges from 0.3 to 0.9 with an average around 0.6. Most absorption tests and field mountings utilize #7 mounting which has a 16-inch air space plenum. Thus they can also act as diaphragmatic absorbers increasing low frequency absorption as opposed to the results of direct mounting of panels to existing ceilings. If lead sheet or other barrier materials are included, acoustical ceilings can handle the full range of acoustical properties if chosen correctly. To aid in this choice several test methods are utilized. ASTM C 635 and 636 standards for "Composition and Installation of Metal Ceiling Suspension Systems for Acoustical Tile and Lay-In Panels" as well as AMA 1-11 Ceiling Sound Ratings which gives the attenuation of sound from one room to another through the ceiling and via the plenum path above the ceiling are extremely useful.

Vibration

A detailed discussion of vibration isolation is beyond the scope of this publication, yet since the idea here is to prepare the small budget studio builder/owner for handling the adjustments needed to alleviate problems as they arise, a general knowledge of the theory and terminology on the subject of vibration isolation and a basic understanding of the mechanics involved would prove helpful. There are only two ways of treating vibrating structures: vibration isolation and damping. Isolation separates a vibrating structure from the receiver while damping reduces the intensity of vibration, thereby quieting down the noise problem.

Damping converts the mechanical vibrations into heat. The qualities of damping include internal friction, viscosity, turbulence, and acoustic radiation. Once the energy is dissipated it can no longer be radiated. Thus damping reduces the amplitude or the impact energy of the noise. Damping materials are generally viscoelastic such as rubber or elastomeric plastics. To help in understanding the effects of damping, imagine hitting a piece of metal with a hammer. Now cover all or part of that piece of metal with a thick, heavy piece of rubber and then hit it again. The sound produced will not only last a shorter length of time, but will have less amplitude. Damping is primarily effective only at or below the dampener's resonance so that a vibration correction plan will depend on whether the excitation frequency is less than, greater than, or equal to that resonant frequency. Damping ratio is also a key factor since, should the system be overdamped, the response will be slow, causing its return to equilibrium to be sluggish. Yet if underdamped, the system response will be fast, but would continually overshoot the equilibrium position until friction finally caused the oscillation to die out. In between these two amounts of damping lies what is called "critical damping" and here the system returns to equilibrium as fast as possible without the repeated overshooting.

As stated vibration isolation separates noise sources from structures, thus reducing the vibrations radiating from the source. Basically all you are doing here is placing some form of flexible material between that which is causing the vibration and that which is receiving and transmitting the vibration. Simply put, something is causing the disturbing vibration and its frequency in either cycles per minute (CPM) or per second (Hz) is one of the most important factors in determining what kind of vibration isolator to use. The source of the vibration is generally easy to locate as the resonant frequency of the vibration is directly related to it. For instance, if the cause of the vibration is a running fan's eccentrics, then the vibration's frequency will be the same as that of the fan's rotating speed. This is called harmonic excitation.

It is extremely important to insure that the lowest vibrating frequency is used when determining the correct isolator. Another important factor is the natural frequency of the isolator itself which is determined by its own spring rate along with the weight of the load that's being supported. No need to be alarmed here as most manufacturers of vibration isolators list the natural frequencies of their products. You should look for an isolator that has a natural frequency that is at least one-third that of the vibration you are trying to isolate. For that matter, the lower the natural frequency of the isolator, the more pronounced becomes that isolator's effectiveness.

Your first step in dealing with correcting a vibration problem (just as with airborne noise control) is starting your approach by first considering the three main points of attack, right at the source, along the path, or at the receiving end. Unlike a situation on a noisy factory floor, isolating the receiver that is in a studio's control room is not a practicality. The path may still be a possibility, but if it is through the building's foundation that may be way out of the question cost-effectiveness-wise. So as always it's best to go right to the source of the problem. First, see if you can alter this source by making it more structurally rigid, try balancing it, possibly changing its mass or "de-tuning" it in any other way. For instance, a small vibrating motor that causes a maddening amount of noise when it's placed on a wobbly wooden stand, produces an almost tolerable amount of noise simply by affixing it to a thick concrete floor. The improvement is obvious even before any vibration isolation mounting scheme is utilized. Think of a washing machine whose foot lengths are adjusted incorrectly. Can you say "Hello, spin cycle!"

VIBRATION ISOLATORS

Decoupling the noise source from the path via the use of vibration isolators is the usual chosen method in dealing with correcting a vibration problem. In general, vibration isolators can be broken down into three categories: metal springs, elastomeric mounts, and resilient pads. The advantages of metal springs are as follows: They are resistant to environmental factors such as temperature, corrosion, and solvents; they do not creep or drift out of place; they are designed to handle maximum deflection; and they work well at low frequencies. The disadvantages include their transmission of high frequency vibrations as well as those at their resonant frequency, and that they can be most troublesome if allowed to take on a rocking motion. You can get around these problems fairly easily. The damping lacking in springs can be augmented by using dampers along with them. Rocking can be avoided by the use of an inertia block that weighs one to two times that of what the springs are isolating. High frequency transmission can be blocked by simply selecting and adding rubber pads in series with the springs.

Elastomeric mountings are most often made of natural rubber or synthetic rubberlike materials such as neoprene, which has the added advantage of being more desirable in harsher environments. In a controlled environment natural rubber is one of the best and most economical isolators as it has an inherent damping and is perfect for machines that operate near resonance or pass through resonance when they start and stop. Rubber used for isolation is rated in durometers (30 being soft while 80 is considered hard). Often, when dealing with very specific or complicated problems, multiple isolators of differing hardnesses can be used to take advantage of the desirable properties of each.

Isolation pads include materials like cork, felt, and compressed fiberglass. These can be obtained in sheet form, cut to size, and stacked to achieve varying degrees of isolation. Cork, like fiberglass, is not susceptible to solvents and can handle a wide temperature range. On the other hand, felt, being made of organic material, can be very sensitive to solvents. Fiberglass isolators are not very useful at frequencies below 20 Hz and cork materials do have a tendency to crumble with age.

Inertia blocks such as isolated concrete slabs are very useful in vibration isolation. They, along with steel structures of sufficient mass, limit motion by overcoming the inertia forces that are generated by the equipment mounted on them. They lower the center of gravity, thus adding stability, and increase mass, thus decreasing the vibration's amplitude along with minimizing any rocking motion. In addition they act as noise barriers between that which is mounted on them and the floor below them. They must be mounted on isolators and they can be fairly cumbersome and costly to construct, especially after the studio has been completed, unless as used in my favorite place outside the studio complex and under your HVAC equipment.

HOW THE PROs DO IT

Professional vibration engineers study and deal with a vibration problem starting with a very simple model called "a single degree of freedom" which takes into consideration only the vertical displacement of the system. This very basic approach is used to point them in the right direction as far as seeking a cure to the problem. In order to actually accomplish this, they also take into account: system mass, damping constant, spring constant, position, velocity, acceleration, inertia force, viscous damping force, linear elastic force, and external forces, including impulse and sine wavelike repeated and random forces. If all this is much more than you want or need to get involved with, take heart, you've got a whole slew of allies at the ready to help you out. Start with the offending machinery. It's no easy task to weigh a several-hundred-pound AC unit or figure out its compressor pump's rotation speed and the same goes for a fan mounted in a wall just below the ceiling. Yet, copying down the manufacturer's name along with the unit's model number should be a lot easier and don't be surprised to find that they stock isolators for that particular device as part of their product line, let alone their having at the ready the exact weight or rotation speed information you seek.

MORE HELP

Even if the manufacturer has been out of business for several years or might as well be because the help they are able to give you is nonexistent, you are still not alone as there are all kinds of trade organizations just falling all over themselves trying to give you very helpful information. I kid you not! It is important to take advantage of this wealth of information because it's often based on many years of study plus practical experience, and it is readily available. As an example, when dealing with air conditioning, vents, or any HVAC (heating, ventilating, and air conditioning) -related noise it's wise to start by contacting an organization such as the Air Movement and Control Association International, Inc., 30 W. University Drive, Arlington Heights, IL 60004, (847) 394-0150, (847) 253-0088, *amca@amca.org*. Believe me, there is an almost unlimited number of trade organizations which make available information having to do with the control of noise. You can prove this to yourself simply by spending a little time in the local public library, especially if you get the help of the librarian on staff, and most of it is free for the asking! For a start, try the National Council of Acoustical Consultants (NCAC) at *www.ncac.com*, the Acoustical Society of America (ASA) at *www.acoustics.org*, Institute of Noise Control Engineering (INCE) at *www.allenpress.com/inceusa/*, and the American Society of Mechanical Engineers (ASME) at *www.asme.org*.

The manufacturers of vibration isolators themselves are also a great source of help in the search for the right product to alleviate a particular problem as they are quite frank and honest about their products' capabilities. Let's say for instance you need isolators suitable to use in isolating your large, in-wall, soffit-mounted speakers from that wall itself, because otherwise you'd end up having vibration-induced low frequency noise being picked up by microphones in the studio whenever you monitored at loud levels while recording (it happens). A manufacturer's catalog such as the one from Barry Controls would steer you away from their series 1000 through 4000 "Cupmount" mounts. While offering good protection against shock and structure-borne vibrations, the manufacturer plainly states that these units provide good vibration isolation characteristics at frequencies above 40 Hz. This is not exactly your best choice if your speakers' low frequency response can handle 32 Hz or extend down to even below 20 Hz.

Not convinced? Okay. How about a manufacturer steering a customer away from one product to another less expensive solution because the latter would do the job better?

Well, that actually happens. I had called the Airspring division of Firestone Industrial Products Company to inquire about one of their "airmount" vibration isolators which reduce structurally transmitted noise very well. The plan was to float a mobil control room on top of computer room type of flooring. The size of the floor was just under 150 square feet, almost 10 feet × 15 feet. The floor panels were 2 square feet each and totaled 36 in number. In order to place an isolator underneath each support 48 mounts would be needed.

Next came a little time going through that manufacturer's product line catalog, which contains all the mathematical equations and data in the form of graphs, tables, and charts that are needed to make the proper selection. Later when I called their distributor to ask about pricing to get the complete picture, they asked for the technical requirements involved (vibration frequency, load, and size) and came back with a another alternative. You guessed it, the mount they recommended I use did cost more. Almost twice as much in fact. However that wasn't their main point.

What they stressed was that I go with a single common floor, either by supporting the computer room flooring on a continuous platform or on something like steel I-beams. This would not only give the structure more lateral stability but also allow for the use of fewer larger-sized mounts. While the recommended mounts were almost twice as expensive, the alternative plan required only one-third the number of mounts thus bringing the actual cost *down* by a third! The trade-offs? None. The original mounts would support 48,480 pounds inflated at 60 psi and would render room isolation down to 8.28 Hz. The recommended mounts would support the exact same 48,480-pound load filled at the same 60-psi pressure and would isolate down to 7.2 Hz. These folks ended up saving me money. I, too, was skeptical so I double-checked all the above figures. When I recently called again to confirm those figures everything, including the pricing, had not changed.

Oh I should mention that in situations like this I never like to use the Internet, at least after the initial contact. A phone call may cost a little bit more but that one-to-one, easy give and take, with less of a problem explaining yourself and getting feedback from a real live human being, cannot be duplicated via computer screen interface. A hand-written note of thanks may not be quite as fast as E-mail, but which would you prefer getting? All this free information and help that's now available makes a huge difference for the musician or any nonacoustician who is trying to build a recording facility that's capable of delivering a professional quality product sound-wise yet still doesn't break the bank. It's one thing to be aware that something is not sounding right, but it's a whole different thing knowing how to go about achieving the correction. It seems as if all the development that has taken place over the past 25 to 30 years in the audio industry has been leading up to a whole new non-secretive approach to acoustics, right now at the turn of the century.

You'll never understand what a big deal this is unless you once tried, back in the late 1960s, to explain to a bona fide professional acoustician/noise-control engineer you hoped to glean some info from that, no, it would not be possible to "simply" decrease the electric guitarist's amplifier level or "for that matter remove" one of two 4- X 12-inch speaker cabinets that comprised his stack, and "yes, sir," the electric bass guitarist did feel it was appropriate to "thump" his fingers against the instrument's strings thereby causing those low frequency shock waves of vibration. See how the girls dancing in front of the stage over by that dual folder horn bass rig get more active when he plays in that percussive manner?

CONTROL ROOM CONSTRUCTION

To consider a recording studio control room an accurate tool for judging the integrity of music reproduction or the sufficiency of dialogue intelligibility, several requirements must be met. Among these are its having a short reverb time, high sound isolation from the outside as well as from other rooms in the facility, and a minimum amount of ambient noise emanating from within the control room itself. Going about achieving adequate vibration and noise control is no mystery, but early planning and close attention to detail will help avoid those complicated and costly down-the-road construction revamps.

All one has to do is follow a few simple procedures: always use proprietary acoustical doors that have cam-lift hinges and extra heavy-duty compression seals around their full perimeters with solid latching hardware to tightly compress those gaskets. Hydraulic and electrically actuated compression, or even assemblies utilizing magnetic closures, are also a possibility though more expensive, yet, without positive latching, the gaskets may not be fully sealed and the resulting sound transmission loss may not match the manufacturer's published ratings. From the top down, your room's roof should be constructed of 6-inch hard-rock concrete with a double-layer gypsum board subceiling. Walls will be adequate of 10-inch-thick grout-filled masonry with two to three layers of gypsum board mounted on furring strips which render a minimum of 4 inches (and up to 16 inches) of air space from the masonry.

This box-in-a-box construction is mounted on a steel-frame flooring base that is itself mounted on spring isolators. This structure also requires an isolation joint around its perimeter with *no* rigid structural connections between it and the rest of the world. This means that no electrical conduit, ventilation duct, or wire-way tubing may be firmly clamped or affixed to the structure unless those clamps themselves have resilient bodies or utilize compliant rubber inserts. Ventilation intake and return lines must be of a flexible ductwork type with its outside properly wrapped with sufficient batting and sealed to preclude any sound from the outside leaking into them, while internal baffles should be employed to stop any wind, fan, or other HVAC-caused noise from being transmitted through them into the room.

Unfortunately, the scope and budget constraints of small studios generally precludes a total sound and vibration isolation scheme. The limited ceiling heights commonly available also hamper the usage of both floating floor and dropped ceilings. The weight of the equipment that is normally found in the average recording studio control room (including console, tape decks, outboard equipment racks, and large reference monitors) can easily come to thousands of pounds, requiring the use of sophisticated isolation mounts. Costs add up very quickly, yet, by using a little ingenuity, along with some of today's modern materials and construction methods, and by taking advantage of the huge number of new suppliers of acoustical products which now include prefabricated kits, putting together a professional-quality recording facility for under six figures is possible especially if you are willing to put in some old-fashioned elbow grease on your part.

A DIFFERENT APPROACH

Let's start with noise reduction and vibration isolation. This tends to be the hard part and it seems it's what is most often ignored. Taking the above box-in-a-box construction method which works just as it is supposed to, let's see what can be done using a modern approach to reduce the costs, thus making it more feasible for the budget studio builder. Remember that isolation-mounted computer flooring example? Well, it used to be that thinking about using computer-room flooring in a budget facility was simply out of the question due to the expense. Today it seems companies start up and fold like photo-development huts and almost everyone has to have a computer room. Come auction time, there's not a lot of interest in the flooring. Nobody including the auction house wants to get involved with the complicated disassembly. Keep it under your hat, but the process is not actually that bad at all. Now it's best to talk to the auction people before the bidding starts and explain that you'll be bringing in a professional crew (friends, musicians, and relatives) that will disassemble and *pack up* the flooring in an *orderly* fashion, *clean up* after themselves, and *completely* remove *all* debris, without causing *any* damage to the existing walls and flooring. Believe me, this approach has been known to help win bids.

Once you have it, assemble this flooring on top of a series of panels or I-beams that have been joined together to form a solid structure that's been separated from the existing building floor by springs, pads, or actual vibration isolation mounts. You'll not only end up with an isolated floating floor, but also the greatest wire-way system I know of. Simply by lifting up the 2-foot-square panels that make up computer-room flooring you'll be able to add, subtract, or replace your wire runs at any time without any difficulty. You can also use this space for incoming air conditioning duct runs (remember heat rises) and be able to feed those directly to problem areas such as near (but not directly into, because of condensation) racks that contain heat-producing power supplies and amplifiers when positioning them outside of your control room is just not possible.

The surface of computer-room flooring is fairly reflective at higher frequencies so you might want to add carpeting to the top of the square panels, maybe in a pattern or even randomly, to help in trimming your control room's midrange and highs. Here common carpet samples from salesman's books work perfectly. This flooring unit as a whole makes an efficient diaphragmatic absorber to help fight low frequency buildup and the muddy sound that results from it. You can add plastic bags, loosely filled with fiberglass batting, to help increase this absorption.

Once you've gotten to this point you have the makings of a giant floor-sized Helmholtz resonator. You will need to surround the perimeter to seal or cap this under the floor chamber. Here lead-filled vinyl sheeting would be the best choice, but the budget approach would more than likely have you sawing up discarded automobile tires into strips using them as layers and building up the sufficient thickness for isolation. Even if you stack up those computer flooring panels, you're still looking at a lot of hole drilling, not to mention the fact that sawing steel-belted automotive tires is no small undertaking, but this is all part of that elbow grease stuff that was mentioned earlier.

RESONANT FREQUENCY VIBRATIONS

Vibrations that are at resonance produce a very high amount of vibration which can often cause damage to very expensive machinery, possibly leading to a complete system failure. Should this happen the result would be manufacturing downtime. Therefore, much study has been devoted to this problem and it behooves us in the field of audio, where large-scale funding of these types of studies is not available, to take advantage of this outside work. In this case it was found that the effective mass of a vibrating object as well as its stiffness determines its resonant frequency. Now stiffness can be changed by the use of damping and vibration isolators but should this prove impractical cost-wise, you now have a second often less-expensive and easier-to-facilitate method of reducing the vibrations simply by experimentally adding mass to the offending part of the system until the vibrations have subsided sufficiently. Thank you very much.

Ambient control room noise isn't due only to your HVAC system, but can also emanate from within your electronic equipment as well. These problems can be often cured with a little damping material. Here, a foam of a nonsilicon formulation will avoid off-gassing and breakdown where the foam turns to dust. I once, during a control room checkout, hunted down an ambient noise problem to a vibrating vacuum fluorescent display which was vibrating up against the back of an outboard processor's front panel. Four small pieces of a product called "confor" from EAR Specialty Composites added to the display's corners completely ended a problem which everyone had been sure was an electrical ground loop hum. Another time the same type of problem ended up being caused by an internal "whisper" fan that was rattling from a combination of a missing mounting bolt and its fan having one blade snapped off. Same cure but this time I cut small pieces from a sample of a closed-cell "poron" foam from Rogers Corp. Besides having low outgassing, and being resistant to temperatures, chemicals, and abrasion, these urethane foams are no problem to cut and they have bondable surfaces making them easy to glue. Like most closed-cell foams they display good resistance to compression set, high energy absorption, and excellent all-around dimensional stability, making them suitable for use as gaskets, seals, shock and vibration mounts, isolation pads, and, as in this case, vibration isolators for PC boards or other internal electronic parts that might cause vibrations. You can also cast your own small rubber vibration isolator parts. A manufacturer named Devcon produces an easily pourable castable liquid urethane that cures to a rubber with a varying rigidity depending on the additives you choose.

A company named Raxxess has started manufacturing a sound isolation equipment rack. It has a clear plastic front door that allows you to view metering, power-on status lights, etc., blower-aided intake air vents, vibration-isolated mounting rails, and an interior lined with 1-inch wedge-shaped foam. I haven't personally heard the results of parking offensive equipment inside, but it's a great idea for noisemakers that cannot be removed from the studio. If you were to open up the back and position it in a corner it might even function well as a bass trap. Another method of utilizing equipment racks as bass traps is to mount the rails in those rear or side control room walls which have some space behind them (the more the better). If you leave some of the rack spaces open or better yet cover some of them with blank panels that are perforated or are actually made up of screen material and also line the rear space behind with absorptive material, you'll not only have a very effective bass trap, but one which, due to the equipment's front panels, defuses high frequencies and yet is still useful at its intended equipment-mounting function.

SOUND ISOLATION AND DUAL PANEL WALL PARTITIONS

In 1950, Mr. A. London published in *Journal of The Acoustic Society of America* the findings derived from his scientific studies of sound transmission through double-wall partitions, here meaning walls made up of two layers of gypsum board with an air gap between.

The results showed that depending on the resonant frequencies involved, the two panels along with the air space between them could in fact act much like a spring. This is now referred to as the mass-spring-mass effect and it was found to be not as pronounced at high and midfrequencies. Yet the results of his tests did show that dual-wall partitions could actually be less effective in providing low frequency sound transmission loss than if only one of the two panels were to be used independently. This does not seem logical until you stop to think of that mass-spring-mass model. At and below their resonant frequency, two similar panels, although separated by an air space, will still move in response to each other and this vibrationlike movement is actually augmented by the springlike effect of the air space between them. While not actually amplified by this effect, it certainly could cause frequencies at and below resonance to be transmitted almost as if unimpeded.

Low frequency transmission through walls can be due to mass law, mass/spring resonances, and coincidence effect. No, not coincidence as in chance happenings, but certain angles of incidence at which plane waves excite a wall in such a manner as to allow that sound wave to be transmitted through a wall with a reduced transmission loss. This happens when the wavelength of the flexing or bending wall itself is equal to the wavelength of the sound wave that is being projected onto it. This ends up tightly coupling the two so as to make them act almost as one. Wave coincidence can cause wall movement and therefore the transmission of sound through that wall to be much greater than a wall of that particular mass would normally allow or in other words, result in transmission losses lower than the mass law would normally indicate. For every frequency above the critical frequency there is a certain angle of incidence at which this occurs, but that's enough of the theoretical. Instead, you should seek out a set of curves given by Magrab which plots the thickness of common building materials versus critical frequencies. Here you'll find that the critical frequency of $\frac{1}{2}$-inch common plywood is about 2 kHz while with glass of the same thickness it's closer to 1250 Hz. Of course you can also look up the equation for critical frequency and do all the math yourself, most enlightening, but very heavy on the time consumption.

High frequencies are normally pretty easily impeded by partitions, it's the stopping of low frequency sound waves that is difficult to achieve. Yet, even lightweight double partitions can yield high sound isolation if properly designed and many now are thanks in large part to Mr. London's work. This is exactly why I stress gathering in as much information on past acoustical work as you can because 99 percent of all the work needed has already been done and it's all pretty much free for the asking. You'll find that sound absorptive materials can be installed in the air gap to lessen the resonance. Hey, you can even lower the resonant frequency by making the air space wider. What you're accomplishing is lowering the mass-spring-mass resonant frequency which in turn reduces direct sound transmission. You'll discover that increasing the density of the panels will also yield this same effect and the same kind of results will occur even if only one of the two panels is changed by making it thicker or heavier. If in fact you have two panels of different densities, the partition will perform more effectively as sympathetic vibrations between the panels will decrease.

You'll discover that increasing the density of a panel can be accomplished in many ways, by simply using a denser material or by the building up or adding of additional layers of the same material, and that making sure each layer's seams do not line up with the ones under it will also cut down on the amount of sound transmitted. If, between those builtup layers, you use a viscoelastic core coupling such as NDS damping sheets from Noise Suppression Technologies, Inc., the partition's low frequency isolation capabilities will be further increased. This product even comes with pressure-sensitive backing and is easily cut making installation uncomplicated. You do not have to cover the complete area between the layers of paneling with elastomeric damping sheeting so adding this type of material in between panels in a kind of patchwork design proves more cost-effective. Do the research and pretty soon you'll see structures that leak sound with new eyes that easily lead you to what will become obvious design and construction flaws.

As with some thermal insulated window glazing in which the pane's edges are glued around the whole perimeter to the frame, rigidly affixed boundaries create a sound bridge that allows mid- and high frequency sound waves to pass through in an almost unimpeded manner, pretty much the way mass-spring-mass resonances do low frequencies. This is a form of nondirect transmission which allows sound to go right through the panel layers. Other routes can be via structural links such as the wall joists and studs. Bridging transmission due to these structural links can be reduced by using panels made up of multiple layers as this makes the panel's coincidence frequency higher. By using premade elastomeric stud mounts like those from Sorbothane, Inc., you'll isolate the sound energy by lowering the resonant frequency of the whole wall system and this damping occurs over a broad frequency range. Laminates, or two layers coupled with a viscoelastic core material between them, provide higher transmission isolation than given by the same material of equivalent thickness. Again, not only are you using two layers instead of one (added mass), but the addition of the core material provides considerably more isolation by reducing structural linkage and therefore the passage of vibration from one panel to another. Vinyl material added to the outside or right on top of your walls can yield an improvement. For example, a vinyl barrier product called Prospec from Illbruck, Inc., when installed over a wall, can add up to 26 dB in increased noise transmission loss.

Don't forget any of the other sound leakage pathways often found in studio walls either, including internal wire-ways, connector panels, electrical outlet boxes, windows, door frames, doors, through-wall hanger lags, and bolts, not to mention any flanking paths. With flanking, airborne sound does not have to pass through a wall because it has been left a nice easy pathway above (ceiling), below (floor), or around (adjacent walls) the wall in question.

Recycled plastic sheeting while not any less expensive, is heavier and denser than sheet rock/gypsum, plywood, and particleboard, especially when built up in layers. The same holds true for the recycled plastic two-by-fours. You can't drive nails into them but screws and all kinds of glues handle the fastening without a problem and the whole thing can be disassembled and reassembled over and over without the normal cracking or crumbled edges. If you put things together in the form of preassembled panels, you can end up with a completely movable control room or at least one made of materials that you can reuse over and over again. Experiment with different glues, especially those in the silicon family, as they also add resilience and your disassembly will only require the use of a putty knife, some alcohol, and a screw gun.

I'd also suggest that you find the location of nearby junk yards and even dumps, you won't believe the stuff you'll come up with. An old 50-gallon drum with a speaker mounted just inside of, and facing away from, the lid, whose screw-off opening has been lined with sponge rubber to make a cushion surround for the microphone that will be inserted there, makes an outstanding reverb chamber. Wrap the drum with a sufficient amount of batting and it can, with some internal acoustic deadening via bagged fiberglass, double as an isolation-chamber for a small guitar amp. The idea is not only to be resourceful and imaginative but to base your experimentation on hard and fast acoustical theory.

A great source for fiberglass bagging is your local dry-cleaner. Those suit, or better yet the longer overcoat-sized, bags will only run you about a nickel apiece and the plastic used is super thin, meaning it'll offer no real impedance to any sound passing through it. They come on a handy roll which makes the work of sliding the fiberglass into the bottom and taping that end along with the smaller top hanger hole go more smoothly. Your best bet is to perform this task *outside*, *before* you ever bring that itchy stuff inside your studio. This may seem a bit excessive on my part but I've seen some pretty bad skin reactions to fibers of various types over the years. Believe me it's best to *never* find out that you are susceptible to an allergic reaction. I once had the job of testing battleship P.A. speakers for a government contractor. This manufacturer had its own "anechoic chamber" that was a small 6-foot x 10-foot x 5-foot affair in a room that was actually three to four feet larger than that in every dimension. All the extra room was completely taken up (even under a raised screen floor) by raw uncovered fiberglass. The most accurate microphones at the time were still tube driven, making not only the test microphone very hot but the interior of that chamber about 120°F as sound absorption materials *do* make great thermal insulators. Many days were spent bringing speakers in, hooking them up, testing them, and then carrying stacks of them out of that room. I found out all about exposed fiberglass and its effect on sweaty open pores and it's no joke. To this day I still spec that *all* fiberglass whenever it's used to be bagged, even when it is to be placed in air spaces inside of walls. Hey, someone has to put it in there and somebody might later have to drill, cut, pull cable through, or remove that wall at some point. I think it's only fair to everyone to assume that you might be that person.

Foam works as well if not better as anechoic chamber wedges and it retains that wedge shape infinitely longer than fiberglass (even the board type). Foam comes in a vast array of densities, pore sizes, thicknesses, and colors. The stuff is also easy as pie to work with. Don't believe me? Then have a go at cutting it with one of those electric Thanksgiving dinner turkey meat-carving knives that you can pick up for a couple of bucks the day after Thanksgiving in second-hand stores. You'll find it enlightening. The only problem I've ever had with the use of foam is its ability to capture, retain, and constantly give off odors such as those from cigarette smoke. You know the scent, it's fairly common in all studios that utilize foam for noise abatement. You can get around this problem by painting the foam panels, but that method also diminishes the absorption properties of the foam. You could try and wash it out (it is possible) but foam retains water almost as well as a sponge and the size of most panels makes squeezing the water out close to impossible. As with the square foam screens used for air filtering, once wet they collect funk even more efficiently.

The Dow Chemical Company has recently announced the availability of the Quash line of odorless "sound management" foam planks. These polylefin-based (polyethylene with an interpolymer) materials offer uniform cell size along with accurate control over density. They inhibit accidental ignition from high-heat and even small fires, handle harsh environments, are impervious to most chemicals, and are 100 percent recyclable. All very nice but buried in the literature is the astounding fact that this foam has a rigid, yet soft, structure that does not absorb or retain water allowing it to be readily cleaned. In fact the literature claims that it "can be washed without affecting acoustical performance" and "because these foams will not deteriorate or promote corrosion they can be removed, cleaned, and then reused." Take it outside, hose it down then either let it air dry or aim a fan at it.

The ever-increasing move toward the use of foam, as well as carpet and other fuzzy (but hopefully nonitchy) material in recording studios over the past couple of decades has simplified the process of manufacturing acoustical correction products. At the very same time, the explosion in the number of professional and private recording facilities has made this manufacturing economically viable. As a result a great many companies now offer acoustical products where a couple of decades ago there were only a handful. Of late more and more have also started offering products that are designed to alleviate very specific problems like standing waves, uneven response due to frequency peaks, and excessive low frequency buildup. You can also now purchase complete acoustical combinations of products as a system or kit for your control room. These can be had from several manufacturers, are normally designed to be very easily mounted onto or in front of an existing room's walls, and render, hopefully, fairly predictable results.

However in this modern, monetized world of fetch-it icons and buy-me buttons you do have to be more vigilant than ever before about performance claims from overzealous advertising or media "aren't I a whiz-kid" types that are just out-and-out too good to be true. For example how about these actual quotations from an editorial blurb about an acoustical correction panel system that ended up becoming "a complete solution to great sound" with which "users can convert *any* ordinary room into a *high performance* recording/audio work-space" (my italics). The real sad part is that this product comes from a very reputable company that manufactures one of the better bass traps available along with other intelligently thought out solutions to various acoustical difficulties.

I know this didn't come from the manufacturer because foam or any other material for that matter cannot make up for serious isolation or dimensional design flaws. If any acoustician were to make claims like those in front of a gathering of knowledgeable acousticians such as those in the Acoustic Society of America he might as well wear a pink electrocuted-hair wig, a large red stick-on rubber nose, and a pair of size 28 shoes since they're going to be rolling in the aisles anyway.

While we're at it, let's clear up the all too common inaccurate assumption that materials, construction methods, and systems that prove effective for thermal insulation also function similarly with airborne sound. As an example, triple-glazed windows work wonders thermally and if the widths of the air gaps within were wide enough they would do well with sound, but even then their equal-sized thickness and parallel-spaced panes are the exact opposite of the proper approach to the design for sound isolation.

So if it's truth in your listening environment that you want, you've got to pay close attention to all those not so small details such as smeared stereo imaging, standing waves, flutter echos, low frequency buildup problems, plus that disconcerting ambient noise, long reverberation times, and excessively differing reverb time lengths at different frequencies. You'll have to put proven acoustical correction methods to use such as corner bass traps, diffusers, Helmholtz resonators, or maybe a diaphragmatic absorpter utilizing a closet space behind it. But how do you know exactly what it is that you need? By getting involved with electro-acoustics.

19

ELECTRONICS AND ACOUSTICS

Today, especially with the wide range usage of microprocessor based electronic equipment, many devices are available to help design and measure studio acoustics. While artificial reverberation is no help in controlling room acoustics, it can readily lend itself to usage on the studio side of the glass. Here the use of close microphone placement techniques and very dead studios (with reverberation times that are very short to the point of nonexistence) yield a very dry sound. The addition of artificial reverberation to the signal fed back to the musicians through the cue or headphone system allows them to hear their instrument as it sounds in a more musical environment. Since this environment is electronically produced it can be varied to match both the instrument and the tempo of the composition being performed. Later during mixdown it can be varied again to either yield better separation of instruments in the mix or to compensate for the addition of subsequent instruments recorded as overdubbs. Again, not only do normal acoustics still figure into the construction of the control room, but now the addition of a great deal of porous absorption material in the studio area is required.

Equalizers allow for the flattening out of any speaker response deviations for a flatter frequency response. One-third of octave equalizers generally do this job very well, but if the problem area is either of a very narrow or a very wide bandwidth, parametric equalizers with their tunable Q or bandwidth controls are utilized. Almost every equalizer introduces some amount of delay to the bands it affects. This results in altered phase response. If equalization is excessive as when used to try to compensate for the speaker/room interface or for the acoustics of the room itself, the resultant phase shift often causes a blurring of the sound as a result of phase distortion. Therefore, while equalization can work wonders for speaker anomalies, it should not be used to compensate for inadequate or incorrect room acoustics.

While the choice of the flattest or truest speakers is best for recording studio work, other factors figure in as well. Mounting is very important, as often larger speaker specifications can only be met when soffited or placed behind the wall's surface. Here the front panel of the speaker cabinet and the wall in which it is mounted form a flat plane. Not only must the mounting be strong enough to support the speaker, but it should have sufficient mass so as not to be easily vibrated by the speaker when reproducing higher levels. Often a box-type structure is first built under the speaker and filled with sand to help achieve this mass.

Point source speaker positioning is also common. Here, picture a triangle. The point is the listening position with the two angles at the base of the triangle representing the two speakers or sources. A problem occurs when moving out of the listening position: The ability to hear the mix as an even stereo blend is lost. As most recording work requires the operator to be constantly moving back and forth across the mixing console, this becomes a major problem. It is therefore better to aim the speakers to some point behind the optimum mix position, allowing the operator to hear both sides of a stereo mix across the width of the console. Another problem with the point-source aiming method is that most speakers are not point sources. Any cabinet made up of two or more drivers cannot be considered a point source as different frequencies are reproduced by different drivers moving the actual source across the speaker's face. A popular misconception occurs with the term Time Alignment. Many people feel that this compensates for the effect of multiple drivers and thus renders a point-source speaker that has perfect phase response. This method introduces a delay to one speaker at a crossover point where two drivers would be reproducing the same frequency, thus rendering the speaker a point source at this one crossover frequency. At frequencies above and below this, however, the source moves across the face of the speaker with frequency. The result is a very slight difference in the time it takes for different frequencies to reach the listener. This "time smear," while very slight can blur the sound and since it is a function of time is also classed as phase distortion.

There is one method of getting around this which lends itself to small studio usage. Speaker cabinets housing single dual-concentric drivers can be considered point-source speakers. A dual-concentric driver is a low frequency driver with a high frequency driver mounted right in its centers so the two drivers have the same center points. Most often single concentric driver cabinets are small by monitor standards. They are often used in small studios as near field speakers (positioned close to the operator, usually on stands). This has two additional benefits besides providing point sources; because they are placed closer to the listener, the level at which they are played is reduced thus lessening that room's needed amount of noise reduction or transmission loss materials. Furthermore, since this lower level and direct path to the listener need not fill the room up with sound, the acoustical properties of the control room figure less in the overall speaker/room interface. The result is that many of the almost insurmountable problems associated with control room acoustics can be aided by the use of near field concentric speakers (especially if stands are utilized which if placed correctly can diminish the amount of early reflections bouncing off the mixing consoles control surfaces).

If you spent any time at all working with the acoustical equations in Chapter 17 you have an idea of how tedious and time-consuming as well as how cumbersome they can be. Testing for level vs. frequency and reverberation time vs. frequency is not only more accurate with the use of microprocessor-based test equipment, but often as simple as pushing one button. Obviously each test involves some set-up procedure, yet these are more often than not very simple and also aided by the equipment's on-board computational electronics. While small studios (and for that matter, many acousticians) cannot afford the purchase price of this sophisticated equipment, renting on a per-day basis is not only affordable, but the information gained during even this short-term usage is invaluable. In order to understand the way in which these devices operate as well as the accuracy and value of the information they render, the following examples will be helpful. The Neutrik 3300 Audiograph System distributed by Sennheiser Electronic Corp., 48 W. 38th St., New York, N.Y. 10018-6297, (212) 944-9440, is a digitally controlled audio analysis system. The 3300 is modular in design and although only the mainframe, input, and output modules are discussed herein additional modules are available which will make the Audiograph a complete audio measurement system. Data output is in the form of permanent hard copy graphic plots. This is in fact a laboratory-quality system, yet its user friendly operation along with many automatic, digitally controlled functioning allows for simplified measurement of many significant audio parameters.

The 3302 mainframe is the base module consisting of a mechanical writing unit, power supply, control logic, and system bus interface. The writing mechanism utilizes 68 mm wide by 50-mm scale single charts or continuous paper rolls. Indication is via a fiber-tipped pen driven by a dc motor whose servo positioning system is controlled by the system bus. This system's ease of operation along with accurate measurement results are due to the logic control circuitry of the bus. The studio generator is digitally controlled over its frequency range of 20 Hz to 40 kHz providing a sine wave voltage to the modules that is accurately related to the markings on the chart paper. Therefore, the need for manual or mechanical syncronization is eliminated and multiple recordings on a single chart produce no frequency deviations. The bus also allows for remote control of the 3300 system.

The plotter itself holds a single pen (threaded for interchange of colors) and is elongated and pointed into a bar marked in scale corresponding to the grid of the chart paper. Therefore, referencing the graph to, say, 0 dB at 1 kHz can be accomplished before actually running a plot. A six-step rotary switch selects paper speeds of 0.1 mm/sec up to 30 mm/sec. Start/stop and fast-forward buttons control recording, pen lift, and advancement of paper. A three-position toggle switch selects modes. "Continuous" provides a registration of level vs. time. Here a single frequency (selected at the output module) is fed at a constant level to the device under test and its output level plotted over a period of time (up to 27.5 min.). "Single" is the setting for normal frequency response plots. The pen automatically drops once the start frequency (again selected at the output module) is reached. Selecting the "reverb time" mode plots decay curves. Upon reaching the frequency selected the audio generator automatically switches off, the pen drops and the paper starts thus yielding a very accurate decay curve picture giving reverberation time vs. frequency.

The 3322 professional output module controls the mainframe's oscillator sweep functions, ½, ⅓, or ⅙ octave warble tones, start frequency for response graphs and the choice of a single frequency for continuous and reverberation decay curve plots. The 3312 input module has a five-step sensitivity switch (off, +20, 0, −20, and −40 dB) with a continuously adjustable uncalibrated vernier. The input module also sets the plot scale to be logarithmic (10, 25, or 50 dB) or linear and writing speed (10, 20, 50, 100, or 200 mm/sec). As can be seen there's a great amount of variability built into the 3300 which makes it applicable to a variety of uses. Its ability to obtain very accurate results of many measurement parameters is very important, yet it is the ease of operation due to automatic digital control along with its confidence of accuracy that opens the door to its usage by those with even little or no technical background.

While the Neutrik 3300 basically utilizes pure tones for measurement it can give readout in the one-third octave format as well. On the other hand, the DN 60 spectrum analyser from Klark Teknik Electronics, Inc. (262a Eastern Parkway, Farmingdale, N.Y. 11735, 516-249-3660) performs ⅓ octave measurements exclusively. The sound source here is pink noise. White noise is the sound you hear when a television set is tuned to a channel that is either not broadcasting, or does not exist. This noise is randomly made up of all frequencies, phases, and amplitudes. Because the widths of successive one-third octave bands have an increasing number of individual frequencies (from only 15 Hz at 31 Hz center frequency all the way up to 8000 Hz at 16 kHz center frequency), a compensation filtering network is added which effectively produces the opposite results (increased level at low frequencies and decreased level at high frequencies). The combination of the two ends up being flat when used in one-third octave filtering and is called pink noise.

The nature of pink noise is random, but via true rms detection and averaging over a period of time it is flat on a one-third octave display. The Klark Teknik DN 60 utilizes three response time settings as well as average and peak detection. It performs the mathematical averaging operation required via a built-in microprocessor and the logarithmic output with the slowest response time results in a visually very flat display.

Once the pink noise produced by the DN 60 has passed through the system under test it is picked up either by electronic connection or by a low impedance microphone input. An internal preamp gain control allows for calibration to microphones with sensitivity between 0.25 m V/ubar (or 2.5 mV/Pa) to 1 mV/ubar (or 10 mV/Pa). Other input preamplifiers are available for mics not in this range, however, because it is not only important for the mic used to be flat with frequency but also have a polar or pick up pattern with the same flatness of frequency off axis. It is more advantageous to use the mic provided by Klark Teknik which has such a pattern and comes with the DN 60's preamp calibrated to it.

An eight-position input level switch adjusts the input to the display's optimum range. Since the mic/preamp combination is calibrated and the display and switch settings are marked, accurate readings in either dBm or SPL are given. The display's resolution can be set for either 1 or 2 dB per graduation (here a graduation is a single LED illumination out of the 15 per each of the 30 bands). A 31st band actually shows the additive effect of all the bands and thus is a full audio bandwidth display. To round up the level setting ability, Klark Teknik has also added the ability to switch in the A-weighing curve to the DN 60.

Three response speeds affect the attack and decay (the speed up or slowing down of the displays changes with time) in the average mode and the decay only in the peak mode for easier visability. The peak hold function is very helpful for the measurement of short duration signals or transients as it leaves a line of single LEDs across the display corresponding to the maximum levels even after the signal is removed. For normal frequency response measurement, the slowest response is not only the flattest appearing display, but makes for easier adjustment of either the whole system, individual components such as equalizers, and allows for more accurate judgments as to the acoustical treatment called for.

Three internal memory banks store the full display plus response speed, average, or peak, weighing if used, line or mic input, maximum hold and peak hold switch setting indications for accurate comparisons and judgment after measurements are computed. Thus using the DN 60 is quite a learning experience. All this is due basically to the built-in microprocessor control. Interface is also possible to options such as a reverberation analyser, graphic plotter, and computer. Furthermore error codes are part of the programming showing malfunctions and in frequency response testing, in overloaded input conditions. In overload, too much input level is sent to either a single band of the display or across the full bandwidth. No damage is incurred or distortion produced due to clipping. The unit is simply warning of possible inaccurate readings. Every time this device is turned on it goes into self-test and if no problems are found the display winks at you, gives a readout of "O.K.," and is ready for use.

As stated the DN 60 can be outfitted with a reverberation measurement option. The RT 60 unit measures reverb via either impulse or steady state pink noise methods. By impulse is meant any sharp transient signal such as a gunshot or even hand claps. The RT 60 simply waits for the impulse and then begins measurement of the decay. The pink noise source in the DN 60 is driven by a line amplifier which is gatable. The gating action is controlled by the microprocessor and enables the output to be turned on and off very precisely. Reverberation decay time is determined by first bringing the signal to a steady state level in the room, removing that signal and then measuring the time it takes for the reverberant level to drop 60 dB. Thus the RT 60 shows level vs. time while the DN 60 alone shows level vs. frequency.

To accomplish reverb measurement normally means running the system at very high levels to improve accuracy. With limited level the RT 60's decay time readout is calculated from a smaller level decay. The DN 60/RT 60 combination samples the decaying sound's energy every 500 microseconds, averages 16 of these samples, and stores them in memory until the test cycle is finished at which point it will have 790 times 8 millisecond samples. This system can measure a decay rate from 0 dB down to −20 dB or any range in between and with as little as a 2 dB drop. The start of measurement is set at 0 dB upon turn-on but can be lowered. While accurate measurements result from a 2 dB drop this also can be extended to over a 20 dB drop although in some cases this may reduce accuracy. As an example, in environments with high background noise such as an auditorium full of heavy metal enthusiasts, the level after signal turn-off may never drop 20 dB.

Whatever setting is used the unit extrapolates for a 60 dB drop in level. Additionally the window or the time frame and other parameters displayed can be changed before or after measurement. Three settings (16 mS, 64 mS, or 208 m S per LED) enlarge or shorten the time span of the display. The resultant decay curve can also be stored in memory. When recalled, the unit shows frequency band measured, if any accumulation was used, how many accumulations were used, and the window and time parameters can still be changed. It is necessary to take many reverb and frequency response measurements and use an average of each for more precise results. This dual unit averages or accumulates up to 32 readings and measures and accumulates reverb at full audio bandwidth or at any individual band as well as average up to 32 real-time displays of frequency response with the adjustment of a single switch. The result of averaging is not only given on the RT 60 numerical time readout but also across the display, in the form of an accumulative graph.

Here the windows are very enlightening. By changing the time frame of the display the graph can be enlarged thus giving a better picture of short decay curves. The main parameter one looks for outside of straight reverb time are irregularities or unevenness of the curve. Bumps or soft "knees" in the curve indicate a problem in the room often caused by some structure reinforcing certain frequencies which causes a decay that is not smooth. All controls are optimized during reverberation timing therefore the user has no control over resolution or speed of response.

The peak/average control handles pink noise turn-on. This is important because it enables the start of the noise just before measurement saving the operator's ears a lot of abuse. The RT 60 is also very user-friendly in that it tells via readout when the unit is properly set up for measurement. It indicates running and calculating as well as accumulating modes and more. Because the onset of the noise source must be of a high enough level to trigger the RT 60 into operation you must drive the noise at what, for frequency response measurement, would be an overload condition. With this unit the level is simply raised until every other pair of LEDs is off. Then pushing "run" shuts the noise off and the measurement begins. It's as simple as that. Once measurement is completed the display shows a graph of the decay curve. Here the one-third octave frequency columns are converted into time columns. All in all, this is a very powerful tool to work with.

Acoustical Testing and Microphones

Microphones play a crucial role in acoustical testing. Not only are they the interfaces that convert energy in the form of sound pressure to that of electrical voltage, but they are used at the very initial point of data gathering and therefore affect every subsequent step down the measurement instrumentation system's path. Any judgments made or action taken as a result of the outcome of these tests must be predicated on the known characteristics of the particular microphone that is being used. So it is important to know the basics regarding microphones used for testing.

In general, the types of microphones used for measurement are condenser, ceramic, and dynamic. Condensers have a diaphragm that is actually one side of a capacitor. When a change in sound pressure causes that diaphragm to be deflected, the resulting movement causes a change in capacitance and thus the electrical output signal. They themselves have a low capacitance and a delicate diaphragm. It's the tension of the diaphragm which determines both the microphones' high frequency response and its sensitivity. When they are shocked as by dropping, or stretched from temperature changes or finger poking, diaphragms loosen. This causes the resonant frequency of the microphone to become lower and the sensitivity to become higher which makes them untrustworthy measurement tools. So all microphones, measurement or not, should be handled with care. Diaphragms can also be harmed by heat and high humidity. I've also heard of cases of older nickel-based diaphragms developing holes (possibly from acid in the atmosphere) which resulted in the reduction of low frequency response. Condenser or capacitor microphones require a relatively high DC polarizing voltage but they have good high frequency response, and are fairly stable in terms of your normal vibrations, pressures, and temperatures as far as damage is concerned.

Electret condensers have a self-polarized plastic diaphragm (usually mylar coated or sputtered with metal) that does not require a high-voltage power supply. However, their frequency response is a bit lacking and they do have temperature limitations. Years ago, a measurement microphone manufacturer told me that because of their susceptibility to temperature and humidity damage, his company could not use sputtered mylar diaphragms in the measurement microphones they manufactured as the unpredictability of the resultant frequency response after the diaphragm had warped or wrinkled from heat and humidity changes made them unreliable.

Ceramic microphones can also be known as piezos if they utilize a piezoelectric crystal behind the diaphragm which, when acted upon by sound pressure, causes strain to the crystal, producing changes in the generated electrical signal. They do not necessarily need a power supply, have a good dynamic range, and are fairly robust, but their frequency response is poor.

Dynamic microphones use a diaphragm with a coil connected to it which moves through a magnetic field much as a speaker does, but here the signal is picked up. The inherent low impedance of these microphones allows them to be used with longer cable runs but these cables must be kept away from magnetic fields such as those produced by transformers or elevator motors as these could affect the accuracy of the measurement readings.

Piezoelectric crystals have fair stability in terms of temperature and humidity and a frequency response range of 10 Hz to 10 kHz. Piezoelectric ceramics have better stability but their response is a little poorer at 20 Hz to 10 kHz, while moving coil dynamics which offer good stability have a response range of 25 Hz to 15 kHz. But it's the capacitor condensers, with fair stability in terms of being able to handle hostile environments such as the heat from factory furnaces, that have a response that extends from 2 Hz to over 20 kHz, and these are readily available and usable in the temperature-controlled environment of recording facilities.

In the past you still had to deal with diaphragm size versus frequency response, special external power supplies, calibration equipment, and very high pricing when considering a measurement microphone. It used to be the case that large diaphragm microphones were more sensitive while small diaphragms gave a better frequency response. This resulted from the fact that the diaphragm itself could cause inaccuracies in free-field sound pressure measurements: A 1-inch-diameter diaphragm will not be accurate at frequencies above 13 kHz since a 1-inch wavelength corresponds to a frequency of 13,536 Hz (see Chapter 17). This can still be the case for those dealing with frequencies above 100 kHz, but for our purposes a common, high-quality ½-inch diaphragm microphone suits the needs of recording studio acoustic measurement just fine.

FREE FIELD VERSUS RANDOM INCIDENCE MICROPHONES

The choice between these two types of microphones depends on both the application and the requirements of the test standards that are being followed. When measuring a single sound source (such as a speaker) while all other sound sources or reflections are at a much lower sound level, the best frequency response can be obtained by aiming a free field microphone on axis right at and close to the source so as to have the sound arrive at the diaphragm at a perpendicular angle (0 degrees incidence). Since free field microphones are less omni-directional at higher frequencies than random incidence microphones they can have more of a high frequency error when measuring off-axis sound sources. Therefore when the measurement is to be made of multiple sound sources and/or many reflections, as in architectural acoustic measurements, more than likely you will be better off with a random incident microphone.

Today most free field and random incidence microphones that are designed specifically to measure sound pressure in a sound field are also designed to compensate automatically for the influence that the microphone itself introduces into the field. The end result is hopefully a measurement of the sound field that appears the same as it existed before the microphone was added, as even these specialty microphones do have some physical effect on the sound field. Here, the presence of the microphone in the sound field will actually change the sound pressure as right around the microphone itself the sound pressure will increase due to added reflections and diffractions caused by the microphone. Additionally, this increase will be different at different incidence angles. As long as the sound waves do in fact arrive from random directions, random incidence microphones are better in diffused sound fields. You can also use a random incidence microphone to measure isolated or single sound sources such as a section of a wall or even a corner by aiming its diaphragm at an 85-degree angle.

At the same time, should you be following a particular standard's protocol, you'll find that ANSI standards call for random incidence, while IEC testing utilizes free field microphones. This is important to your later judgment and comparison of different test results. You can by-pass this choice in a pinch simply by changing the angle of the microphone's positioning (for free field microphones, when pointed at and close to a sound source, a 0-degree angle of incidence, and for random incidence microphones, an 85-degree angle).

Pressure microphones are generally mounted directly into a wall, thus becoming part of the acoustical space, which lessens the effect of their presence. Due to the fact that these microphones do not require any kind of aiming at all, they can still yield accurate sound pressure readings even with this type of mounting. Of course, these are one of the most special and single purpose of all the microphones and they cannot be used for normal audio, including recording purposes. So it's free field when close, random for whole room, and pressure for strictly scientific applications.

The question of whether you're measuring a free field or a diffused field is important, since using the wrong type of microphone can cause substantial test errors at frequencies above 4 or 5 kHz. To find out which you are dealing with, simply make a measurement, reposition the microphone by turning it 90 degrees, and measure again. If there is a drop-off in the high frequency content then you're in a free field. If there is no real drop-off (less than a couple of dB at 5 kHz) then it's a diffused field. Now, the only things the test engineer has to worry about are microphone positioning, possible electro/mechanical interference, and the distance to the preamp's location.

Advances have been made in microphone manufacturing methods to the extent that today you can purchase a microphone stable enough that the graph it is shipped with holds true to that microphone for many years unless some form of damage occurs. Now, not only has routine calibration become unnecessary, but powering via the phantom supplies found in most every recording console is commonplace. Some of the best microphones presently available have prices under 1000 (instead of several thousand) dollars and are usable for more than just measurement purposes. Most of these microphones are of the capacitor/condenser type, instead of the older pressure or force pickups. They work wonderfully as music recording tools, as well as giving you a much more cost-effective solution to your measurement needs. There is currently available from a company called Earthworks a high-quality, extended (9 Hz to 30 kHz), flat (± 1 dB) frequency response omnidirectional test/recording microphone (the M30BX) that is powered by a single AA battery that will last through more than 1500 hours of operation and it carries a list price of 600 dollars. Crown International, Inc. offers a recording/acoustical measurement microphone (the CM-150) that's a precision (having a virtually flat frequency response from 20 Hz to 20 kHz) omnidirectional condenser microphone that has a metal diaphragm, which gives it stability over a very wide range of environmental situations, yet it was designed specifically to be an audio recording microphone as well. Its suggested retail price is 795 dollars. Gold Line, long known as a manufacturer of easy-to-operate, inexpensive acoustic test and measurement equipment now sells a quality $\frac{1}{2}$-inch diaphragm condenser omnidirectional microphone (the MK 10) that renders a frequency response from 20 Hz to 20 kHz with a maximum ± 2 dB variation from dead flat, handles sound pressure levels up to 128 dB, yielding a dynamic range of over 100 dB all at a suggested retail price of 250 dollars. This is at best a small sampling, and, best of all, since all of these microphones are more than just adequate budget measurement tools, the choice of which one to purchase now comes down to your subjective audio listening and recording tastes. Now that's progress!

Want cutting edge? Bruel & Kjoer, a long-time manufacturer of top-of-the-line acoustic measurement instruments including microphones, is reportedly working on applying the methods and materials used in the manufacture of integrated circuits (IC chips), including micromachining and microelectronics. They have come up with a microelectro/mechanical microphone transducer having the accuracy and repeatability inherent to microprocessor manufacturing to the extent of yielding a microphone diaphragm of new materials and with a mounting method that is reported to result in a "multi-field" microphone. This device promises to have a flat frequency response in all sound fields, including free and diffused, no matter what the angle of incidence.

Acoustic Testing and the PC

Due to the vast computational power of computers it is now possible for the measurement system to automatically take into consideration any of the anomalies of your more common, but still quality, recording microphones calibrating themselves to the microphone's response and giving test results that are very accurate. This was unheard of even a mere 10 years ago, but it is now commonplace with systems manufacturers such as Linear X, who perform "microphone data file" calibrations "to support any third-party microphone."

Personal computer and even laptop-based acoustic and vibration measurement via PCMCIA cards as opposed to the use of very expensive, traditional, single-purpose test hardware is now a reality. However, in order to acquire quality data, poor test procedures which might cause distortion in the recorded data must, as always, be avoided. The test engineer must pay close attention to the basics of signal processing such as correct impedance matching and level setting. Proper level setting, also necessary for repeatability, may require an external conventional voltmeter to augment the often seemingly arbitrary on-screen signal generator output level setting via fader icon or even dB scaling, which can be of questionable referencing. An example of this problem would be the sound system under test being fed a swept frequency signal while a calibrated microphone picks up the system's output within your control room and feeds it back to the PC's processor for recording. The end result then displayed in a nice, easily understood graphic form helps you to see that your room/system has a nasty resonant peak in its frequency response that's both narrow band and centered around 4 kHz. So you decide, "Hey let's bring in those foam square things with the pyramid shapes and stick them up and see what that does to the sound." Unless you can reset the signal generator to the exact same output level (as in 500 mV into 600 ohms) you're going to be left only slightly better off than just guessing or depending on some casual subjective opinion in your judgment and decision making. Not really a problem as you trust your ears? Well, factor in that on the ride into work that day you were subjected to a repeated onslaught of a 5 kHz screech from a faulty bus or train braking system. "Well, I thought it sounded better at the time, maybe we can return those cut up foam sheets."

It's always best to go with a complete testing system either by renting or one that comes along with its hired acoustical consultant/operator and you then won't even have to worry about things like low frequency discrepancies due to possible impedance mismatch.

PC sound card–based testing is most often not adequate for testing vibration as structural response measurement must extend down to only a couple of hertz. True laboratory data acquisition systems record frequency data down to DC. Sound cards are not typically capable of handling much below 20 Hz (even though some can resolve the signal down to 10 Hz). But it's below this is where structural vibrations often occur and that's usually beyond these systems' capabilities. Again if you go with a complete test package there's at least someone such as its manufacturer, a rental company's technical staff, or the owner/operator you've hired that can answer any questions should they arise.

The Internet could end up being one of the technological drivers in the area of computer control in the testing of both structural dynamics and acoustics. This is due to the fact that it already has the capability to deliver not only the software needed but also the data storage and signal processing hook-up which would allow acoustic measurement instruments outfitted with Web-server capabilities to themselves be less complex and thus be more affordable. Additionally, the needed (spelled $) technological developments in consumer digital DVD music electronics required small cost-effective analog-to-digital converters (ADCs) that had both high dynamic and frequency ranges, making them useful in vibration and acoustical test instrumentation. They also have lower power requirements, smaller size, and yet cost only a few dollars. Most important to data acquisition, these 24-bit ADCs need far less anti-aliasing electronic real estate, and their expanded dynamic and frequency ranges can in fact be fully transmitted digitally.

Everything seems to be falling into place that would allow the studio owner to purchase if not rent a fairly inexpensive (as far as test equipment goes) data acquisition peripheral; hook it up over ISDN phone lines to manufacturers of, or consultants in, acoustical products; and have their room evaluated. They could then receive the expected results and any recommendations not only graphically over the Web, but audibly as well via digital signal processing and through ISDN phone lines. By this I mean the ability to actually listen to real aural demonstrations of the resulting changes that would be expected to take place when utilizing various preconstructed absorption, reflection, or diffraction aids such as panels, tiles, or traps mounted at different positions in the actual space being tested. This is not as much of a futuristic pipe dream as it may seem as all of the individual components needed to make this possible already exist. On the other hand just look what happened to the promise of what was to be the information (but now "buy-me button" potholed) highway and the horror of companding that is at present considered "music" being transmitted over the Web, so it may not pay to hold your breath waiting for someone else to put it all together. For now you will just have to rent the test gear yourself or get the help of an acoustician and use ISDN to ship and receive audio signal.

READING BLUEPRINTS

Blueprints are drawn to show the details of a design. Mechanical blueprints have very rigid directions pertaining to details of size and shape. They are drawn this way so as to yield the exacting results required of machined parts whose fit and finish often demand the meeting of tolerances within plus or minus one-thousandth of an inch (±0.001"). Representative blueprints such as the ones used in this publication are drawn to be a little more flexible in showing sizing, construction methods, and positioning details. This is to allow for variation in the implementation and layout of the design. They also tend to be more enlightening than mechanical blueprints in terms of revealing the philosophy or theory behind a design. Therefore they are more instructional and useful in applications such as acoustical designs where it is necessary to have the ability to modify and even completely revise designs in order for them to fit in with the needs of a particular situation.

Representative blueprints while very helpful, do however demand a little more from the thought process. Some are almost sketchlike, such as Figs. 2-4 and 2-10. These are used more to present ideas in a not overly restrictive way. Yet others, like Fig. 10-11 can be quite detailed, giving specific dimensions, layout positioning, materials used, and often even have sections with more meticulous details as in this figure where it is shown how to calculate the perforation percentage of pegboard. We find from the text, that the very first figure or blueprint, Fig. 2-1, was drawn to be representative because "sound locks come in all shapes and sizes." Using this drawing as an example and focusing on the text's description of the lower wall section, which is referred to as "wainscoting" it is explained that a 1-inch-thick layer of Tectum or perforated peg-board could be used over a fiber-glass-filled airspace obtained by furring out 2×2s for good midband absorption. The wainscot detail of Fig. 2-1 also shows these components of the lower wall (glass fiber-filled air space between 2×2 furring covered by "ordinary pegboard"). This detailed section is additionally helpful as it demonstrates the design's internal layout. On the other hand it could be confusing if you had not looked at the full ceiling-to-floor drawing above it which shows this detailed section (marked as "see below") to actually be standing upright with the view being from above it, looking down toward what would be the floor.

As a representative drawing, one of its important benefits lies in its flexibility. Illustrated here is the fact that if your lower wall or wainscoting is 4-feet high and runs in 8-foot lengths, you could, without it being detrimental to the design's functioning, utilize a full-sized sheet of pegboard (which comes in 4' × 8' sheets) simply by rotating this wall section 90 degrees, thereby easing up on both your material and labor costs. Budget acoustics demand the flexibility of representative blueprints.

You could run into some additional confusion if you didn't know from checking another drawing such as say, Fig. 3-4 to find that the open-sided areas in Fig. 2-1, which had been filled with slanted lines, represent the existing wall. More than likely you would have had to search out other examples of this open-sided slanted-line fill besides Fig. 3-4, such as the plastered walls in Figs. 5-4 and 5-5 or as in Fig. 12-5, where they represent the existing wall structures "in large Studio A." In this figure they are used within the small drawings which show the view of the panel layout from above. These are located below each of the four wall drawings that give the same panel layout but from the facing or front view. You would have found that they represent both the ceiling and wall in Fig. 11-6 and both of those plus the floor in Fig. 11-5.

It might have taken someone all of this investigation to finally come to the conclusion that this fill pattern represents an existing structure. At this point you'd comprehend that this slanted-line fill with one open side represented the existing wall and floor in Fig. 10-7 and that Fig. 11-6 is actually two drawings. Here the left-hand side depicts the air space behind the louver (as seen from above) with the open-ended, slanted-line-filled area representing both the wall behind the panel and the small section of wall that abuts against its side. The right-hand section of the drawing depicts the actual floor-to-ceiling layout of the louver panels, here rotated 90 degrees so as to lay on its left side, showing the framework and cross-bracing by depicting it with broken lines.

In this search you would also have run across slightly different looking and enclosed line fills such as those used in Fig. 4-1 or Fig. 4-2, explained to be brick in the detail section of Fig. 4-1 or perhaps the very similar gypsum of Fig. 14-8. From both the text and drawing of Fig. 14-12 we find that the somewhat similar fill found here is 4 inches of 703 fiber glass. The fills that are pointed out to be chipboard in Figs. 6-7 and 6-9 or plywood in Fig. 8-9 have diagonal lines that are slightly curved, much like wood grain.

I do not think it is possible to overstress the importance of fully reading the text in order to understand some of the design ideas, especially those which develop across several chapters. Even grasping a concept that's covered in only a single chapter requires thorough reading in order to discern its many subtle details so as to perceive the often diverse problem-solving applications it can address. Take for example the seemingly simple design of the louver-type absorber of Fig. 11-5. The text from pages 144 through 155 covers this topic in detail concerning aspects of this design such as its ability to incorporate adjustable absorption control. You would not have known the correct panel layout and placement, nor the changes you could expect in both the frequency response and reverberation time (per frequency) in the location where you utilized any of these low frequency, broad band,

and/or reflective panels. To leave out all of this helpful and long-term educational information doesn't make sense, plus as Mr. Everest put it, such "procedures could result in some weird unbalanced acoustical conditions."

Some of the difficulty in one's fully understanding a drawing may be due in part to book sizing. For this reason, even though the drawing of Fig. 10-7 points out the fact that in the construction of this polycylindrical panel, a foam vibration isolation strip is included, due to the constraints of the book's sizing, you may not have understood that it is to be attached along the edge of the plywood bulkhead underneath the $^3/_{16}$-inch hardboard or plywood covering skin unless you had read the text. You also would have needed to have read the text in order to understand that the block-shaped object that's shown as "M" in the detail of Fig. 3-4 and which is described in the text as "a strip running the length of a wall" is in fact molding that is fastened to the existing wall and allows for the panel units to "easily be positioned laterally anywhere" across the wall. This perfect as-suggested mounting solution, which is simple to install, versatile in usage, and inexpensive, might have remained a complete mystery. Finally, if the garage door in Fig. 7-3 were not explicitly pointed out to be such in both the drawing and the text there would have been no way to fathom this fact.

Once you've discovered the material that the pattern or fill in question represents, your job is still only half done. Now you must replace that fill pattern, if necessary, with one that corresponds to the actual material you're dealing with, and use this pattern in your drawings, that is, the blueprints you have to draw to document your own construction. This endeavor is well worth the effort if only just to have the hard-copy record, let alone the whole host of additional benefits. Not only will you be able to adapt existing designs to your own particular needs more readily with every drawing you do but you'll find it easier to explain to workers those parts of your own drawings that illustrate any particularly important construction detail or method they are required to perform. Down the road should things change or you find that your original thinking was somehow flawed, you'll be able to readapt the design and blueprint you had ended up with more readily into one more suitable to your new re-thought-out situation.

Having accurate blueprints gives you the ability, for instance, to precisely relocate an installed rattling mounting fixture long after all construction has been completed (usually when the rattling starts) and then modify or completely change out that hardware without unnecessarily destroying any of your already finished work. Believe me, this is big. If you're still not convinced that doing your own blueprint drawings is worth the trouble how about the financial benefits you'll be getting? As an example let's say you actually built some of those adjustable absorption louver-type panels. Time passes and you're ready for whatever the reason to sell your studio. The blueprints you drew not only aid in your explanation of these acoustical panels' adjustable functioning to a prospective purchaser, but just the fact that they exist indicates that all your hard work and expense did not result in something that will be perceived as some hair-brained fluff.

You may find that the number of panels needed to achieve the desired frequency and reverberant room response requires additional ones to be constructed. On the other hand let's say things are going so well businesswise that you've decided to add on a second studio which will require more panels to be constructed. In either case you'll find that the building side of the task at hand is not only eased but the hardest part, that of the actual design and layout work having been already completed, saves you money in both labor time and lack of mistakes in the purchasing of construction materials. Indicating any modification you desire such as changes to facilitate the use of bagged (less unpleasant) fiberglass will also be a cinch. You'll find that customers, be they recording engineers, producers, musicians, or even guests, will often ask about these types of structures. Hopefully you've added a little wood stain, used colored and intricately shaped foam in place of the fiberglass, and utilized currently available hardware for easier and smoother mechanical operation of the variable sound reflection/absorption function. If this ends up being the case you may be in a position to market professional acoustical panels yourself. Don't take this suggestion lightly as these variable absorption panes end up functioning perfectly, in addition to being good looking, in the home theater environment.

It doesn't matter if you do not intend to take studio building to anywhere close to an occupational endeavor, everyone will inevitably have to change some of the materials utilized or at least the original designs dimensions to suit his or her own purposes. This more than likely will not be a matter of simply using photocopier enlargements to achieve larger scale drawings. Therefore to make going through the drawings easier and to aid you in doing your own drawings, when this is the case, all the representative material fills have now been consolidated to form an easy-to-use reference and guide. Some other fill patterns commonly used in acoustic designs have also been added just in case should you either run into them elsewhere or need to show that material in your drawing to correctly depict the fills which correspond to the actual materials that exist in your studio.

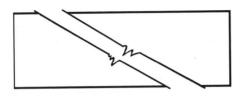

Drawing split for illustration
But is continuous, with size
or length often given

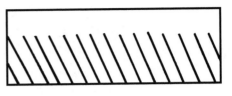

Existing structure

Fiberglass

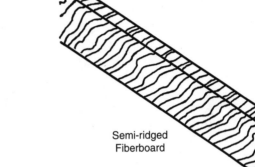

Semi-ridged
Fiberboard

Fiberboard,
Lengthwise view as
used under poured
concrete flooring

Plate glass

Figure 20-1

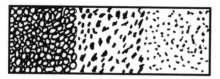

Foam, Sponge, Neoprene
or Cement as specified

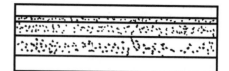

Gypsum Wallboard
three layers at 1/4,
1/2 and 3/4 inch thickness

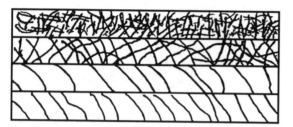

Plywood, Chipboard, Particle
or Presswoood as specified

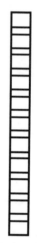

Pegboard

Plywood

Figure 20-1 *(Continued)*

Lumber:

Running Board

1 × 1

1 × 2

1 × 3

1 × 4

1 × 6

1 × 8

Blocked

Continuous

Continuous

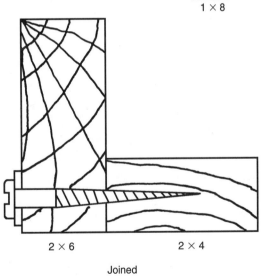

2 × 6   2 × 4

Joined

Figure 20-1 *(Continued)*

Finish plywood,
facing view

Lumber:

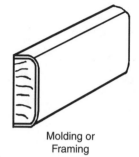

Molding or
Framing

Masonry:

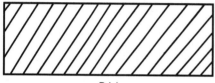

Brick

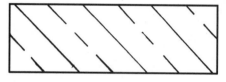

Stone

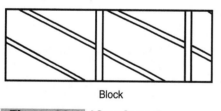

Block

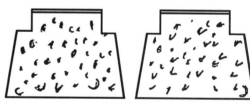

Concrete risers,
Piers or Mounts

Figure 20-1 *(Continued)*

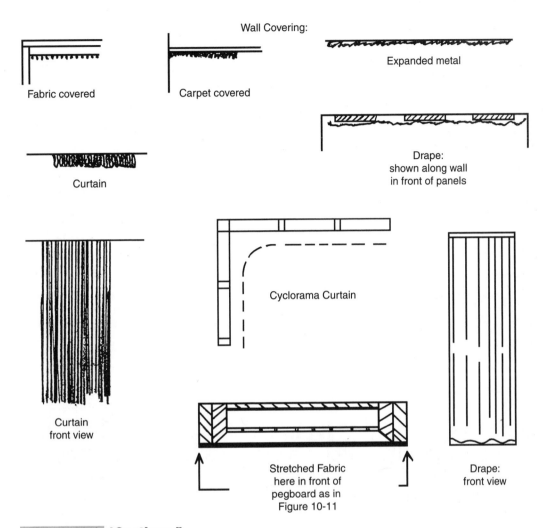

Wall Covering:

Fabric covered

Carpet covered

Expanded metal

Curtain

Drape:
shown along wall
in front of panels

Cyclorama Curtain

Curtain
front view

Stretched Fabric
here in front of
pegboard as in
Figure 10-11

Drape:
front view

Figure 20-1 (*Continued*)

Drywall

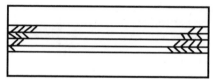

Plywood

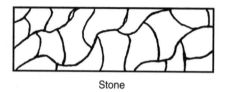

Stone

Fieldstone

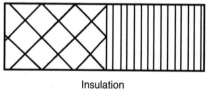

Insulation

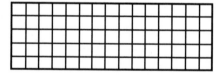

Rigid Insulation

Figure 20-1 (*Continued*)

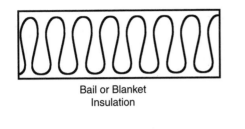

Bail or Blanket
Insulation

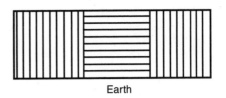

Earth

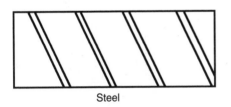

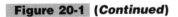

Marble

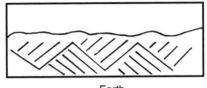

Earth
Rough Grade

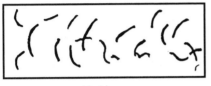

Steel

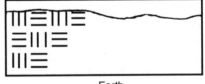

Earth
Finish Grade

Figure 20-1 (*Continued*)

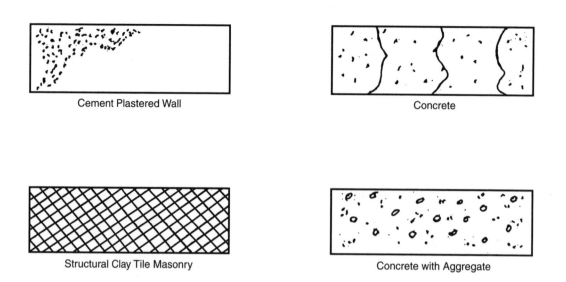

Cement Plastered Wall

Concrete

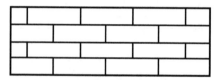

Structural Clay Tile Masonry

Concrete with Aggregate

Brick or Block Masonry
"Elevation" (front on) view

Concrete Block

Figure 20-1 (*Continued*)

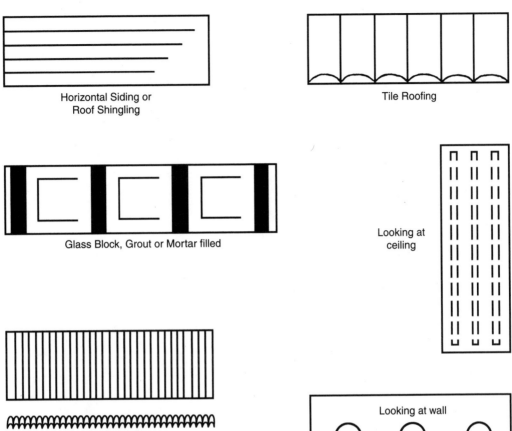

Horizontal Siding or
Roof Shingling

Tile Roofing

Glass Block, Grout or Mortar filled

Looking at
ceiling

Sheet Metal Flashing
curved or triangled at
top and bottom for
Corrugated Sheet Metal

Looking at wall

Fluorescent light fixtures

Figure 20-1 *(Continued)*

Appendix A

SOUND ABSORPTION COEFFICIENTS FOR GENERAL BUILDING MATERIALS AND FURNISHINGS

Complete tables of coefficients of the various materials that normally constitute the interior finish of rooms can be found in the various books on architectural acoustics. The following shown list of materials give approximate values which will be useful in making simple calculations of the reverberation in rooms.

Materials	125 Hz		500 Hz	1000 Hz	2000 Hz	4000 Hz
				Coefficients		
Brick, unglazed	.03	.03	.03	.04	.05	.07
Brick, unglazed painted	.01	.01	.02	.02	.02	.03
Carpet						
⅛ " pile height	.05	.05	.10	.20	.30	.40
¼ pile height	.05	.10	.15	.30	.50	.55
³⁄₁₆ " combined pile and foam	.05	.10	.10	.30	.40	.50
⁵⁄₁₆ " combined pile and foam	.05	.15	.30	.40	.50	.60
Concrete block, painted	.10	.05	.06	.07	.09	.08
Fabrics						
Light velour, 10 oz. per sq. yd., hung straight, in contact with wall	.03	.04	.11	.17	.24	.35
Medium velour, 14 oz. per sq. yd., draped to half area	.07	.31	.49	.75	.70	.60
Heavy velour, 18 oz. per sq. yd., draped to half area	.14	.35	.55	.72	.70	.65
Floors						
Concrete or terrazzo	.01	.01	.01	.02	.02	.02
Linoleum, asphalt, rubber or cork tile on concrete	.02	.03	.03	.03	.03	.02
Wood	.15	.11	.10	.07	.06	.07
Wood parquet in asphalt on concrete	.04	.04	.07	.06	.06	.07
Glass						
¼ ", sealed, large panes	.05	.03	.02	.02	.03	.02
24 oz., operable windows (in closed condition)	.10	.05	.04	.03	.03	.03
Gypsum board, ½ " nailed to 2 × 4's 16" o.c., painted	.10	.08	.05	.03	.03	.03
Marble or glazed tile	.01	.01	.01	.01	.02	.02
Plaster, gypsum or lime, rough finish on lathe	.02	.03	.04	.05	.04	.03
Same, with smooth finish	.02	.02	.03	.04	.04	.03
Hardwood plywood paneling ¼ " thick, wood frame	.48	.22	.07	.04	.03	.07
Water surface, as in a swimming pool	.01	.01	.01	.01	.02	0.2
Wood roof decking, tongue-and-groove cedar	.24	.19	.14	.08	.13	.10
Air, sabins per 1000 cubic feet at 50%RH				.9	2.3	7.2

Reprinted from ACOUSTICAL CEILINGS—USE AND PRACTICE
by permission of the Ceilings & Interior Systems Contractors Association.

Appendix B

SOUND ABSORPTION

COEFFICIENTS OF OWENS

CORNING PRODUCTS

OWENS-CORNING FIBERGLAS CORPORATION
Absorption Coefficients For Type 703 Fiberglas
Semi-rigid Boards (3 lbs/cu. ft. density) and Allied Materials

Facing Material	Core Material	OCTAVE BAND CENTER FREQUENCIES, Hz						NRC
		125	250	500	1000	2000	4000	
None[1]	1" 703	.06	.20	.65	.90	.95	.98	.70
¼ " Pegboard[2]	1" 703	.08	.32	.99	.76	.34	.12	.60
⅛ " Pegboard[3]	1" 703	.09	.35	.99	.58	.24	.10	.55
None	1" TIW Type I	.11	.33	.70	.80	.86	.85	.65
¼ " Pegboard	1" TIW Type I	.08	.41	.99	.82	.26	.32	.60
1" Nubby Glass Cloth Board	None	.04	.21	.73	.99	.99	.90	.75
1" Textured Glass Cloth Board[5]	None	.05	.22	.67	.93	.99	.85	.70
1" Painted Linear Glass Cloth Board	None	.03	.17	.63	.87	.96	.96	.65

Facing Material	Core Material	125	250	500	1000	2000	4000	NRC
None	2" 703	.18	.76	.99	.99	.99	.99	.95
¼ " Pegboard	2" 703	.26	.97	.99	.66	.34	.14	.75
Perforated Metals[4]	2" 703	.18	.73	.99	.99	.97	.93	.95
1" Painted Linear Glass Cloth Board	1 " 703	.18	.71	.99	.99	.99	.99	.90
1 " Nubby Glass Cloth Board	1" 703	.25	.76	.99	.99	.99	.97	.95
None	2" TIW Type I	.25	.75	.99	.99	.99	.99	.95
¼ " Pegboard	2" TIW Type I	.26	.89	.99	.58	.26	.17	.70
Perforated Metal	2 " TIW Type I	.25	.64	.99	.97	.88	.92	.90
1 " Linear Glass Cloth Board	1 " TIW Type I	.23	.72	.99	.99	.99	.99	.90
1 " Nubby Glass Cloth Board	1 " TIW Type I	.26	.75	.99	.99	.99	.99	.95
1 " Linear Glass Cloth Board	1 " Air space	.04	.26	.78	.99	.99	.98	.75

NRC = Noise Reduction Coefficient. It is the average of absorption Coefficients for 250, 500, 1000, and 2000 Hz rounded to the nearest multiple of 0.05.

Facing Material	Core Material	125	250	500	1000	2000	4000	NRC
None	3" 703	.53	.99	.99	.99	.99	.99	.95
¼" Pegboard	3" 703	.49	.99	.99	.69	.37	.15	.75
1" Painted Linear Glass Cloth Board	2" 703	.59	.99	.99	.99	.99	.99	.95
1" Nubby Glass Cloth Board	2" 703	.50	.99	.99	.99	.99	.97	.95
None	3" TIW Type I	.46	.99	.99	.99	.99	.99	.95
¼ Pegboard	3" TIW Type I	.53	.99	.97	.51	.32	.16	.70
1" Painted Linear Glass Cloth Board	2" TIW Type I	.48	.99	.99	.99	.99	.99	.95
1" Nubby Glass Cloth Board	2" TIW Type I	.51	.99	.99	.99	.97	.95	.95
1" Painted Linear Glass Cloth Board	2" Air space	.17	.40	.94	.99	.97	.99	.85

Facing Material	Core Material	125	250	500	1000	2000	4000	NRC
None	4" 703	.99	.99	.99	.99	.98	.98	.95
¼ Pegboard	4" 703	.80	.99	.99	.71	.38	.13	.75
1" Painted Linear Glass Cloth Board	3" 703	.88	.99	.99	.99	.93	.98	.95
1" Nubby Glass Cloth Board	3" 703	.75	.99	.99	.99	.99	.97	.95
None	4" TIW Type I	.57	.99	.99	.99	.99	.99	.95
¼" Pegboard	4" TIW Type I	.70	.99	.94	.58	.37	.19	.70
1" Painted Linear Glass Cloth Board	3" TIW Type I	.77	.99	.99	.99	.99	.99	.95
1" Nubby Glass Cloth Board	3" TIW Type I	.71	.99	.99	.99	.99	.92	.95
1" Painted Linear Glass Cloth Board	3" Air space	.19	.53	.99	.99	.92	.99	.85

[1]Absorption values would be unchanged for open facings such as wire mesh metal lathe, or light fabric.
[2]Perforated ¼" holes, 1" o.c.
[3]Perforated ⅛" holes, 1" o.c.
[4]24 gauge, ³⁄₃₂" holes, 13% open area.
[5]Absorption values of textured glass cloth may be interpolated to lie between linear and nubby glass cloth for all other thickness of wall treatment.
TIW = Thermal insulating wool

Facing Material	Core Material	125	250	500	1000	2000	4000	NRC
None	5" 703	.95	.99	.99	.99	.99	.99	.95
¼" Pegboard	5" 703	.98	.99	.99	.71	.40	.20	.75
1" Painted Linear Glass Cloth Board	4" 703	.87	.99	.99	.99	.99	.99	.95
1" Nubby Glass Cloth Board	4" 703	.88	.99	.99	.99	.99	.96	.95
None	5" TIW Type I	.83	.99	.99	.99	.99	.99	.95
¼" Pegboard	5" TIW Type I	.78	.99	.89	.63	.34	.14	.70
1" Painted Linear Glass Cloth Board	4" TIW Type I	.77	.99	.99	.99	.99	.99	.95
1" Nubby Glass Cloth Board	4" TIW Type I	.79	.99	.99	.99	.99	.98	.95

Facing Material	Core Material	125	250	500	1000	2000	4000
None	6 " 703	.99	.99	.99	.99	.99	.99
¼ " Pegboard	6 " 703	.95	.99	.98	.69	.36	.18
1 " Painted Linear Glass Cloth Board	5 " 703	.99	.99	.99	.99	.99	.99
1 " Nubby Glass Cloth Board	5 " 703	.92	.99	.99	.99	.99	.99
None	6 " TIW Type I	.93	.99	.99	.99	.99	.99
¼ " Pegboard	6 " TIW Type I	.95	.99	.88	.64	.36	.17
1 " Painted Linear Glass Cloth Board	5 " TIW Type I	.87	.99	.99	.99	.99	.99
1 " Nubby Glass Cloth Board	5 " TIW Type I	.92	.99	.99	.99	.99	.93
1 " Painted Linear Glass Cloth Board	5 " Air space	.41	.73	.99	.98	.94	.97

Facing Material	Core Material	125	250	500	1000	2000	4000
1 " Painted Linear Glass Cloth Board	6 " 703	.86	.99	.99	.99	.99	.99
1 " Nubby Glass Cloth Board	6" 703	.85	.99	.99	.99	.99	.99
1" Painted Linear Glass Cloth Board	6 " TIW Type I	.95	.99	.99	.99	.99	.99
1 " Nubby Glass Cloth Board	6" TIW Type I	.95	.99	.99	.99	.99	.94

All material combinations installed and tested against a solid wall (i.e., # 4) mounting) TIW = Thermal Insulating Wool

16" O.C.

Wood Studs	Facing	Insulation	125	250	500	1000	2000	4000	NRC
2 × 2's	None	2¼ " Fiberglas	.30	.69	.94	.92	.92	.98	.85
2 × 4's	None	3½ " Fiberglas	.34	.80	.99	.97	.97	.92	.95
2 × 4's	1 " Painted Linear Glass Cloth Board	3½ " Fiberglas	.66	.99	.99	.99	.99	.97	.95
2 × 4's	1 " Nubby Glass Cloth Board	3½ " Fiberglas	.67	.99	.99	.99	.99	.90	.95
2 × 4's	¼" Pegboard	3½ " Fiberglas Paper faced*	.45	.99	.87	.41	.30	.14	.70
2 × 4's	None 1 " Painted	3½ " Fiberglas	.38	.96	.99	.68	.47	.35	.80
2× 4's	Linear Glass Cloth Board	3½ " Fiberglas	.66	.99	.99	.96	.99	.99	.95
2 × 4's	1 " Nubby Glass Cloth Board	Paper faced* 3 ½" Fiberglas	66	.99	.99	.98	.99	.95	.95
2 × 4's	¼" Pegboard	Paper faced* 3 ½" Fiberglas	.50	.99	.70	.41	.38	.27	.60
2 × 6's	None 1" Painted	6" Fiberglas	.67	.99	.99	.99	.99	.98	.95
2 × 6's	Linear Glass Cloth Board	6" Fiberglas	.89	.99	.99	.99	.99	.99	.95

Reprinted by permission from "Industrial Noise Control," publication No. 5-BMG-8277 (1978), Owens-Corning Fiberglas Corporation, Fiberglas Tower, Toledo, Ohio 43659

Appendix C

SOUND ABSORPTION COEFFICIENTS FOR TECTUM*

Absorption Coefficients Hz

Thick,	Mfg.	125	250	500	1000	2000	4000
	2	.08	.14	.27	.57	.59	.63
	4	.07	.12	.24	.44	.70	.54
1″	7	.45	.43	.31	.42	.56	.79
	8	.18	.53	.96	.90	.71	.90
	2	.09	.16	.36	.70	.49	.78
1½″	4	.09	.14	.31	.65	.66	.64
	7	.44	.43	.33	.49	.66	.77
	2	.13	.20	.50	.70	.58	.72
2″	4	.12	.20	.48	.80	.62	.94
	7	.48	.46	.36	.55	.74	.79
	2	.14	.29	.77	.67	.80	.85
2½″	4	.18	.28	.63	.87	.62	.80
	2	.14	.32	.78	.60	.84	.91
3″	4	.18	.35	.82	.76	.75	.88

*Gold Bond Building Products

Appendix D

MANUFACTURERS
OF ACOUSTICAL MATERIALS

This listing should not be considered complete, as many manufacturers will have been inadvertently left out and even those listed will undoubtedly not have all of their products and services indicated. The straight-out best bet is to jump into the Thomas Register Web site. There you'll find most U.S. manufacturers listed, along with their local distributor's addresses and phone numbers.

Accessible Products, Co. Inc. 2122 W. 5th Place, Tempe, AZ 8528, (408)967-888   Absorptive materials, acoustical insulation, ducts, barrier

Acon, Inc. 4600 Webster St., P.O. Box 1324, Dayton, OH 45404, (937) 276-2111   Panels, barrier panels, walls, enclosures, and composite panels

Acoustical Solutions 3603 Mayland Ct., Richmond, VA 23233, (800) 782-5742,
Web site: www.acousticalsolutions.com   Absorption panels, sheet foam, vinyl barrier materials, diffusion panels

Acoustical Surfaces 123 Columbia Court N., Chaska, MN 55316, (952)448-5300, Web site: www.acousticalsurfaces.com Felts, foams, glass and mineral fiber, perforated sheet metal, wall treatments, ceilings, panels, absorbers, pipe laggings, loaded plastics, sheet glass and plastics, curtains, panels, windows, composites, enclosures, vibration damping materials and isolation systems

Acoustical Interiors 574 Industry Lane, Frederick, MD 21704, (800) 221-8975 Web site: www.acoustical interiors.com Absorption panels and ceiling tiles

Acoustics First Corp. 2247 Tomlyn St., Richmond VA 23230-3334, (8888) 765-2900
Web site: www.acousticsfirst.com   Foam, baffles, barriers, vibration damping, composites, bass traps, absorption and diffusion panels

Acoustic Sciences Corp. P.O.Box 1189, Eugene, OR 97440, (800) 272-8823, www.asc-soundproof.com
Absorption and diffusion panels, bass traps and complete control-room sound shaping kits.

Acoustic Systems 415 E. St. Elmo Rd., Austin, TX 78745, (800) 749-1460, www.acousticsystems.com
Modular soundproof rooms and booths, panels, doors, windows, enclosures.

Advanced Component Technologies, Inc. 91 16th St., South, Northwood, IA 50459, (641)324-2231
Web site: www.goact.net   Foams, perforated sheet metal, spray-on coatings, foam composites, vibration damping materials

The Aeroacoustic Corp. 169-197 Highland Parkway, Roselle, NJ 07203-2690 (908)241-9718   Panels, barrier panels, walls, windows, enclosures, composite panels, ducts, silencers

Aetna Felt Corp. 2401 W. Emaus Ave., Allentown, PA 18103, (800) 526-4451   Vibration damping material, mass-loaded plastic barriers

Air-Loc Products 5 Fisher St., P.O. Box 269, Franklin, MA 02038 (508)528-0022 or (800)528-7555

Airsan Corp. 4554 W. Woolworth Ave., Milwaukee, WI 53218, (414) 353-5800   Duct silencers

Airtex Inductries, Inc. 3558 82nd St. N., Minneapolis, MN 55412, (612) 522-3643,   Foams, wall treatments, mass-loaded plastics barriers, sealants, foam barrier composites, vibration damping materials

Albany International 1373 Broadway, Albany, NY 12204, (518) 447-6400   Felts, acoustical materials

All Felt Products, Inc. 2010 S. Vineyard Ave., Ontario, CA 91761, (714) 947-1774   Felts, acoustical materials

All Force Acoustics, Lord Cooperation 2001 Peninsula Dr., P.O. Box 1067, Erie, PA 16512
(814) 838-7691   Panels, curtain, barrier panels

Allied Felt Group Star Lake Ave., Bloomingdale, NJ 07403, (201) 838-1616   Felts, acoustical materials

Allied Witan Co. 13805 Progress Parkway, Cleveland, OH 44133, (216) 237-0630   Fan silencers, general industrial silencers

Alpha Dyne Inc. 727 Harding St., NE, Minneapolis, MN 55413, (612) 378-1080   Panels

Alpha-One Mfg. Co., Ltd. P.O. Box 299, Copiagne, NY 11726, (516) 822-4141   Acoustical materials

Amber/Booth Co., Inc. 1403 N. Post Oak, Houston, TX 77055 (713) 688-1228 E-mail: sales@amberbooth.com

American Acoustical Products 6 October Hill Rd., Holliston, MA 01746 (508)429-1165
E-mail: americanacoustic@aol.com   Felts, foam barrier composites, fiber barrier composites, quilted composites

American Enka Co. Enka, NC 28729, (714) 667-7713 Floating floors

American Air Filter Co. Box 35690, Louisville, KY 40232, (502) 637-0011   Enclosures, filters

American Boa Inc. P.O. Box 1301, Cumming, GA 30130, (404) 889-9400   Absorbers, ducts, absorptive materials

American Felt & Filter Co. 34-1D John St., P.O. Box 95, Newburg, NY 12550, (914) 561-3560   Felts, acoustical materials

American Industrial Contracting Inc. 264 E. Beaver St., Sawickley, PA 15143, (412) 761-7806   Panels, acoustical materials

American National Rubber Co. P.O. Box 7338, Huntington, WV 25778, (800) 624-3410   Acoustical materials

American Star Cork Co., Inc. 35-53 62nd St., Woodside, NY 11377, (718) 335-3000   Acoustical materials

Anaconda 698 South Main St., P.O. Box 2618, Waterbury, CT 06723, (203) 574-3500   Ducts, vibration isolation materials

Anamet Inc. 698 S. Main St., P.O. Box 2618, Waterbury, CT 06725 (203) 574-3500   Vibration isolation materials

Antipon Inc. 62 Omega Dr., Newark, DE 19713, (302) 454-7666   Felts, foams, glass fiber, mass-loaded plastic barriers, sealants, fiber barrier composites, foam barrier composites, vibration damping materials

Architectural Design Products, P.O. Box 575 American Fork, UT 84003, (801)756-6046   Acoustical materials

Armstrong World Inductries, Inc. P.O. Box 3001, Lancaster, PA 17604 (717) 397-0611   Mineral fiber, wall treatments, ceiling systems, panels, vibration damping materials

Air Crest Products Co., Inc. 500 W. Cermak, Chicago, IL 60616   (312) 733-7117   Acoustical products

Aspen Plastics Inc. 1425 W. Fullerton, Addison, IL 60101, (312)627-4430   Enclosures, acoustical materials

Auralex Acoustics Inc. 8851 Hague Rd., Indianapolis, IN 46256, (800) 959-3343 Web site: www.auralex.com Foam sheet, absorption and diffusion panels, clear vinyl barriers, bass traps, complete control-room kits

Automation Devices, Inc. P.O. Box AD, Automation Park, Fairview, PA 16415 (814) 474-5561   Enclosures

Barray Products, Inc. 237 25th St., Brooklyn, NY 11232, (718) 965-7000   Acoustical materials

Barley Earhart Co. 233 Divine Hwy., Portland, MI 48875, (517) 647-4117   Felts, foams, mass-loaded plastic barriers, curtain, fiber barrier composites, foam barrier composites, composite curtains

Barry Controls 40 Guess St., Brighton, MA 02135, (617)787-1555; 254-7381
Website: www.barrymounts.com   Vibration isolation mounts and materials

Barrier Corp. 9908 SW Tigard St., Tigard, OR 97223, (503) 639-4192   Foams, mass-loaded plastic barriers, foam barrier composites

Beaird Industries, Inc. 601 Benton Kelly St., Shreveport, LA 71106 (318)868-1701
Web site: www.beairdindustries.com   Duct silencers, fan silencers, exhaust silencers

Blachford, Inc. 1400 Nuclear Drive, West Chicago, IL 60185 (630)231-8300 Web site: www.blchford.com Foams, glass fiber, loaded plastics, pipe lagging, composites, enclosures, vibration damping materials

Bostik Division Of Emhart Boston St., Middleton, MA 01949, (617) 777-0100 Vibration damping materials

Boyde Corp. 6630 Owens Dr., Pleasanton, CA 94566 (415) 463-1760   Acoustical materials

BRD Noise and Vibration Control 112 Fairview Ave., P.O. Box 127, Wind Gap, PA 18091 (610)863-6300 Foam, glass fiber, perforated sheet metal, spray-on coatings, panels absorptive materials, wall treatments, mass-loaded plastic barriers, curtains, doors, operable partitions, barrier panels, walls, windows, fiber barrier composites, foam barrier composites, composite curtains, open plan partitions, vibration damping materials, vibration isolation materials, ducts, duct silencers, electric motor silencers, fan silencers, general industrial silencers

Brand Industries Inc. 1420 Renaissance Dr., Park Ridge, IL 60068 (312) 298-1200   Acoustical materials

Barwley Felt Co., Inc. 160 Fifth Ave., New York, NY 10010 (212) 243-7779   Felt, acoustical materials

The Brewster Corp. Old Saybrook, CT 06247, (203) 388-4441   Panels, partitions

Brunswick Corp. Technetics Div., 2000 Brunswick Lane, Deland, FL 32724, (904) 736-1700   Felt, duct silencers, fan silencers

Buckey Rubber Product Inc. N. Jackson and Raber Sts., Linda, OH 45801, (419) 228-4441   Acoustical materials

Burgess-Manning, Inc. 27 Thomas Ave., Orchard Park, NY 14127
(716)662-6540   Barrier panels, walls, windows, ducts, duct silencers, fan silencers, general industrial silencers

Burnett Processing   Court St. Rd., Syracuse, NY 13206 (315) 437-1131   Acoustical materials

Canada Metal Co., Ltd.   Noise Control Division, 721 Eastern Ave., Toronto, Ontario, Canada M4M 1E6 (416) 465-4684   Barriers

Capaul Ceilings   210 W. 33rd St., Suite 129, Oakbrook, IL 60521 (312) 325-9242   Panels, ceilings

L.E. Carpenter   880 Third Ave., New York, NY 10022 (212) 751-3565   Panels

Cary Corp.   9950 Rittman Rd., Wadsworth, OH 44281 (216) 334-1524   Various acoustical material

Certainteed   5587 New Peachtree Rd., Atlanta, GA 30366 (404) 457-1171   Ducts

The Celotex Corp.   Building Product Div., P.O. Box 22602, Tampa, FL 33622 (813)871-4543   Ceilings

Chemfab   Materials Technology Div., Water St., P.O. Box 476, North Bennington, VT 05257 (802) 447-1131 Absorptive fabric

Charles W. House & Sons, Inc.   19 Perry St., P.O. Box 158-1, Unionville, CT 06085 (800) 243-7064   Felt, acoustical material

Chase Ind. Inc.   8104 Reading Rd., Cincinnati, OH 45222 (513) 821-3939   Acoustical material

Chemlex Div. Raychem Corp.   2555 Bay Rd., Box 8036, Redwood City, CA 94063 (415) 361-4900 Absorbers, ducts, absorptive materials

Chemical Coating and Engineering Co., Inc. Brook St. & Baltimore Pike, Media, PA 19063 (215) 566-7470   Acoustical material

Chestnut Ridge Foam Inc.   P.O. Box 781, Latrobe, PA 15650 (412) 537-9000 Acoustical material

Childers Products Co.   23350 Mercantile Rd., Cleveland, OH 44122 (216) 464-8020   Panels

Cid Assoc. Inc.   P.O. Box 445, Dept. 11, Oakmont, PA 15136 (412) 828-2495   Acoustical materials

Claremenont Co.   175 State St., Meriden, CT 06450, (203) 238-2384   Acoustical materials

Clark Cutler McDermott Co.   Franklin, MA 02038 (617) 528-1200   Felts, acoustical materials

Clark Foam Products Corp.   4632 W. 53rd St., Chicago, IL 60632 (312) 284-6900   Foam

ClearSonic Mfg. Inc.   1223-B Norton Rd., Hudson, OH 44236 (330) 650-1420, Web site: www.clearsonic.com Clear vinyl barriers, diffusion and absorption panels

Clevaflex Clevepak Corp.   2500 Westchester Ave., Purchase, NY 10577 (914) 251-2726   Ducts, absorptive material

Commercial Acoustics   5960 W. Washington St., Phoenix, AZ 85043 (602)233-2322, Web site: www.mfmca.com   Perforated sheet metal, panels, doors, windows, enclosures, silencers

Comp-Aire Systems, Inc.   4185 44th St., Grand Rapids, MI 49508 (616) 698-9660   Enclosures, acoustical materials

Compo Industries, Inc.   Panel Chemical Division, 21 River Dr., Cartersville, GA 30120   Vinyl & urethane absorptive materials

Concrete Plank Co., Inc.   2 Porete Ave., North Arlington, NJ 07032 (201) 996-7600   Acoustical materials

Continental Felt Co.   22 W. 15th St., New York, NY 10011   (212) 929-5262   Felt

Control Elect. Co., Inc.   107-8 Allen Blvd., Farmingdale, NY 11735 (516) 694-0125   Barrier, absorptive materials

Controlled Acoustics Corp.   12 Wilson St., Hartsdale, NY 10530 (014) 428-7740   Felts, perforated sheet metal, wall treatments, ceiling systems, wall treatment systems, mass-loaded plastic barriers, curtains, doors, operable partitions, barrier panels, walls, windows, fiber barrier composites, foam barrier composites, composite curtains, enclosures, open plan partitions, composite panels, quilted composites, vibration isolation materials, ducts, duct silencers, fan silencers, general industrial silencers

Conweb Ceiling Products Div.   332 Minnesota St., P.O. Box 43237, Saint Paul, MN 55447 (612) 221-2284 Ceiling and wall panels

Craxton Acoustical Products Ind.   2838 Vicksburg Lane, Plymouth, MN 55447 (612) 553-9890   Ceiling systems, absorptive materials, wall treatment systems

Crest Foam Industries Inc.   100 Carol Place, Moonachie, NJ 07074 (201)807-1113, Web site: www.crestfoam.com   Reticulated (absorption) and compressed (damping), polyurethane foam

CTA Acoustics, Inc.   560 Kirts Blvd., Suite 120, Troy, MI 48084 (248)362-9550, Web site: www.ctaacoustics.com   Foams, glass and mineral fiber, panels, composites, silencers

Cupples Products Division   Robertson Co., H. H., 25050 S. Hanley Rd., St. Louis, MO 63144 (314) 781-6729   Enclosures

Cyro Ind.   155 Tice Blvd, Box 488, Westwood, NJ 07675 (201) 930-0100   Acoustical material, panels

E. J. Davis Co.   10 Dodge Ave., P.O. Box 326, North Haven, CT 06473 (203)239-5391, Web site: www.ejdavis.com   Felts, foams, glass and mineral fiber, panels, curtains, composites

Dayco Corp.   333 W. First St., Dayton, OH 45402 (513) 226-7000   Barriers, absorptive materials

DBA Inc.   P.O. Box 413, Buford, GA 30518 (404) 945-2929   Acoustical materials

Delaware Valley Corp.   500 Broadway, Lawrence, MA 01841 (617) 688-6995   Felt, fabric, insulations

DeMarco MAX VAC Corp.   1412 Riverview Dr., McHenry, IL 60050 (815)344-2222, Web site: www.maxvac.com   Silencers

Designer Acoustics, Inc. 2070 Five Mile Line Rd., Penfield, NY 14526 (716) 385-3320   Ceiling systems, panels, absorptive materials, composite panels

Devac Inc. 10130 State Highway 55, Minneapolis, MN 55441 (800) 328-5717   Windows

Devcon (800)933-8266, Web site: www.devcon.com Castable (non-shrinking) liquid urethane (vibration damping material) with selectable hardnesses

Diamond Mfg. Co. 243 W. 8th St., Wyoming, PA 18644 (570)693-0300   Perforated sheet metal

Discrete Technology (516) 678-2254   Sound-sorber panels

Dodge Cork Co., Inc. 13 Laurel St., Lancaster, PA 17604 (717) 295-3400   Cork

Donray Co. 8500 So. M. Center Rd., Cleveland, OH 44143 (216) 449-6450   Barriers, absorptive materials

Door-Man Manufacturing Co. 721 E. Elmwood, Troy, MI 48083 (313) 583-6000   Sheet glass metal & plastic, doors, barrier panels, fiber barrier composites, enclosures, duct silencers

Dupont Company 1004 Market St., Wilmington, DE 19898 (302) 774-1000   Mass-loaded plastic barriers

Duracote Corp. 350 Diamond St., Revenna, OH 44266 (330) 296-9600, Web site: www.duracote.com Felt, wall treatment, panels, absorptive materials, wall treatment systems, curtain operable partitions, barrier panels, fiber barrier composites, foam barrier composites, composite curtains, open plan partitions, composite panels, vibration damping materials, mass loaded plastics

E.A.R. Specialty Composites 7911 Zionsville Rd., Indianapolis, IN 46268 (317)962-1111, Web site:www.earsc.com   Foams, barriers, composites, Damping, vibration isolation, grommets, mounts, gasketing seals, moldable damping material, pipe lagging, loaded plastics, panels, composites, vibration damping materials, vibration isolation systems

Eagle Ind. 34300 Lakeland Blvd., Eastlake, OH 44094 (216) 951-0600

Eckel Industries Acoustics Division, 155 Fawcett St., Cambridge, MA 02138 (617) 491-3221, Web site: www.eckelacoustic.com   Felt, perforated sheet metal, wall treatments, ceiling systems, panels, absorptive materials, mass-loaded plastic barriers, curtains, operable partitions, barrier panels, walls, windows, foam barrier composites, composite curtains, enclosures, open plan partitions, composite panels, vibration damping materials, ducts, duct silencers, fan silencers

Empire Acoustical Systems 36744 Constitution Dr., Trinidad, CO 81082 (719)846-2300, Web site: www.empireacoustical.com   Mineral fiber, perforated sheet metal, panels, wall treatments, curtains, doors, composites, enclosures, silencers

Enatech Corp. 961 Shallow Ford Rd., Kenesaw, GA 30144 (404) 926-3422   Ducts, duct silencers, fan silencers

Engineering Sales and Service P.O. Box 558, Dayton, OH 45405 (513) 277-2047   Foams, spray-on coatings, wall treatments, panels, absorptive materials, wall treatment systems, mass-loaded plastic barriers, curtains, operable partitions, barrier panels, walls, windows, foam barrier composites, composite curtains, enclosures, open-plan partitions, composite panels, vibration damping materials, ducts, duct silencers, fan silencers

Enviro Acoustics Co. 6150 Olsen Memorial Hwy, Minneapolis, MN 55422 (612) 542-8665   Acoustical materials

Environmental Elements Corp. 3700 Koppers St., P.O. Box 1318, Baltimore, MD 21203 (301) 368-7000 Panels, absorptive materials, barrier panels, walls, ducts, duct silencers, electric motor silencers, fan silencers, general industrial silencers

Environmental Health System 105 Irving Box 1006, Framingham, MA 01701 (617)872-6042   Acoustical materials

Essi Acoustical Products Co. 11750 Berea Rd., Cleveland, OH 44111 (216)251-9933, Web site: www.essiacoustical.com   Wall treatments, ceiling systems, panels, absorptive materials, wall treatment systems, composite panels

Exxair Corp. 1250 Century Ct. N., Cincinnati, OH 45246 (513)671-3363, E-mail: techelp@xair.com General industrial silencers

Fabreeka Products Co., Inc. 1023 Turnpike St., P.O. Box 210, Stoughton, MA 02072 (781)341-3983 Web site: www.fabreeka.com   Foam, fiber barrier composites, foam barrier composites, vibration isolation materials, ducts

Faddis Concrete Products 3515 Kings Highway, Downingtown, PA 19335 (610)269-4685 E-mail: info@faddis.com   Spray-on coatings

Federal Metals P.O. Box 959, Newark, NJ 07101 (210) 589-0500   Barriers

Federated Fry Metals Inc. 6th Ave. and 41st St., Altoona, PA 16602 (814) 946-1611   Glass metal and plastic sheet, doors, walls

Ferguson Perforating and Wire 130 Ernest St., Providence, RI 02905 (401)941-8876, E-mail: sales@fergusonperf.com   Glass, metal and fiber sheet

Ferro Corp. Composites Division, 34 Smith St., Norwalk, CT 06852 (203) 853-2123   Foam and barriers

Fiberflex of Georgia Inc. Unit 15B, 690 Murphey Ave. SW, Atlanta, GA 30310 (404) 753-1360   Acoustical materials

Fiberex Inc. P.O. Box 1148, Aurora, IL 60507 (312) 896-4800   Acoustical materials

Fidelity Felt & Mfg. Co. Front and Venango Sts., Philadelphia, PA 19140   Felt and cork products

Firestone Industrial Products Co. 12650 Hamilton Crossing Blvd., Carmel, IN 46032 (317)818-8600 Web site: www.firestoneindustrial.com Vibration isolation materials

Flaregas Corp. 100 Executive Pk., Spring Valley, NY 10977 (914) 352-8877   Acoustical materials

Fluid Kinetic Corp. P.O. Box CE, Ventura, CA 93002 (805) 644-5587, Fan silencers, general industrial silencers

FM Tubecraft Support Systems, Inc. (516) 567-8588   Acoustafoam

Foam Design Inc. 444 Transport Court, Dept. US, Lexington, KY 40581 (606) 231-7006   Foam, acoustical materials

Foam Form Inc. 6703 Governor Printz Blvd., Wilmington, DE 19809 (302) 798-6843   Foam, acoustical materials

Foam-Lite Plastics, Inc. P.O. Box 5, Knoxville, TN 37901 (615) 577-2635   Foam, acoustical materials

Foam Packaging Ltd. 281-A Halstead Ave., Harrison, NY 10525 (914) 835-2230   Foam, acoustical materials

Foam Products Corp. 2525 Adie Rd., Maryland Heights, MO 63043 (314) 739-8100   Foam, acoustical materials

Foam Products Div. 122 Manton Ave., Providence, RI 02909 (401) 421-8962   Foam, acoustical materials

Foam Products Inc. 4747 Bronx Blvd., Bronx, NY 10470 (212) 324-8000   Foam, acoustical materials

Foam Products Inc. 900A 77th Ave., Oakland, CA 94621 (415)569-9681   Foam, acoustical materials

Foam Seal Inc. 2716 E. 79th St., Cleveland, OH 44104 (216) 881-8111   Foam, acoustical materials

Foam Technology Div. Lance Industries, Inc., 50 Aleppo St. Providence, RI 02909 (401)222-1710   Foam

Foam Tek, Inc. 11650 Emerald St., Dallas, TX 75229 (214) 241-0096   Foam

Foamade Industries P.O. Box 4494, 2550 Auburn Court Auburn Hills, MI 48057 (800) 221-7388   Foam

Foamair Packaging Co., Inc. 4043 Ridge Ave., Bldg. 24, Philadelphia, PA 19129 (215) 849-3545   Foam

Foams Converters, Inc. 111 Park Dr., Montgomeryville, PA 18936 (215) 641-1686   Foam

Foamedge Corp. 3850 Granger Rd., Akron, OH 44313 (216) 666-0280   Foam, acoustical materials

Foamfab Div. Brick Assoc., 5 Lawrence St., Bloomfield, NJ 07003 (201) 429-8942   Foam, acoustical materials

Frommelt Industries, Inc. 4343 Chavenelle Rd., Dubuque, IA 52004 (800) 553-5560   Curtains, doors, barrier panels, fiber barrier composites, foam barrier composites, composite curtains, enclosures, composite panels, quilted composites, vibration isolation materials

Gale Noise Control Div. Norwood Mfg., P.O. Box 6750, Bridgewater, NJ 08857 (201) 725-9500   Panels, doors, walls, enclosures, composite panels, duct silencers, fan silencers

General Acoustics Corp. 12248 Santa Monica Blvd., Los Angeles, CA 90025 (213) 820-1531   Perforated sheet metal, wall treatments, ceiling systems, panels, wall treatment systems,mass-loaded plastic barriers, curtains, doors, operable partitions, barrier panels, walls, windows, fiber barrier composites, composite curtains, enclosures, composite panels, ducts, duct silencers, electric motor silencers, fan silencers, general industrial silencers, absorptive materials

General Rubber Corp. 9 Empire Blvd., Hackensack, NJ 07606 (201) 641-4700   Absorptive materials, ducts

Carroll George Inc. 15th Street S., P.O. Box 144, Nortwood,IA 50459 (515) 324-2231   Foam barrier panel composites

Georgia Pacific Corp. 133 Peachtree St., NE, Atlanta, GA 30303 (404) 521-4728   Acoustical materials

Gerb Vibration Control Systems, Inc. 1950 Ohio St., Lisle, IL 60532 (630)724-1660, Web site: www.gerb.com Enclosures, rooms, vibration isolation systems, silencers

Gilmane Brother Co. 102 Main St., Gilman, CT Foam, acoustical materials

Globe Industries Inc. 2638 East 126th St., Chicago, IL 60633 (312) 646-1300   Felt, ceiling systems, sealants, barriers, panels, fiber barrier composites, foam barrier composites, vibration damping materials

A. Goethal Sheet Metal Inc. 6209 Industrial Court, P.O. Box 159, Greendale, WI 53129 (414) 423-0500 Absorptive materials and ducts

Gold Bond Building 2001 US Rexford Sand Rd., Charlotte, NC 28211 (704) 365-0950   Acoustical materials

Goodfriend A & Ass. Lewis S. 7 Saddle Rd., Cedar Knolls, NJ 07927 (201) 540-8811   Acoustical materials

The Goodyear Tire & Rubber Co. P.O. Box 185, Greenburg, OH 44232 (216) 896-5000   Vibration isolation

W.R. Grace and Co. Construction Product Div., 62 Whittemore Ave., Cambridge, MA 02140 (617) 876-1400 Acoustical materials

Great Lakes Industrial Assoc. 1221 S. Bowen St., P.O. Box 628, Jackson, MI 49204 (517) 784-7146   Foams, glass fiber, ceiling systems, panels, wall treatment systems, mass-loaded plastic barriers, curtains, operable partitions, barrier panels, walls, foam barrier composites, composite curtains, enclosures, composite panels, quilted composite, vibration isolation materials, fan silencers, absorptive materials

Greene Rubber Co., Inc. 160 Second St., Cambridge, MA 02142 (617) 547-7655   Foam, acoustical materials

GT Acoustical Technologies P.O. Box 9527, Providence, RI 02940 (401) 724-1900 Ceiling systems, curtains, foam barrier composites, composite curtains, enclosures

Hamfab Inc. Bridge & 9th Sts., P.O. Box 353, Leighton, PA 18235 (215) 377-4120 Acoustical materials

Harrington & King Perforating Co., Inc. 5655 W. Fillmore St., Chicago, IL 60644 (312) 626-1800 Perforated sheet metal

Henges Associates 12100 Prichard Farm Rd., Maryland Heights, MO 63043 (314) 739-2600 Foam, panels, doors, enclosures, open plan partitions, composite panels

Hexcel 11711 Dublin Blvd., Dublin, CA 94566 (415) 828-4200 Panels, absorptive materials

Higgott-Kane Industrial Noise Controls Ltd. 1085 Bellamy Road N., Scarborough, Ontario, Canada M1H 3C7 (416) 431-0641 Ceiling systems, panels, doors, operable partitions, enclosures, open plan partitions

Hilmac Inc. 26004 Atlantic Ave., Wall, NJ 07719 (201) 449-9296 Felt, acoustical materials

Hitco (ARMCO Inc.) 1600 W. 135th St., Garden, CA 90249 (213)321-8080 Felt, acoustical materials

Holtz Rubber Co., Inc. 1129 S. Sacramento St., P.O. Box 109 Lodi, CA 95241 (209) 368-7171 Ducts, absorptive materials

Hufcor P.O. Box 591, 1205 Norwood Rd., Janesville, WI 53545 (608) 756-1241 Doors, barrier panels, walls, operable partitions

Huyck Felt Renessela, NY 12144 (518) 445-2711 Absorptive fabric

Hydra Matic Packing Co. 2992 Frank Rd., Bethayres, PA 19006 (215) 676-2992 Acoustical materials

Hydra-Zorb Corp. 2450 Commercial Dr., Auburn Hills, MI 48057 (313) 373-5151 Vibration isolation materials

IDE Processes Corp. Noise Control Div., 106 81st Ave., Kew Gardens, NY 11415 (713) 544-1177, Web site: www.frontier.net/ide Panles, enclosures, composite panels, duct silencers, fan silencers, general industrial silencers

Illbruck 3800 Washington Ave. N., Minneapolis, MN 55412 (800)662-0032, Web site: www.illbruck-sonex.com Foam, wall treatments, ceiling systems, panels, absorptive materials, wall treatments, sealants, curtains, barrier panels, walls, enclosures, open plan partitions, composite panels, sonex

Imi-Tech Corp. 701 Fargo, Elk Grove Village, IL 80007 (312) 981-7610 Foam, ceiling systems, panels, barrier panels, foam barrier composites

Industrial Acoustics Company 1160 Commerce Ave., Bronx, NY 10462 (718) 931-8000, Web site: www.industrialacoustics.com Foam, wall treatments, ceiling systems, panels, absorptive amterials, wall treatments, sealants, curtains, barrier panels, walls, enclosures, open plan partitions, composite panels, sonex

Industrial Noise Control Inc. 1411 Jeffrey Dr., Addison, IL 60101 (312) 620-1998 Foam, perforated sheet metal, wall treatments, ceiling systems, panels, absorptive materials, wall treatment systems, mass-loaded plastic barriers, curtains, doors, operable partitions, barrier panels, walls, windows, fiber barrier composites, foam barrier composties, composite curtains, enclosures, open plan partitions, compostie panels, vibration damping material, ducts, duct silencers, fan silencers, general industrial silencers

Insta Foam Product, Inc. 1500 Cedarwood Dr., Joliet, IL 60435 (800) 435-9359 Foam

Insul-Art Acoustic Corp. 107 Allen Blvd., Farmingdale, NY 11735 (516) 694-0002 Panels

Insul Coustic Corp. Jernee Mill Rd., Sayreville, NJ 08872 (210) 257-6674 Acoustical materials

Insulation Corp. of America 2571 Mitchell Ave., Allentown, PA 18103 (215) 791-4200 Foam, acoustical materials

Insulation Industries Inc. Hainesport Industrial Park, P.O. Box 485, Hainesport, NH 08036 (609) 261-1600 Acoustical materials

Insulation Materials Distributors Inc. 58-25 Flushing Ave., Maspeth, NY 11378 (718) 456-8863 Acoustical materials

Interior Acoustics Inc. 176 Rt 206 South, Somerville, NJ 08876 (800) 221-0580 Glass fiber, wall treatments, ceiling systems, panels, absorptive materials, wall treatment systems, curtains, barrier panels, fiber barrier composites, composite curtains, composite panels, quilted composites

International Cellulose Corp. 12315 Robin Blvd., Houston, TX 77045 (713)433-6701, E-mail: icc@spray-on.com Spray-on coatings, walltreatments, composites, vibration damping materials

International Steel Wool Corp. P.O. Box 936, Springfield, OH 45501 (513) 323-4651 Acoustic materials

Jamison Door Co. P.O. Box 70, Hagerstown, MD 27140 (800) 532-3667, E-mail: gwh www.jamison-door.com Doors, windows

Johnston Environmental 1502 E. Chestnut, Santa Ana, CA 92701 (714) 547-8288 Foams, panels

Johns-Manville 22 E. 40th St., New York, NY Ceilings, panels

KFE Inc. 1333 10th Ave., Columbus, GA 31901 (404) 324-1189 Enclosures, curtains

The Kennedy Co. P.O. Box 1216, Scottsboro, AL 35768 (205) 259-4436, E-mail: vinylusa www.kennedyvinyl.com Mass-loaded plastic barriers, composite panels

Kimbal-Clark Corp. P.O. Box 2001, Neenah, WI 54956 (414) 729-1212 Acoustical materials

Kinetics Noise Control Inc. 6300 Irelan Pl., Dublin, OH 43016 (614)889-0480, Web site: www.kineticsnoise.com Foams, glass and mineral fiber, perforated sheet metal, wall treatments, ceilings, panels, pipe lagging, loaded plastics, curtains, composites, enclosures, vibration damping materials, vibration isolation systems

King Manufacturing Co. 1500 Spring Lawn Ave., Cincinnati, OH 54223 (513) 541-5440 Curtains, panels

Kirkhill Rubber Co. 300 E. Cypress Blvd., Brea, CA 92621 (714) 529-4901 Foam, acoustical materials

George Koch Sons, Inc. P.O. Box 358, Evansville, IN 47744 (812) 426-9880, E-mail: sales www.kochllc.com Ceiling systems, panels, wall treatment systems, sheet glass, metal & plastic, barrier panels, walls, windows, fiber barrier composites, enclosures, open plan partitions, composite panels

Koppers Co. 1900 Kopper Blvd., Pittsburgh, PA 15219 (412) 227-2000 Acoustical materials

Korfund Dynamics Corp. P.O. Box 235, Westbury, NY 11590 (516) 333-7580 Panels, curtains, barrier panels, composite curtains, enclosures, composite panels, vibration isolation materials, duct silencers, fan silencers, general industrial silencers, floating floors

Krieger Steel Products Co. 4896 Gregg Rd., Pico Rivera, CA 15219 (412) 227-2000 Acoustical materials

L-A-B Division Mechanical Technology Inc., Onondaga, Skaneatoles, NY 13152 (315) 685-5781 Vibration isolation materials

Laubenstein Mfg. Co. 419 So. Blvd., Ashland, PA 17921 Enclosures, acoustical material

Lauren Manufacturing Co. 2228 Reiser Ave., S.E., New Philadelphia, OH 44663-3345 (330)339-3373, (800)683-0676, Web site: www.lauren.com Elastomeric products including soft and dense sponges, weather stripping with pre-applied pressure sensitive adhesive tape, clip-on and other seals, silicones, polymers extrusions, flexible compressed cellular foams and other vibration isolation and sealing materials,

Lead Industries Assoc. Inc. 292 Madison Ave., New York, NY 10017 Lead sheeting

Lewcott Chemicals & Plastics 86 Providence Rd., Millbury, MA 01527 (800) 225-7725 Acoustical materials

Linear Products Corp. P.O. Box 902, Cranford, NJ 07016 (201) 272-2211 Foam, glass fiber, perforated sheet metal, wall treatments, panels, absorptive materials, wall treatment systems, mass-loaded plastic barrier, sealants, curtains, doors, operable partitions, barrier panels, walls, windows, fiber barrier composites, foam barrier composites, composite curtains, enclosures, open plan partitions, composite panels, quilted composites, vibration damping materials, vibration isolation materials, duct silencers

Lorbrook Corp. 730 State St., Hudson, NY 12534 (518) 828-1592 Felt, acoustical materials

Lord Corporation Industrial Div. P.O.Box 10039, Erie, PA 16514 (814) 456-8511 Vibration isolation materials

Leveland Assoc. Inc. P.O. Box 3485, Peabody, MA 01960 (617) 531-7235 Acoustical materials

Lundell Manufacturing Corp. Minneapolis, MN (763) 559-4114, Web site:wwwlundellmfg.com Foams, composite foams some with protective and decorative surfaces.

3M Acoustic Control Systems 3M Center-220-7E-01, St. Paul, MN 55144 (651)733-5245, Web site:www.mmm.com Glass and mineral fiber, perforated sheet metal, curtains, panels, composites, vibration damping materials, vibration isolation systems, silencers

Machinery Mountings, Inc. 11 Constance Ct., Haupauge, NY 11788 (631)851-0480 Machinery mounts, springs, vibration dampers, vibration isolation systems

Manville Corp. P.O. Box 5108, Denver, CO 80217 (303) 978-4900 Felt, glass fiber, mineral fiber, wall treatments, ceiling systems, panels, wall treatment systems, walls, windows, fiber barrier composites, foam barrier composites, enclosures, open plan partitions, quilted composites, vibration isolation materials, ducts, duct silencers, electric motor silencers, fan silencers, general industrial silencers

Markertek (845) 246-3036, (800) 522-2025 Web site: www.Markertek.com Foam sheet, absorption panels and basstraps

Mason Industries Inc. 350 Rabro Dr., Hauppauge, NY 11788 (516) 348-0282, E-mail: info www.mason-ind.com Spray-on coatings, vibration isolation materials, floating floors

MB Dynamics 25865 Richmond Rd., Cleveland, OH 44146 (216) 292-5850 Vibration isolation materials

MBI Metal Building Interior Products Co. 5309 Hamilton Ave., Cleveland, OH 44114 (216) 431-6400 Web site: www.mbiproducts.com Glass fiber, mineral fiber, wall treatments, ceiling systems, panels, wall treatment systems, curtains

McCarty Industries, Inc 81 Drendel Lane, Naperville, IL 60565 (630)773-1721, Web site: www.mccarty.com Felts, foams, glass and mineral fiber, pipe lagging, seals, loaded plastics, composites, vibration damping materials

McGill Air Pressure Corp. 190 E. Broadway Ave., Westerville, OH 43081 (614)797-2106, Web site:www.mcgillairpressure.com Foams, perforated sheet metal, spray-on coatings, wall treatments, panels, pipe lagging, loaded plastics, curtains, doors, operable partitions, seals, windows, composites, enclosures

McNichols Co. 5501 Gray St., Tampa, FL 33609 (800) 237-3820 Absorptive materials, ducts

Mearl Corp. The Foam and Chemicals Div. 220 W. Wastfield Ave., P.O. Box 208, Roselle Park, NJ 07204 (201) 245-9500   Foam, acoustical materials

Mechanical Felt and Textiles, Co., Inc. 1075 Louson Rd., Union, NJ 07023 (201) 688-0690   Barriers, absorptive materials

Mikco Mfg. Inc. 1063 W. Washington Ave., P.O. Box 126, Cleveland, WI 53015 (414) 693-8163   Enclosures

Metalastik Vibration Control Systems (Dunlop) 3 Pullman Court, Scarborough, Ontario, Canada M1X 1E5 (416) 297-0565   Vibration isolation materials

Metal Industries Inc. 1310 N. Hercules Ave., Clearwater, FL 33518 (813) 441-2651   Absorptive materials, ducts

Met-L-Wood 6755 W. 65th St., Chicago, IL 60638 (312) 458-5900   Panels

Metric Felt Co. 35 S. Peoria, Dept. A, Chicago, IL 60607 (312) 733-0040   Felt

Middle Atlantic Products North Corperate Dr., Riverdale, NJ 07457 (973)839-1976, Web site: www.middleatlantic.com   Absorption panels

Midwest Acoust-A-Fiber Inc. 7790 Marysville Rd., Ostrander, OH 43061 (614) 666-2231   Foams, mineral fiber, wall treatments, ceiling systems, panels, wall treatment systems, curtains, fiber barrier composites, foam barrier composites, open plan partitions, composite panels, quilted composites, vibration damping materials

Milcut Inc. 4837 W. Woolworth Ave., Milwaukee, WI 53218 (414) 353-1300   Foam, ceiling systems, panels, wall treatment systems, curtains, fiber barrier composites, foam barrier composites, composite curtains, composite panels

Minerals Pigments and Metal Division Pfizer Inc. E. 42nd St., New York, NY 10017 (212) 573-2816 Acoustical materials

Minnesota Flexible Corp. 14944 Minnetonka Industrial Rd., Minnetonka, MN 55345   Absorptive materials, ducts

Minor Rubber Company Inc. 49 Ackerman St., Bloomfield, NJ 07003 (800)433-6886, Web site: www.minorrubber.com   Natural rubbers, compounds, polymers, urethanes, neoprene and silicone elastomer products in raw block or extrusion, grommet, vibration isolator, bumper, boot and bellow form, with custom molding available.

Modernfpold P.O. Box 310, New Castle, IN 47363 (317) 529-1450   Panels

Molded Acoustical Product, Inc. One Danforth Dr., Easton, PA 18042 (215) 253-7135, E-mail: psbjab@nowonline.com   Absorptive materials, ducts

Morgan Electro Ceramics 232 Forbes Rd., Bedford, OH 44146 (440)232-8600   Vibration isolation, active dampers

MPC Inc. Noise Control Products Div., 835 Canterbury Rd., Westlake, OH 44145 (440) 835-1405, Web site: www.mpcsilentwale.com   Glass fiber, wall treatments, ceiling systems, panels, absorptive material, wall treatment systems, barrier panels, walls, enclosures, open plan partitions, composite panels

M/RAD Corp. 71 Pine St., Woburn, MA 01801 (617) 935-5940   Vibration isolation materials, floating floors

MSC Laminates & Composites, Inc. 2300 E. Pratt Blvd., Elk Grove Village, IL 60007 (847)806-2100 Web site: www.laminatesandcomposites.com Constrained-layer composites, vibration damping materials.

MTS Systems 14000 Technology Dr., Minneapolis, MN 55344 (800)933-4617, Web site: www.mts.com Vibration dampers, vibration damping materials, vibration isolation systems

E.N. Murray Co., Inc. 707 Umatilla St., Denver, CO 80204 (303) 892-1106   Felts, glass fiber, wall treatments, absorptive materials, mass-loaded plastic barriers, curtains, fiber barrier composites, foam barrier composites, composite curtains, vibration damping material, vibration isolation material, duct silencers, general industrial silencers

Muth Co., K.W. 2821 North 19th St., Sheboygan, WI 53081 (414) 458-9181   Acoustical materials

NABCO Glazed Products Div. Trenwyth Industries, Inc., One Connelly Rd., Emigsville, PA 17318 (717) 767-6868   Acoustical materials

National Cellulose Corp. 12315 Robin Blvd., Houston, TX 77245 (713) 433-6701   Spray-on coatings, wall treatments, ceiling systems, wall treatment systems, walls, enclosures

National Felt Co. 25 Mechanic St., Easthampton, MA 01027 (413) 527-3445   Felt

National Perforating Corp. Parker St., Clinton, MA 01510 (800) 225-9050   Perforated sheet metal

Neiss Corp. P.O. Box 478, Rockville, CT 06066 (203) 872-8528   Panels, wall treatment systems, operable partitions, barrier panels, walls, enclosures, open plan partitions, composite panels

Netherland Rubber Co. Boone and Burbank Sts., Cincinnati, OH 45206 (513) 221-4800   Felt, acoustical materials

NetWell Noise Control 6125 Blue Circle Dr., Minnetonka, MN 55343 (800) 638-9355 Web site: www.controlnoise.com   Foam sheet, absorption panels, vinyl barrier and composite panels, bass traps

The Noble Company P.O. Box 350, Grand Haven, MI 49417 (800)878-5788, Web site: www.noblecompany.com   Floating floors, vibration isolation systems

Noise Control Assoc., Inc. 32 Park St., Montclair, NJ 07042 (201) 746-5181 Glass fiber, wall treatments, ceiling systems, wall treatment systems, mass-loaded plastic barriers, sealants, glass, metal and plastic sheet, curtains, barrier panels, fiber barrier composites, foam barrier composites, composite curtains, enclosures, open plan partitions, composite panels, quilted composites, vibration damping material, ducts, duct silencers

Noise Control Co. 6217 E. Husking Peg Rd., Chana, IL 61015 (815)453-2462, Web site: www.noisecontrolco.com Foam, glass fiber, wall treatments, ceiling systems, panels, absorptive materials, wall treatment systems, mass-loaded plastic barriers, glass, metal & plastic sheet, curtains, doors, operable partitions, barrier panels, walls, windows, fiber barrier composites, foam barrier composites, composite curtains, enclosures, composite panels, quilted composites, vibration damping materials, ducts

Noise Control Products, Inc. 800 Chettic Ave., Copiague, NY 11726 (516) 226-6100 Foams, panels, barrier panels, walls, foam barrier composites, enclosures, composite panels, duct silencers, fan silencers

Noiseknock Industrials P.O. Box 1400-L Northridge Branch, Dayton, OH 45414 (513) 237-8300 Barriers, panels

Noise Reduction Corp. 32321 Country Hwy. 25, Redwood Falls MN 56283 (507) 644-3067 E-mail: noisered@means.net Foam, glass fiber, sealants, curtains, foam barrier, composite curtains, enclosures, vibration damping materials

Noise Suppression Technologies, Inc. 4182 Fisher Rd., Columbus, OH 43228 (614)258-4455, Web site: www.noisesuppression.com Foams, glass fiber, spray-on coatings, wall treatments, ceilings, panels, pipe lagging, loaded plastics, curtains, composites, enclosures, vibration damping materials

North American Vinyl Ind. 6632 W. Wrightwood Ave., Chicago, IL 60635 (312) 237-2235 Acoustical materials

Norwood Industries Sub Seton, Co., 100 N. Morehall Rd., Malvern, PA 19355 (800) 523-0336 Foams

Nott Co., Rubber Fabrication Div., 1724 New Brighton Blvd., P.O. Box 1422, Minneapolis, MN 55440 (612) 781-9561 Acoustical materials

Nuclead Inc. 110 Pacific St., Cambridge, MA 02138 (617) 876-7217 Lead sheeting

O'Brien Partition Co., Inc. 5301 E. 59th St., Kansas City, MO 64130 (800)821-2595 Panels, absorptive materials, wall treatment systems, walls

Ohio Rubber Co. 3911 Ben Hur Ave., Willoughby, OH 44094 (216) 942-0500 Insulating materials, ducts, vinyl-clad rubber sheet

Overly Manufacturing Co. P.O. Box 70, Greensburg, PA 15601 (800) 979-7300, (412) 834-7300 Doors, windows

Owens Corning Fiberglas Corp. Fiberglas Tower, Toledo, OH 43659 (419) 248-8000 Felt, mineral fiber, spray-on coatings, wall treatments, ceiling systems, panels absorptive materials, wall treatment systems, barrier panels, fiber barrier composites, foam barrier composites, vibration isolation materials, vibration damping material, ducts

Ozlite Corp. 1755 U.S. Butterfield Rd., Libertyville, IL 60048 (312) 362-8210 Acoustical materials

Panelford Inc. P.O. Box 680130, Miami, FL 33168 (305) 688-3501 Panels

Paramont Inc. 150 U.S. Fieldcrestone, Edison, NJ 08817 (201) 755-1400 Foam, acoustical materials

Parsons Mfg. Corp. 1055 O'Brien Dr., Menlo Park, CA 94025 (415) 324-4726 Barrier, absorptive materials

Patterson Asc. 935 Summer St., Lynnfield, MA 01940 (781)334-5777, E-mail: patersoncomp@aol.com Foam, wall treatments, panels, absorptiveoperative partitions, barrier panels, walls, foam barrier composites, composite curtains, enclosures, open plan partitions, composite panels, duct silencers, electric motor silencers

Peabody Noise Control Inc. 6300 Irelan Place, Dublin, OH 43017 (614) 889-0480 Foams, glass fiber, wall treatments, panels, wall treatment systems, curtains, barrier panels, fiber barrier composites, foam barrier composites, composite curtains, enclosures, composite panels, quilted composites, vibration damping materials, vibration isolation materials, floating floors

Phoenix Fabrication, Inc. 1559 Pine St., Oxnard, CA 93030 (805) 483-1143 Duct silencers

Pioneer Industries 401 Washington Ave., Carlstadt, NJ 07072 (201) 933-1900 Panels, doors, operable partitions, barrier panels, windows

Pittsburgh Corning Corp. 800 Presque Isle Dr., Pittsburgh, PA 15239 (412) 327-6100 Panels, ceilings

Plastomer Corp. 37819 Schoolcraft Rd., Livonia, MI 48150 (313) 464-0700 Enclosures, barriers

Gordon J. Ploock & Assoc. Inc. 19120 Old Detroit Rd., Rocky River, OH 44114 (216) 333-8710 Wall treatments, ceiling systems, panels, absorptive materials, wall treatment systems, curtains, doors, operable partitions, barrier panels, walls, composite curtains, enclosures, open plan partitions, composite panels, quilted composites, ducts, duct silencers, electric motor silencers, fan silencers

Premetco International Inc. P.O. Box 1426, Shreveport, LA 71164 (318) 797-1111 Acoustical materials

Presto Manufacturing Co., Inc. 2 Franklin Ave., Brooklyn, NY 11211 (718) 852-0187 Fiber barrier composites, foam barrier composites, composite curtains, quilted composite, vibration damping material

Primacoustic (a division of JP CableTek Electronics Ltd.) 114-1585 Broadway, Port Coquitlam, British Columbia, Canada V3C 2M7 (604) 942-1001, Web site: www.primacoustic.com   Panels, bass traps and complete control-room kits

Proco Product Inc.
3274 Tomahawk Dr., P.O. Box 590, Stockton, CA 95201 (800) 344-3246   Ducts, absorptive materials

Prota FAb Corp. 2319 Grissom Dr., St. Louis, MO 63146 (314) 567-4444   Panels, barrier panels, walls, enclosures, composite panels

Proudfoot Company Inc. P.O. Box 1537, Greenwich, CT 06836 (203) 869-9031, E-mail: info www.sound-blox.com   Glass fiber, wall treatments, absorptive materials, wall treatment systems, glass metal & plastic sheeting, barrier panels, walls, fiber barrier composites, masonry, composite curtains, quilted composites

Pyrok, Inc. 121 Sunset Rd., Mamaroneck, NY 10543 (914)777-7070, Web site: www.pyrokinc.com Spray-on coatings, wall treaments, ceilings

Quaker State Oil Refining Co. Sound Off Div., Oil City, PA 16301   Ceilings, panels, barriers

Quietflo Noise control Div. 103 Airport Executive Pk., Nanuet, NY 10954 (914) 352-8877   Ducts, duct silencers, electric motor silencers, fan silencers, general industrial silencers, dampers, louvre silencers

Raxxess 261 Buffalo Ave., Paterson,NJ, 07503 (800) 398-7299, E-mail: sales@raxxess.com   Noise isolating equipment racks

Rectical Foam Corp. P.O. Box 369, Laport, IN 46350 (21() 326-7977   Foam, foam barrier composites

Rectical Foam Corp. Hambelin Ave., Morristown, TN 37814 (615) 581-8350   Foam, acoustical materials

Reddaway Mfg. Co., Inc. 32 Euclid Ave., Newark, NJ 07105 (201) 589-1410   Acoustical materials

Reliable Plastic, Inc. Chimney Rock Rd., Boundbrook, NJ 08805 (201) 469-8686   Foam, acoustical materials

Rempag Foam Corp. 61 Kuller Rd., P.O.Box 2585, Clifton, NJ 07015 (973)88-18880, Web site: www.rem-pac.com   Industrial suppliers of most types of open and closed cell foams including those with skin-like surfaces and pre-applied pressure sensitive adhesives for noise and vibration control

Richards Parents & Murry Inc. 206 S. 14th Ave., Mount Vernon, NY 10550 (914) 664-3464   Acoustic materials

Richards Willcox Mfg. Co. 339 Third St., Aurora, IL 60507 (312) 897-6951   Doors, operable partitions, barrier panels

Riley-Beaird, Inc. P.O. Box 31115, Shreveport, LA 71130 (318) 865-6351   Fan silencers, general industrial silencers

Rink Sound Control Group Div. Philips Industries P.O. Box 5486, Tucson, AZ 85703 (602) 622-7601 Perforated sheet metal, wall treatments, panels, absorptive materials, wall treatments, enclosures, composite panels, vibration isolation materials, duct silencers, fan silencers

Roblin Building Product Inc. 786 Terrace Blvd., Depew, NY 14043 (716) 683-0227   Acoustical materials

Rogers Corp. Poron Materials Div., 245 Woodstock Rd., Woodstock, CT 06281 (860)928-3622, Web site: www.rogers-corp.com/bmu   Silicones and cellular urethane foam gaskets, seals or shock and vibration isolation

Rogers Foam Corp. 20A Vernon St., Somerville, MA 02145 (617) 623-3010   Foam, acoustical materials

Roush Industries 11953 Market St., Livonia, MI 48150 (800)486-3637, Web site: www.roushind.com   Foams, vibration damping materials, vibration isolation systems

RPG Diffusor Systems 651-C Commerce Drive, Upper Marlboro, MD 20774 (301) 249-0044, Web site: www.rpginc.com   Diffusion and absorption panels, bass traps, control-room kits, room sizing and modeling software

M.W. Sausse & Co., Inc. 25590 W. Ave. Stanford, Valencia, CA 91355 (818) 367-2211   Vibration isolation materials

Scotfoam Corp. 1500 E. 2nd St., Eddystone, PA 19013 (215) 876-2551   Vibration isolation materials

Sealwall Products, Inc. 36310 Lakeland Blvd., Eastlake, OH 44094 (216) 951-3445   Acoustical materials

Semco Mfg. Inc. 409 Vandiver, P.O. Box 1797, Columbia, MO 65205 (314) 443-1481   Ducts, absorptive materials

Sentinel Foam Products Inc. 130 North St., Hyannis, MA 02601 (800) 323-5005   Foam

Shelmark Industries Inc. 1100 Steelwood Dr., Columbus, OH 43212 (614) 486-5234   Enclosures, barriers

Silent Source (800)583-7174, (413)584-7944, Web site: www.silentsourse.com   Foam sheet, absorption panels

Simplex Mfg. Co., Inc. Linene Pl., Auburn, NY 13022 (315) 252-7524   Acoustical materials

Singer Partitions 444 N. Lake Shore Dr., Chicago, IL 60611 (312) 222-1860   Barriers, curtains

Singer Safety Co. 3800 N. Milwaukee Ave., Chicago, IL 60641 (800) 621-0089, Web site: www.singersafety.thomasregister.com   Foam, ceiling systems, panels, absorptive materials, wall treatment systems, mass-loaded plastic barriers, curtains, doors, fiber barrier composites, composite curtains, composite panels, quilted composites, vibration damping materials

Smart-Aim Co. P.O. Box 1006, Framingham, MA 01701 (617) 877-4308   Spray-on coatings, panels, wall treatment systems, curtains, doors, operable partitions, barrier panels, walls, composite curtains, enclosures, open plan partitions

Sommer Cork Co. 6342-C West Irving Park Rd., Chicago, IL 60634 (312) 283-5340   Acoustical materials

Sonic Barrier Sound Products 3400 Lysander Lane, Richmond, British Columbia, Canada U7B 1C3 (604) 273-5171   Panels

Sorbothane, Inc. 2144 State Rte. 59, Kent, OH 44240 (330)678-9444, Web site: www.sorbothane.com   Vibration isolation, mounts, wall treatments, vibration damping materials, vibration isolation systems

Soundcoat Co., Inc. One Burt Drive, Deer Park, NY 11729 (516) 242-2200, E-mail: info www.soundcoat.com   Felt, wall treatments, wall treatment systems, mass-loaded plastic barriers, curtains, fiber barrier composites, foam barrier composites, composite curtains, quilted composites, vibration damping materials, fan silencers, absorptive materials

Sound Construction & Engineering Co. 522 Cottage Grove Rd., Bloomfield, CT 06002 (203) 243-1428   Panels, barrier panels, walls, ducts, fan silencers, duct silencers

Sound Control Material & Instrument Co. 734 E. Hyde Park Blvd., Inglewood, CA 90302 (213) 673-5195   Acoustical materials

Sound Fighter Systems, Inc. 6135 Linwood Ave., Shreveport, LA 71106 (318) 868-3626, Web site: www.soundfighter.com   Acoustical materials

Soundseal 50 HP Almgren Dr., Agawam, MA 01001 (413)789-4444, Web site: www.soundseal.com   Foams, glass fiber, spray-on coatings, wall treatments, pipe lagging, loaded plastics, curtains, doors, panels, composites, enclosures, vibration isolation systems

Spaulding Fiber Co. Inc. 310 Wheeler St., Tonawanda, NY 14150 (716) 692-2000   Barrier, absorptive materials

Specialty Composites Corp. Delaware Industrial Park, Newark, DE 19713 (800) 544-5180   Mass-loaded plastic barriers, sealants, composite panels, vibration damping materials

Specialty Doors, Inc. 269 W. 154th St., Holland, IL 60473 (312) 339-4331   Door, barrier panels, windows

The Spencer Turbine Co. 600 Day Hill Rd., Windsor, CT 06095 (860)688-0098, Web site: www.spencerturbine.com   Fan silencers

Spray Tech Corp.
1457 N. Main St., Delophos, OH 45833 (800) 272-7772   Sound insulation

Standard Felt Co.
115 S. Palm Ave., Alhambra, CA 91802 (213) 283-1106   Felt, acoustical materials

StanPro 2401 South Gulley Rd., Dearborn MI 48124 (800)333-3265, (313)562-2352, Web site: www.stanpro.com   Aluminum core push-on self-sealing weatherstripping, decorative and protective sealing and trims, bumpers, window channels, latex foam seals and rubber glass gripping and setting seal strips

Stark Ceramics Inc. P.O. Box 8880, Canton, OH 44711 (800) 321-0662   Absorptive materials, wall treatment systems, masonry

Starrco Co. Inc. 1515 Fairview Ave., St. Louis, MO 63132 (800) 325-5259   Panels, barrier panels, walls, open plan partitions, composite panels

State Insulation 525 Johnstone Ave., Perth Amboy, NJ 08861 (201) 442-5656   Acoustical materials

Streetcraft Corp. 2700 Jackson Ave., P.O. Box 12748, Memphis, TN 38182-0748 (901) 452-5200   Absorptive materials, ducts

Steelite Inc. East First Ave., Hazkhurst, GA 31539 (912) 375-2575   Acoustic materials

Stephenson & Lawyer Inc.
3831 Patterson Ave. SE, Grand Rapids, MI 49506 (616) 949-8100   Acoustical material

Stiles Rubber Co. P.O. Box 92, Rockaway, NJ 07886 (201) 625-9660   Vibration damping materials, vibration isolation materials

Stock Drive Products Division Designatronics Inc., 2101 Jericho Turnpike, New Hyde Park, NY 11040 (516) 328-3300, E-mail: hvarum www.sdp-si.com   Vibration damping materials, vibration isolation materials

Stoddard Silencers Inc. P.O. Box 397, Grayslake, IL 60030 (312) 223-8636   Duct silencers, fan silencers, general industrial silencers

Stretchwall A Division St. Joel Berman Assoc., Inc. 42-03 35th St., Long Island City, NY 11101 (718) 271-4400   Wall treatments, panels, absorptive materials, wall treatment systems, operable partitions, barrier panels, walls

Sullivan Co. 251 S. Van Ness Ave., San Francisco, CA 94103 (415) 861-4455   Acoustical materials

Sunnex Inc. 3 Huron Drive, Natick, MA 01760-1314 (800)445-7869, Web site: www.sunnex.com   Vibration isolation materials, mounts, systems

Superior Energies Inc. 3115 Main Ave., P.O. Box 386, Groves, TX 77619 (409) 962-8549   Acoustical materials

Supreme Supply Co. 734 E. Hyde Park Blvd., Inglewood, CA 90302 (213) 678-3491   Acoustical materials

Tamer Industries 185 Riverside Ave., Somerset, MA 02725 (508)677-0900, Web site: www.tamerind.com   Foam, glass and mineral fiber, perforated sheet metal, wall treatments, panels, curtains, doors, composites, enclosures

Target Enterprises Inc. Roxbury, VT 05669 (802) 728-3081   Absorptive materials, vibration isolation materials

Taylor Merchant Corp. 212 W. 35th St., New York, NY 10001 (212) 757-7700   Barriers, absorptive materials

Taytrix 134 Bay St., Jersey City, NJ, 07302 (800)829-8749, Web site: www.taytrix.com   Movable and stackable reflective, absorptive and see-through combination panel room dividers (go-bos)

Technical Foam Products 222 Rampart St., Charlotte, NC 28203 (704) 333-1131   Foam glass fiber, wall treatments, sealants, fiber barrier composites, foam barrier composites, composite curtains, quilted composites, vibration damping material

Technical Manufacturing Corp. 15 Centennial Dr., Peabody, MA 01960 (978) 532-6330, Web site: www.techmfg.com   Active vibration isolation systems, bases, floating floors, pneumatic systems, vibration dampers

Technical Specialities Co. Inc. 400B Airport Executive Park, Spring Valley, NY 10977 (800) 874-0877   Acoustical materials

Tech Products Corp. 5030 Linden Ave., Dayton, OH 45432 (513) 252-3661   Felt, glass fiber, spray-on coatings, wall treatments, ceiling systems, panels, absorptive materials, wall treatment systems, sealants, curtains, operable partitions, barrier panels, walls, windows, fiber barrier composites, foam barrier composites, composite curtains, enclosures, open plan partitions, composite panels, quilted composites, vibraion damping materials, vibration isolation materials, ducts, duct silencers, fan silencers, general industrial silencers, floating floors

Tecnifoam, Inc. 7145 Boone Aveenue North, Minneapolis, MN 55428 (763)537-7000   Foam, composites, barriers, ceiling tiles

Tectum, Inc. P.O. Box 3002, Newark, OH 43058 (888)977-9691, Web site: www.tectum.com   Wall treatments, ceilings, panels, composites, enclosures

Tempmaster Corp. 1222 Ozark St., N. Kansas City, MO 64116 (816) 421-0723   Absorptive materials, ducts

Tenneco Chemical Inc. Foam Division, W 100 Century Rd., Paramus, NJ 07652 (201) 262-7500   Foam, acoustical materials

Tex-Techn Ind. Inc. 150 Industrial Park Rd., Middletown, CT 06457 (203) 632-2211   Barriers, absorptive materials

Thunderline Corp. 8707 Samuel Barton Dr., Belleville, MI 48111 (313) 397-5000   Ducts, absorptive materials

Titus Products Div. Phillips Industries, 990 Security Row, Richardson, TX 75081 (214) 699-1030   Ducts, absorptive materials

Traconstics Inc. Box 316, Austin, TX 78764 (800) 531-5412   Acoustical materials, enclosures, panels

Transco Products, Inc. 55 E. Jackson Blvd., Chicago, IL 60604 (312)427-2828, Web site: www.transcoproducts.com   Panels, curtain, barrier panels, enclosures, composite panels, ducts

Uniroyal Inc. Middlebury, CT 06749 (800) 243-3535   General acoustical materials

Unisorb Machinery Installation Systems, P.O. Box 1000 Jackson, MI 49204 (517) 764-6060   Vibration isolation material

US Laminating Corp. 100 Wilbur Place, Bohemia, NY 11716 (516) 567-0300   Foam, wall treatments, wall treatment systems, foam barrier composites

United McGill Corp. 1501 Kalamazoo Dr., P.O. Box 909, Griffin, GA 30224 (404) 228-9864   Foam, mineral fiber, spray-on coatings, wall treatments, panels, ceiling systems, sealants, glass, metal and plastic sheet, operable partitions, barrier panels, walls, windows, foam barrier composites, enclosures, composite panels, quilted composites, vibration damping materials, ducts, duct silencers, fan silencers, general industrial silencers, absorptive materials

US Mineral Products, Co. Furnaca St., Stanhope, NJ 07874 (201) 347-1200   Acoustical materials

United Process Inc. 279 Silber St., Agawam, MA 01001 (413) 789-1770   Foam, glass fiber, spray-on coatings, wall treatments, ceiling systems, panels, wall treatment systems, mass-loaded plastic barriers, curtains, walls, windows, fiber barrier composites, foam barrier composites, quilted composites, vibration damping materials, absorptive materials

United States Gypsum Co. 101 S. Wacker Dr., Chicago, IL 60606 (312) 321-4000   Mineral fiber, panels, sealants, barrier panels, walls, open plan partitions

Universal Foam System Inc. 6001 S. Pennsylvania Ave., P.O. Box 548, Chudahy, WI (414) 744-6066   Foam, acoustical materials

Universal Silencer Division Nelson Industries, Inc., P.O. Box 411, Highway 51 West, Stoughton, WI 53589 (608) 873-4272   Duct silencers, fan silencers, general industrial silencers

Vanec Metro Square 2655 Villa Creek Dr., Dallas, TX 75234 (214) 243-1951   Panels, barriers

VAW Systems 1300 Inkster Blvd., Winnipeg, Manitoba, Canada R2X 1P5 (204)697-7770, Web site: www.vawsystems.com   Enclosures, rooms, silencers

Verilon Product Co. Box 997, Wheeling, IL 60090 (800) 323-1056   Acoustical materials

Vibra Check Inc. 22 Linden St., Boston, MA 02134 (617) 782-2808   Vibration damping materials, vibration isolation materials

Vibra Sciences Inc. 234 Front St., W. Haven, CT 06516 (203) 934-6113   Vibration isolation materials

VibraSystems, Inc. 310 Rayette Rd., Concord, Ontario, Canada L4K 2G5 (905)738-7810, Web site: www.vibrasystems.com   Vibration damping materials

Vibration & Noise Engineering Corp. 2655 Villa Creek Dr., Dallas, TX 75234 (214) 243-1951 Panels, enclosures, vibration damping materials, vibration isolation materials

Vibration Control Engineering 582 Cambrain Way, San Ramon, CA 94583 (415) 838-2482   Vibration damping materials, vibration isolation materials

Vibration Eliminator Co., Inc. 15 Dixon Ave., Copoague, NY 11726 (631)841-4000 Vibration isolation materials, floating floors

Vibration Isolation Co. 225 Grand St., Paterson, NJ 07501 (973) 345-8282, Web site: www.vibrationiso.com Vibration isolation materials, floating floors, springs

Vibration Isolation Product Inc. 11275 San Fernando Rd., San Fernando, CA 91340 (818) 896-1191 Vibration damping materials, vibration isolation materials

Vibration Mountings & Controls 113 Main St., P.O. Box 37, Bloomingdale, NJ 07403 (973)492-8430 Web site: www.vmc-kdc.com   Foam, glass fiber, wall treatments, ceiling systems, wall treatment systems, sealants, curtains, barrier panels, fiber barrier composites, foam barrier composites, composite curtains, enclosures, quilted composites, vibration damping material, vibration isolation material, fan silencers, floating floors

Vibro-Acoustics/Aseptic Systems Div. 6150 Olson Memorial Hwy., Minneapolis, MN 55422 (612) 542-8261, E-mail: tcharlton www.vibro-acoustics.com Ceiling systems, panels, wall treatments, doors, barrier panel walls, windows, enclosures, composite panels, ducts, duct silencers, fan silencers, general industrial silencers, electric motor silencers, absorptive materials

Vibro/Dynamics Corp. 2443 Braga Dr., Broadview, IL 60155 (708)345-2225, E-mail: vibro@worldnet.att.net Vibration isolation materials

Vicracoustic Division 290 Canal Rd., Fairless Hills, PA 19030 (800) 523-6671   Wall treatment, ceiling systems, wall treatment systems

Vlier Engineering Barry Wright Corp., 2333 Valley St., Burbank, CA 91505 (818) 843-1922   Vibration isolation materials, general industrial silencers

VSM Corp. 7513 North Field Rd., Cleveland, OH 44146 (216) 439-5400   Acoustical materials

Vulcan Lead Product Co. 1400 W. Pierce St., Milwaukee, WI 53204 (414) 645-2040   Acoustical materials

Wendell Fabrics Co. P.O. Box 145, Blacksburg, SC 29702 (803) 839-6341   Acoustical materials

WhisperRoom Inc. 116 S. Sugar Hollow Rd., Morristown, TN 37813 (423)585-5827, (800) 200-8168 Web site: www.whisperroom.com   Sound isolation booths and enclosures, diffusion panels

Wilshire Foam Products Inc. 1240 E. 230th St., Carson, CA 90745 (213) 549-5444   Foams, foam barrier composites

Wilson Sales Co. 732 N. San Gabriel, Box 1265, Rosemead, CA 90745 (213) 388-4222   Acoustical materials, panels

Witco Corp. 520 Madison Ave., New York, NY 10022 (212) 605-3800   Acoustical materials

Zero International Inc. 415 Concord Ave, Bronx, NY 10455-4801 (800) 635-5335, Web site: www.zerointernational.com   Door isolation systems, including jams, cam hinges, gasket sealing, thresholds, ramps and saddles

REFERENCES

1. Everest, F. Alton, *Acoustic Techniques for Home and Studio,* 2nd edition, TAB Books Nl. 1696 (1984), especially Chapter 5 "Resonances in Rooms," and Chapter 6, "Standing Waves in Listening Rooms and Small Studios."
2. Ibid. Fig. 10-4, page 183.
3. Ibid. Pages 164–169.
4. Ibid. Chapter 13, "Adjustable Acoustics."
5. Ibid. Chapter 9, "Acoustical Materials and Sturctures," especially pp. 169–173.
6. Gilford, C.L.S., *The Acoustic Design of Talk Studios and Listening Rooms,* Proc. I.E.E., Vol. 106, Part B, No. 27, May, 1959, pp. 245–258.
7. Sepmeyer, L.W., *Computer Frequency and Angular Distribution of the Normal Modes of Vibration in Rectangular Rooms.* Jour. Acous. Soc. Am., Vol. 37, No. 3, March, 1965, pp. 413–423.
8. Louden, M.M., *Dimension–Rations of Rectangular Rooms with Good Distribution of Eigentone,* Acustic, Vol. 24 (1971) pp. 101–104.
9. Bolt, Richard H. and Manfred Schroeder, Personal communication from Dr. Bolt.
10. Springs, N.F., and K.E. Randall, *Permissible Bass Rise in Talk Studios,* BBC Engineering, No. 83, July, 1970, pp. 29–34.
11. Davis, Don, Editor, *Syn-Aud-Con Newsletter,* Vol. 5., No. 4, July, 1978, page 25.
12. Souther, Howard, *Improved Monitoring with Headphones,* dB The Sound Engineering Magazine, Vol. 3, No. 2, Feb., 1969, pp. 28–29 and Vol. 3, No. 3, March, 1969, pp. 17–20.
13. Runstein, Robert E., *Modern Recording Techniques,* Howard W. Sams & Co., (1974).
14. Eargle, John M., *Sound Recording,* (1976), Van Nostrand Reinhold.
15. Woram, John, *The Recording Studio Handbook,* Sagamore Publishing Co., Plainview, NY (1976).
16. Rettinger, Michael, *Instrument Isolation for Multiple Track Music Recording,* Preprint No. 1119 (J-3), presented at the 54th Audio Engineering Society Convention, May, 1976.
17. Rettinger, Michael, *Sound Insulation for Rock Music Studios,* dB The Sound Engineering Magazine, Vol. 5, Nol. 5, May (1971), page 30.
18. Rettinger, Michael, *Recording Studio Acoustics,* dB The Sound Engineering Magazine, Part 1, Vol. 8, No. 8, Aug. (1974) pp. 34–47.
    Part 2, Vol. 8, No. 10, Oct. (1974) pp. 38–41.
    Part 3, Vol. 8, No. 12, Dec. (1974) pp. 31–33.
    Part 4, Vol. 9, No. 2, Feb. (1975) pp. 34–46.
    Part 5, Vol. 9, No. 4, Pr. (1975) pp. 40–42.
    Part 6, Vol. 9, No. 6, June (1975) pp. 42–44.
19. Rettinger, Michael, *Noise Level Limits in Recording Studios,* dB The Sound Engineering Magazine, Vol. 12, No. 4, April (1978) pp. 41–43.
20. Rettinger, Michael, *Studio Rumbles,* dB The Sound Engineering Magazine, Vol. 7, No. 9, Sept. (1973) pp. 46–48.
21. Cooper, Jeff, *Building a Recording Studio,* Recording Institute of America, New York, (1978).
22. Matel, Juval, *Advanced Room Acoustics,* Preprint No. 1312, presented at the 59th Audio Engineering Society Convention, Feb.–Mar. 1978.
23. Hansen, Robert, *Studio Acoustics,* dB The Sound Engineering Magazine, Vol. 5, No. 5, May (1971) pp. 16–24.

24. Bruce, Robert H., *How to Construct Your Own Studio in One Easy Lesson,* Preprint No. 1245, presented at the 57th Audio Engineering Society Convention, May (1977).

25. Storyk, John and Robert Wolsch, *Solutions to 3 Commonly Encountered Architectural and Acoustic Studio Design Problems,* Recording Engineer/Producer, Vol. 7, No. 1, Feb. (1975) pp. 11–18.

26. Brown, Sandy, *Recording Studios for Popular Music,* 5th International Congress on Acoustics, Liege, 1965, paper G036.

27. Olson, N., *Survey of Motor Vehicle Noise,* Jour. Acous. Soc. Am., Vol. 52, No. 5 (part 1) (1972) pp. 1291–1306.

28. Hudson, R.R. and K.A. Mulholland, *The Measurement of High Transmission Loss (The Brick Wall Experiment),* Acustica, Vol. 24, (1971) pp. 251–261.

29. Burroughs, Lou, *Microphones: Design and Application,* Sagamore Publishing Co., Plainview, NY (1974).

30. —*Performance Data–Architectural Acoustical Materials,* Acoustical and Insulating Materials Association, Bulletin No. 31, 1971–72.

31. —*Performance Data–Acoustical Materials.*

32. Harris, Cyril M., editor, *Handbook of Noise Control,* McGraw-Hill, (1957) pp. 9–10.

33. Randall, K.E. and F.L. Ward, *Diffusion of Sound in Small Rooms,* Proc. Inst. Of Elect. Engrs., Vol. 107-B, pp. 349–350.

34. Wente, E.C., *The Characteristics of Sound Transmission in Rooms,* Jour. Acous. Soc. Am., Vol. 7, Oct. (1935) pp. 123–126.

35. Audio Visual Source Directory, Spring/Summer 1987, published by Motion Picture Enterprises Publications, Inc., Tarrytown, NY 10591, (212) 245-0969.

36. Bolt, R.H. and R.W. Roop, *Frequency Response Fluctuations in Rooms,* Jour. Acous. Soc. Am., Vol. 22, No. 2, March (1950) pp. 280–289.

37. Volkmann, John E., *Polycylindrical Diffusers in Room Acoustic Design,* Jour. Acous. Soc. Am., Vol. 13, Jan. (1942) pp. 234–243, especially Fig. 2.

38. Somerville, T. and F.L. Ward, *Investigation of Sound Diffusion in Rooms by Means of a Model,* Acustica, Vol. 1, No. 1 (1951) pp. 40–48.

39. Head, J.W., *The Effect of Wall Shape on the Decay of Sound in an Enclosure,* Acustica, Vol. 3 (1953) pp. 174–180.

40. Mankovsky, V.S., *Acoustics of Studios and Auditoria,* Hastings House Publishers, New York (1971).

41. Shiraishi, Y., K. Okumura, and F. Fujimoto, *Innovations in Studio Design and Recording in the Victor Record Studios,* Jour. Audio Engr. Soc., Vol. 1, No. 5, May (1971) pp. 405–40.

42. Sabine, Paul E. and L.G. Ramer, *Absorption-Frequency Characteristics of Plywood Panels,* Jour. Acous. Soc. Am., Vol. 20, No. 3, May (1948) pp. 267–270.

43. Young, Robert W., *Sabine Reverberation Equation and Sound Power Calculations,* J. Acous. Soc. Am., Vol. 31, No. 12, Dec. (1959) p. 1681.

44. Schroeder, M.R., *Diffuse Sound Refection by Maximum-Length Sequences,* Jour. Acous. Soc. Am., Vol. 57, No. 1 (1975) pp. 149–150.

45. Schroeder, M.R., *Binaural Dissimilarity and Optimum Ceilings for Concert Halls: More Lateral Sound Diffusion,* Jour. Acous. Soc. Am., Vol. 65, No. 4, April (1979) pp. 958–963.

46. D'Antonio, Peter and John H. Konnert, *The Reflection Phase Grating Diffusor: Design Theory and Application,* Jour. Eudio Engr. Soc., Vol. 32, No. 4, April (1984) pp. 228–238.

47. D'Antonio, Peter and John H. Konnert, *The RPG Reflection Phase Grating Acoustical Diffusor: Applications,* 76th Convention of the Audio Engineering Society, Oct. 1984, preprint 2156.

48. D'Antonio, Peter and John H. Konnert, *The RFZ/RPG Approach to Control Room Monitoring,* 76th Convention of the Audio Engineering Society, Oct. 1984, preprint 2157.

49. D'Antonio, Peter and John H. Konnert, *The RPG Reflection Phase Grating Acoustical Diffusor: Experimental Measurements,* 76th Convention of the Audio Engineering Society, Oct. 1984, preprint 2158.

50. D'Antonio, Peter and John H. Konnert, *The Role of Reflection Phase Grating Diffusors in Critical Listening and Performing Environments,* 78th Convention of the Audio Engineering Society, May 1985, preprint 2255.

51. D'Antonio, Peter and John H. Konnert, *The Acoustical Properties of Sound Diffusing Surfaces: The Time, Frequency, and Directivity Energy Response,* 79th Convention of the Audio Engineering Society, Oct. 1985, preprint 2295.

52. D'Antonio, Peter, John Konnert, and Bill Peterson, *Incorporating Phase Grating Diffusors in Worship Spaces,* 81st Convention of the Audio Engineering Society, Nov. 1986, preprint 2364.

53. D'Antonio, Peter, *New Acoustical Materials and Designs Improve Room Acoustics,* 81st Convention of the Audio Engineering Society, Nov. 1986, preprint 2365.

INDEX